Business of Diving

Business of Diving

John E. Kenny

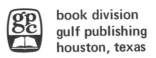

book division
gulf publishing
houston, texas

Library of Congress

Catalog Card Number 74-114693

ISBN-0-87201-183-6

Contents

Part 1
The Last Reentrant

Part 2
Diving Technology

Part 3
Diving Activity

Part 4
The Future

Preface

There have been many books written about the basic physiological concepts and rudimentary procedures of diving technology. Library bookshelves are being expanded to house the rapidly growing number of publications that describe the adventures of treasure diving, the beauty of sport diving, the heroism of underwater military missions, and the monetary rewards of commercial diving. But nowhere has diving been treated as a wide field of activity that plays an important role in exploiting the oceans. This book has been written to fill that gap. We hope to fill it in a new way.

When Gulf Publishing Company and I began discussing the probable content of this volume over a year ago, we realized that the presentation of a highly technical subject to an audience with a wide range of interests could be an impossible task in the context of normal book organization. To solve this problem, the book has been written and assembled in a unique way—the main text is written to be understood by an intelligent reader with little exposure to diving. The content is designed to provide a knowledge of the human physiological characteristics that permit reentry into the ocean, a brief summary of man's past attempts to reenter, an overview of the latest diving equipment and systems, a review of diving activity in all major segments of the field, and an analysis of diving as a business, both from the inside looking out and the business as a customer.

For those with a previous working knowledge of diving, each section relating to technology contains a supplement with in-depth discussion of the latest developments and theory. Appendices are located at the end of the section. They should be considered a must for the experienced, and optional for those who do not require that degree of technical assistance.

Introduction

There are valid hypotheses about our origin among the simple ocean creatures, reinforced by speculation about our early survival on this planet using the bounty of the nearshore environment. Regardless of the accuracy of these postulates, there is no doubt of our present cultural affinity for the shoreline and our interest and curiosity about the undersea environment. In recent years, our growing population, diminishing resources and concomitant awareness of major resource imbalance in future generations has turned our attention to the untapped wealth of resources beneath the sea.

Man has always been a minor ocean predator. He has foraged the shallow waters, hunted air-breathing mammals, ineffectively fished the coastal waters, and most recently, recovered minerals and hydrocarbons from the continental shelves adjacent to all the larger land bodies. Our future needs and a satisfactory return on our investment, however, cannot be realized by a haphazard approach to an area that is twice as large as our planet's land surface and undoubtedly contains a proportional store of food and basic resources. The ecology of our atmosphere, land, and all the major water bodies is so closely interrelated that it is impossible to make a major perturbation in one environment without causing a dramatic

chain reaction in the entire ecosystem. Therefore, the manner and means of extracting and utilizing these vast subsea resources become extremely important. Even a moderately intelligent approach to the technical problems involved will require a vast bank of knowledge—a bank into which we have just made our first deposit.

One of the basic considerations involved in making meaningful advances in ocean technology is the hostility of the underwater environment to human existence. This inability of man to be an effective observer and performer in the ocean has, until recently, impeded progress. The appearance of remote underwater systems in all phases of ocean work is making possible a limited advance. As on land, effective implementation of sophisticated technology requires direct human participation with on-site perception and real time problem solving. To be in the sea, man is forced to become a reentrant—the last reentrant mammal on earth.

All other air-breathing mammals that returned to the sea have done so by natural adaptation and mutation. These changes involved not only structure, but the development of specialized physiological systems that respond to the environment. The research of P. F. Scholander and other investigators at the Scripps Institution of Oceanography shows that diving mammals have a built-in mechanism that reduces their underwater oxygen requirements. These mammals maintain a normal blood flow to the vital organs such as the heart and brain during submergence, but blood flow to the kidneys and most muscles is greatly diminished. The result is a marked lowering of oxygen requirements during a dive and a simultaneous increase in the animal's ability to remain underwater. Unwilling and unable to wait for natural adaptation, man has created an arsenal of mechanical appliances with which to attack this hostile environment.

In recent years, this repertoire of adaptive, mechanical substitutes has enabled man to enter the underwater world, not only as a passer-by, but as a semi-permanent resident. With this new capability, he is able to continue many of his terrestrial activities past the tidal low water mark into the nearshore. In no more than two or three years, he will be an experienced worker at depths to 2,000 feet, and all the earth's continental shelves will be within his domain. His time in residency will be determined mainly by psychological considerations. There may be some physiological barriers similar to those anticipated for extended space flights, but these are only surmised today.

Since 1950, we have seen a series of diving technology events that could only have been fantasied twenty-five years before—colonization of the sea floor for up to two months, deep dives exceeding 1,000 feet in the open sea, and the change of diving from a vertical elevator technology to a mobile capability with diver range of several miles. Diving has become an integral part of nearly every major ocean activity—industry, science, naval mission and recreation. Considered as a single entity in 1970, it is a large business—$135 million a year. By 1980 diving will be a $2 billion per year business.

For all its size and recent public exposure, diving is the most misunderstood of the new ocean technologies. Perhaps this is due to the rather romantic or true adventure type of publicity to which it has been subjected. Whatever the reason, misunderstanding about the business of diving is seated in general concepts. Diving is not an end in itself; it is merely a means to an end. In proper context, diving techniques and technology are a science of transportation. They permit movement from our atmospheric environment to the less hospitable underwater world and then a safe return. What is accomplished underwater is a function of the diver's specialized capability. A commercial diver must first be a rigger, mechanic, welder or some other journeymen specialist. The scientific diver must first be a scientist. The military diver is usually an explosive ordinance specialist. The sport diver is pursuing some interest such as hunting, photography, or simply extending his terrestrial sightseeing.

To My Father

who taught me to love the sea.

Business
of Diving

Part 1

The Last Reentrant

1. Man Underwater

Living in our surface environment, we are unaccustomed to thinking about the physical forces constantly working on our body systems. Only when we are estranged from our normal state do we become conscious of the physics of living, especially when we consider the human body existing in space or under the surface of the water.

High altitude medicine became a reality when nineteenth-century balloonists ascended to heights where the atmosphere would no longer support life and they were affected acutely—sometimes fatally—by a lack of oxygen. Space medicine has established a minimum set of physical parameters for the environment necessary to support life. Those who go into space take their environment with them—either in their capsules or space suits. The limits of how deep a man can go in the ocean is another question—one that is being answered continually, each time differently.

There is no doubt that man can travel to the deepest trenches in the ocean with no ill effects if he is willing, as in space flight, to surround himself with a synthetically created environment. On January 23, 1960, man reached the deepest known spot on the ocean floor. Using the bathyscaphe *Trieste* (Figure 1-1), the equivalent of an underwater balloon with a pressureized gondola, Lt. D.

Walsh and Jacques Piccard touched bottom at 37,800 feet in the Challenger Deep. Throughout the dive, the pressure inside the diving chamber varied little from that of the normal atmosphere. Similar concepts have been used in the design of diving equipment. Maintain man at sea level pressure (1 atmosphere), let him breathe normal atmospheric air and the problem is solved-theoretically. Armored, pressure-proof diving suits were actually built and successfully tested 50 years ago.

In recent years, the idea of a single-atmosphere diving system was again studied by Litton Industries in the United States, but little action followed. If man is to be little more than an observer underwater, with a minimum of articulatory capability, a single-atmosphere diving system is certainly a workable scheme. But, if mere observation is the goal, why not use a tethered bathysphere such as William Beebe did in 1934? The answer to this question is that man can only justify being in the sea by becoming productive. To be productive and creative, he must be mobile and equipped so that he can use his direct sensory perception. Unlike space, no one is willing to pay a billion dollars per pound for recovered rocks and dust from the sea to solve evolutionary postulates. Man in the sea must work.

Figure 1-1. In 1960, the bathyscaph Trieste I took man to the deepest point in the ocean. Lt. D. Walsh and Jacques Piccard descended 37,800 feet into the Challenger Deep. (Photo: U.S. Navy, NELC.)

The future of man in the sea is based on economic exploitation of the subsea resources that are becoming so sorely needed to support our overpopulated planet. The acid test for any underwater system, including man, consists of only two questions: can it effectively and economically perform the required task and can it survive the hostile ocean environment? When these questions are asked of man as an underwater system, the answer must be in the context of time and the changes that are rapidly taking place. Within the past two years, the balance has shifted heavily toward positive answers. Less than 10 years ago, the consensus was definitely negative.

There are many routes to proving that man is really a sea creature who has not completely lost his aquatic ties; the saline solution that is so much a part of human physiology, the analogy of certain stages of human fetal development to development of other sea mammals, the gill rakes in the embryo and the saline solution in the womb are examples. But to reenter the sea, man needs adaptive substitutes such as the helium/oxygen SCUBA and thermal protection suits shown in Figure 1-2.

These analogies are all true, but they are also philosophical observations leading only to philosophical arguments. There is no argument with the fact that man cannot extract life-giving oxygen from the water with his physiological equipment. His underwater vision is so bad that in his normal environment, he would be considered legally blind; his hearing is all but gone in the water; his mobility underwater is ridiculous; his ability to protect himself from a new set of predators might be compared to cavemen surviving in a land of carnivorous dinosaurs. And yet, in 1970 man is found preparing to work at depths of 2,000 feet in the open ocean, with ambient water temperature of 4°C or less, with almost complete absence of light from his warming sun and a pressure of 905 pounds on each square inch of his body. The remarkable aspect is that at 2,000 feet man will not only be existing, but will be a thinking creature capable of creative work, with most of his sensory processes only slightly modified from his surface capabilities. To understand how the human system can operate under these stresses, we must examine the body in this new set of physical conditions.

Figure 1-2. Wearing adaptation substitutes consisting of helium/oxygen SCUBA and a thermal protection suit, Sealab aquanaut Wally Jenkins is joined on an underwater sortie by a natural reentrant— a sea lion. (Photo: U.S. Navy, by Johnson.)

Seeing in the Water

Nearly everyone has had the experience of opening his eyes underwater in a swimming pool or in the sea. Aside from the discomforting irritation, the observable phenomenon is that all objects seem to be out of focus. This loss of visual acuity can be easily understood by studying the structure of the eye as an image receptor both in air and in water.

Refraction

When light travels from one transparent material to another of a different density or physical structure, the light rays are refracted or bent. The amount of change in direction, or refraction, depends upon material structure and is measured by a quantity called the *refractive index*. The simplified cross-sections of the human eye (Figure 1-3) show how this operates. Light passing from an object through the air (Figure 1-3A) comes in contact with the cornea, the outer covering of the eye.

The refractive index of the transparent cornea is different from that of air so the light rays bend or refract, causing them to pass through the eye lens. The amount of bending at the cornea is proportional to the ratio of the refractive index of air to that of the cornea material. When the eye is immersed (Figure 1-3B), the water, having a different index of refraction, causes the impinging light to refract differently. The net effect is that the image is not sharp when it reaches the sensitive retina and is consequently perceived out of focus.

In some reentrant mammals that spend a considerable portion of time both above and in the water, a specially developed mechanism allows them to change the geometry of the eye and to focus on objects beneath the sea surface. Man, not having made such a physical adaptation, must look elsewhere for a solution, and the most obvious one was a long time coming. An early remedy, purportedly used by pearl divers, was to carry a viscous oil in the mouth and excrete it when one wished to see an object clearly underwater. Theoretically, the oil, with a more favorable ratio of refractive index, focused the object when it was in the light

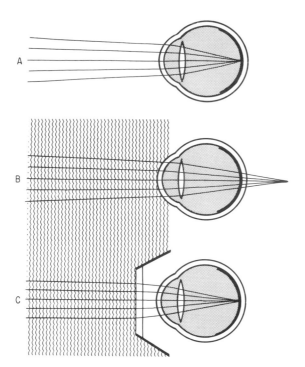

Figure 1-3. A water medium affects vision. Normally, light striking the corneal surface is bent or refracted as shown in A, giving a sharp image on the retina. When the eye covering is in contact with water (B), the light is bent differently, moving the focal point beyond the retina. With a glass interface (C), the image is sharp but the visual field is reduced.

path. Figure 1-3C shows how the correction is now made. When air is maintained in front of the eye by means of a face mask or goggles, the refraction ratio normal in the atmosphere is reestablished, allowing the image to be focused on the retinal surface. However, it is not exactly the same image.

Although we have established an air/corneal interface at the eye, we have still another interface, or meeting of mediums, at the glass/water of the face mask. There is a difference in refractive index, so that even with an optically smooth piece of glass, light rays are bent. This refraction is, however, not as detrimental and does not affect the focus on the retina but it does narrow the field of view. The result is that objects seen underwater through a face mask or goggles seem larger until

perceptual adjustment is made. This "magnification," caused by a 25% reduction in the field of view, actually improves vision in close underwater tasks. Sight is the only sensory perception that is enhanced by the physics of the underwater environment.

Sight in Stress Situations

Although psychology and sensory acuity may not be apparently related in our normal environment, their important interrelationship in a hostile or foreign environment under stress situations has recently come to light. The actual process is not completely understood, but the results show that the responses differ from a layman's expectations. The natural assumption would be that, placed in danger, the sensory system would respond by increasing its performance. Drs. Gershon Weltman and Glen Egstrom of the University of California found that under stress subjects lose their peripheral perception, and if the stress is at a sufficiently intense level, perception is reduced almost to tunnel vision. Considering that diving, especially deep diving, is a stress environment, this phenomenon, added to the loss of visual field, becomes an important factor in estimating underwater performance capability.

Color Perception

The modification of color perception underwater is only of passing interest in the diving business. This modification is a purely physical phenomenon and only present when vision depends on natural light coming from the surface. Sunlight is considered pure or white light, and is composed of high frequency radiations which if viewed separately would appear as "colors." Each of these colors has a characteristic electromagnetic frequency and a specific capability for transmission through water.

When white light strikes the surface of the water, a portion of it is reflected. That which does pass into the water medium is immediately affected by the transmission characteristics of the water. Figure 1-4 shows the relative abilities of the constituent colors in white light to pass through clear sea water. The first to stop, or be absorbed, is the red. The last, near the top of the visible spectrum, are the blue-greens. The effect on vision is obvious. According to our chart, reds disappear at approximately 35 feet in clear sea water. This means that a red object, viewed by light coming from the surface, will no longer look red but will appear as a dull brownish gray. The same holds true for all other colors beyond their transmission depth. Beyond the color penetration depth limit, a red object illuminated with an artificial light source containing energy output in the red frequency spectrum will again exhibit its red color. Since most diving at relatively deeper depth is performed with the aid of artificial light sources, this phenomenon is only of academic interest.

If it were only true that what you can't see or what you don't know can't hurt you, the business of diving would be extremely attractive. Our diver's field of vision has already been reduced through physics and psychology. The physical and psychological factors become minor considerations when discussing turbidity.

According to the *American Heritage Dictionary of the English Language,* turbid means "having sediment or foreign particles stirred up, or suspended, muddy, cloudy." In the business of diving, turbidity means isolation, blindness and stress. Figure 1-5 shows a diver working in very turbid water. Those who are acquainted with diving only through the camera lens never have the opportunity of observing normal diving conditions.

Turbidity is actually particulate matter suspended in the water, and the effect and degree vary from site to site. In New York harbor, for instance, turbidity reduces vision to 3 inches or less. Any ambient light that penetrates to the working depth is so randomly dispersed and reflected that it is of no use. In clean nearshore coastal areas, visibility will vary from 3 to 15 feet. On offshore sites, depending upon currents and wind, visibility may reach up to 200 feet. Turbidity reduces lateral visibility to the equivalent of looking through a heavy atmospheric fog.

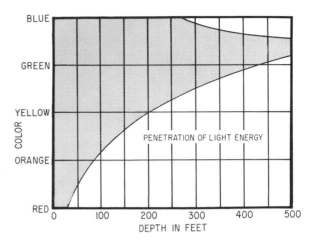

Figure 1-4. The shaded area shows approximate penetration of light energy in coastal sea water. Red disappears from visible light between 30 and 40 feet. The blue-green portion of the spectrum was measured below 1,600 feet.

Hearing in the Water

There is absolutely nothing silent about the sea. To man, the underwater environment is a confused cacophony of omnidirectional sounds. The sound perceptors used by humans in the atmosphere to receive, interpret and classify sounds are nearly inoperative in the water. Once the human ear becomes occluded with water, it ceases to operate as a sensitive neuro- mechanical instrument. The damping effect of water against the eardrum severely reduces its sensitivity. The human sound perception system, however, does not depend solely upon the ear as a receptor. The entire head, particularly the bone structure, is a sound receptor. In the thicker water medium, it takes over as receiver.

Directional Differentiation Loss

One of the most disturbing phenomena associated with underwater hearing is not the loss of audio acuity, but loss of the ability to determine the direction of the sound source. Recent research conducted in Canada indicates that this ability may only be diminished and is recoverable as a learned skill. But it is generally accepted among experimental underwater phoneticians that if this

Figure 1-5. Nearly obscured by turbidity, a diver uses an underwater light with main output in the blue-green portion of the light spectrum. The light, manufactured by Hydro Products, penetrates sea water more effectively than an incandescent output. (Photo: Hydro Products.)

information is available it is at a level requiring sophisticated electronic assistance to be usable.

This loss of directional differentiation is not entirely related to the physiological factors that affect hearing system sensitivity. It is also dependent upon the difference in the physical properties of air and water. Two factors must be dealt with —the comparative speed of sound in water and air, and the mechanics of human head structure. If the human head were 3 feet in diameter, the problem of directional differentiation would not exist. Equipped with an ear separation of less than one-third of that required, the problem resolves itself to one of physics.

Human sound sensors are tuned to the speed of sound in air and determine direction by two phenomena—the relative phase, or difference in time of the arrival of the sound at each ear, and the relative level of sound reaching each ear. Accuracy of directional differentiation varies with individuals and their environmental needs, but even

in individuals having great sensitivity variance of the left and right ears, the basic capability for a high level of differentiation is present. When sound reaching the ear could be ambiguous, that is, comes from a source that could be interpreted on the basis of phase as a specific direction, or one 180° out of phase, the shape of the external ear creates a front to back sound ratio that solves the perception problem.

In water we are faced with different phenomena. First, the speed of sound in water is so much greater than in air that the ear separation does not provide sufficient time lag to interpret direction (Figure 1-6). In addition, the activation of bone conduction reception in the viscous water medium and the concomitant sensitivity loss of the ear reduce the effectiveness of the mechanical separation present.

We find our underwater man then, not only losing a measurable part of his hearing sensitivity, but losing his ability to locate sound sources as well. Combining this with a restricted field of vision and a loss of visual range caused by water turbidity seems to be asking more than should be expected when the problem is complicated by the added effects of pressure.

Pressure

The connotation of pressure in the context of human physiology is normally one of pain and discomfort. To understand the effects of pressure on the total human organism, it is necessary to be aware of the physical forces involved. Again citing swimming pool experience, it is not uncommon to feel pain and discomfort when diving to the bottom of the pool. Usually pain is felt in the ear and the sinus cavities of the head. The actual discomfort comes from differential pressure—a difference of pressure on two sides of a body tissue which causes a physical stress to be placed on the material. This differential pressure or "squeeze," is not a penalty for entering the underwater world; it can be prevented by using proper diving techniques.

Pressure Mechanics

The mechanics of pressure in water are not difficult to understand: The most pertinent to div-

ing is the pressure on an enclosed gas resulting from depth. The effects of pressure on gas and the resultant physiological ramifications are so important to an understanding of diving that we shall look closely at gas under various pressure situations before proceeding.

To describe a gas, some basic terms—common words in everyday jargon—must be defined in a relatively precise manner in order to gain a visceral understanding of the phenomena involved in diving. Our concern with diving physiology and physics is involved with only the major or gross reactions of the human systems to pressure. Particles of matter smaller than the atom will not be of specific concern.

An atom is the smallest unit particle of an element. There are more than one hundred basic substances that have been identified in our physical world. These are normally arranged by physicists in a periodic table of elements according to their atomic weight, and are the constituent parts of all known matter, including man. Some atoms, because of their basic structure, are never found singly. They occur in combination with other atoms of the same substance. Oxygen, for instance, which accounts for half the weight of the earth's crust, normally appears as a combination of two or more atoms. In its normal form, O_2 (two oxygen atoms combined) is used by the body to

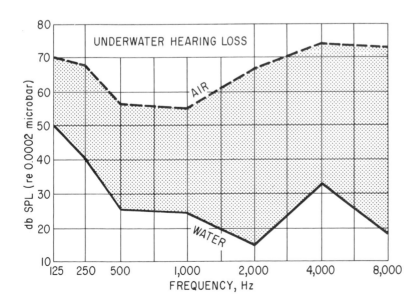

Figure 1-6. Underwater hearing loss is shown by shaded area in plot of sound pressure level thresholds (SPL) for response to single frequency tones in air and water. From tests by Drs. Hollien and Brandt, University of Florida, Communications Science Laboratory.

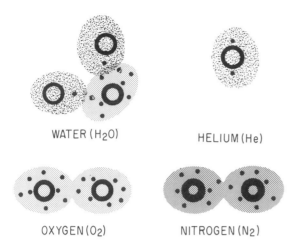

WATER (H$_2$O)   HELIUM (He)

OXYGEN (O$_2$)   NITROGEN (N$_2$)

Figure 1-7. Molecules are the smallest combinations of atoms necessary to form a physical substance. Molecules of water, helium, oxygen and nitrogen are shown as they normally occur in nature. In Figure 1-8, an enclosed gas develops pressure through impacts of molecules on the container walls. Individual molecules have random motions, but all sides of the container receive equal amounts of energy impacts.

sustain life functions. *When in combination with itself or other atoms as the smallest physical unit of a substance, the combination is called a molecule.* Two atoms of oxygen, or O$_2$, is a molecule of oxygen. Two atoms of hydrogen combined with one atom of oxygen, or H$_2$O, is the substance water. Figure 1-7 shows the molecular configuration of water, helium, oxygen and nitrogen.

Atoms and molecules are never at rest except in a state of extreme cold. To see the effects of their motion and to see how pressure is developed, consider a closed container that holds a gas of any molecular structure. The container is sealed, so that gas can neither enter or escape. Molecules of the gas are constantly in motion. As each molecule strikes the side of the container, it strikes with a force that is proportional to its molecular weight (the sum of the weights of the individual atoms that make up the molecule). There is a continual bombardment of the container sides by molecules. The total force exerted on the side of the container is a function of the number of molecules within the container and the speed with which they are moving. Given a fixed number of molecules, the faster they travel, the more collisions they have with the container sides and with other

molecules. The motion and direction of the molecules are random, but statistically, all sides of the container will receive an equal number of impacts on a unit area. This is shown schematically in Figure 1-8.

Pressure Measurement

Although we have not used the term yet, what was described is *pressure*. To use the classical definition, *pressure is the force per unit area.* If a one pound mass were placed upon an area that was one inch on each side, a pressure of one pound per square inch, or one psi would be exerted. Similarly, the force exerted by a gas can be measured in pounds per square inch—each pound of pressure caused by impacts of molecules against a 1 square inch area. Pressure is always measured in force per unit area. In the metric system, the unit of force or weight is the kilogram; the unit of area is the square centimeter. Regardless of the unit of measure, the basic physics involved is identical.

So far, an isolated container that is apparently in a vacuum has been discussed. Take the container out of the vacuum and place it in the atmosphere. Then open the container and let air enter. With the container open, there is an equal density collisions and, therefore, twice as much pressure. Assuming that the pressure of the normal atmosphere is 14.7 pounds per square inch, the pressure within the container would be 29.4 psi. If the volume of our container had been reduced to one-third its original size, again without any of the gas escaping, the pressure within would have increased three times to 44.1 psi.

Pressure was defined as the force per unit area. The pull of gravity on matter also causes force. For instance, a light container approximately one foot square filled with sea water will weigh about 64 pounds. The bottom of a 12 inch square container has an area of 144 square inches. *To find the pressure exerted upon the bottom, we simply divide the force by area* and find that 64 pounds divided by 144 square inches gives us a pressure of .445 pounds per square inch (Figure 1-10). A column of water 1 foot high creates a pressure of .445 psi. If the container had been 2 feet high the pressure on the bottom would have been .890 psi. A column of water 100 feet high creates a pressure of 44.5 psi; a 1,000 foot column creates a pressure of 445.0 psi.

In underwater work, a common measure of pressure is the atmosphere. *One atmosphere of pressure is equivalent to the same force per unit*

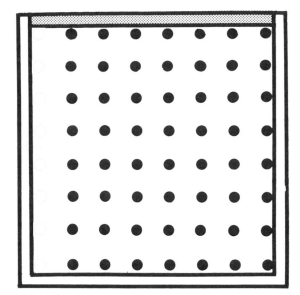
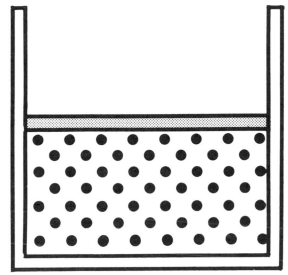

Figure 1-9. Gas density and pressure are doubled when the volume within which a specific number of molecules is contained is reduced to one-half the original volume.

of air inside the container and outside its walls. If the container is now sealed and its size mechanically reduced to one-half, allowing none of the gas inside to escape, the gas density inside the container, or the number of molecules in each cubic inch, is doubled—twice as much as in each cubic inch of the surrounding atmosphere (Figure 1-9). With twice as many molecules, there are twice as many

Figure 1-10. Water pressure is caused by the pull of gravity on the mass of the water. Each cubic foot of sea water weighs 64 pounds. A column of sea water 1 inch square and 1 foot high exerts a pressure of 0.445 psi.

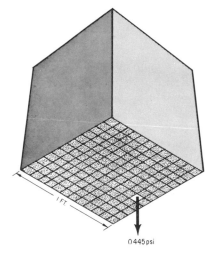

0.445 psi

area as that exerted by the earth's atmosphere at sea level. We live under a constant pressure of 1 atmosphere or approximately 14.7 pounds. This pressure is, of course, transmitted to the surface of the water so that the total or *absolute pressure* exerted on an underwater object is the sum of the pressure due to the weight of the atmosphere and the weight of water above it. Each 33-foot interval in ocean depth is equal to the pressure created by the atmosphere at sea level. An object 33 feet in the ocean is therefore subject to a total or absolute pressure of 29.4 psi or 2 atmospheres (AtA).

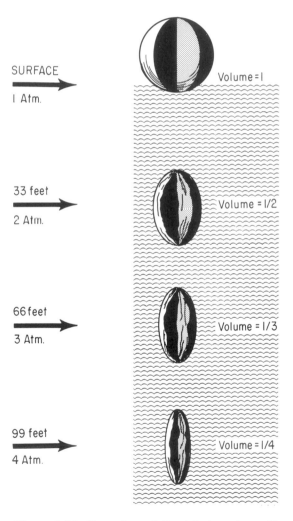

SURFACE
1 Atm.
Volume = 1

33 feet
2 Atm.
Volume = 1/2

66 feet
3 Atm.
Volume = 1/3

99 feet
4 Atm.
Volume = 1/4

Figure 1-11. Operation of Boyle's Law is readily seen when a pliable container is lowered into the water. Volume is inversely proportional to the absolute pressure.

The Effects of Pressure

Relating these physical facts to diving techniques leads into the field of diving physics and physiology. It is not a far step from the closed container to some of the aspects of free diving. Free diving is the process of descending from the surface without the use of any auxiliary breathing apparatus. Pearl Divers of the South Pacific have been performing this feat for thousands of years. The diving women of the Orient, the *ama* have been engaged in free-diving activity for 1,500 years and can make up to thirty dives an hour to depths of 60 feet. On each dive their lungs operate as closed elastic containers that are subjected to the pressure caused by the column of water and atmosphere over them. Figure 1-11 shows how this takes place.

Before considering the structure of the respiratory system and how it survives the stress of pressure, it will be simpler to first examine the physical forces involved. Instead of a rigid container, this time consider a supply of gas contained in a pliable plastic bag. If the volume of the inflated bag is 1 cubic foot at the surface, the physical effects on it will not be unlike a lung subjected to the same forces. With the container sealed, it is gradually lowered into the sea. At the surface, there was an equal density of air in the container and in the atmosphere, but as the bag is lowered into the water, the pressure on the outside of the container begins to increase because of the weight of the surrounding water. At a depth of 10 feet there has been an increase of 4.45 psi in the ambient or surrounding pressure.

To equalize the pressure between the inside and outside of the container, more collisions between the entrapped molecules and the container walls must take place. Two alternatives are available in this constant temperature environment. More molecules of gas must be added, or the molecules must travel a shorter distance between collisions so that more impacts will be received by the container wall. Since we cannot add gas molecules, the container must collapse until the pressure is equalized.

In the case of this pliable container, a volume measurement would show that at a depth of 33 feet, where the pressure is 2 AtA, the container

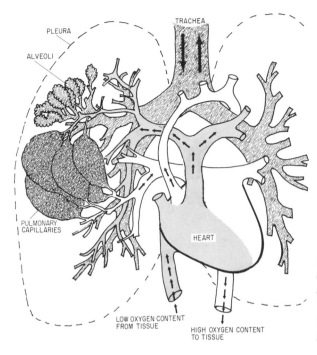

PLEURA

ALVEOLI

TRACHEA

PULMONARY
CAPILLARIES

HEART

LOW OXYGEN CONTENT
FROM TISSUE

HIGH OXYGEN CONTENT
TO TISSUE

Figure 1-12. In this simplified schematic of the cardio-pulmonary system, the lined area shows the air path through the intake system to the alveoli. The dotted area is the path followed by blood from the tissue to the pulmonary capillaries, where gas exchange takes place.

would have exactly one-half its surface volume. At a pressure of 3 atmospheres obtained at 66 feet, the container would have only one-third its original volume. This linear process can be mathematically derived and is described by Boyle's Law. It continues until the molecules begin to crowd one another at a pressure beyond that obtainable in the ocean. Applied to free diving and the lungs, the answer is the same. At a depth of 33 feet, the volume contained within the lungs is approximately one-half what it was on the surface. At a depth of 99 feet, well within the reach of free divers, the volume is only one-fourth of the surface volume.

Human lungs are able to withstand this compression with little or no discomfort. There is, in fact, no discomfort until the elastic limit of the system is reached. This limit varies with individuals. In February, 1967, a U.S. Navy underwater instructor, Robert Croft, made a free dive to a depth of 212.7 feet. His rare ability for compression was due to an abnormally large lung capacity and flexible rib cage resulting from a childhood illness. A normal lung structure would not permit this degree of compression.

The Lungs Underwater

To appreciate the technical problems of diving it is mandatory to understand the basic lung structure and its relationship to the cardiac system. The lungs are the point where life-sustaining oxygen is transferred to the blood stream and carbon dioxide, CO_2, is removed. Figure 1-12 is a schematic drawing of the cardio-respiratory system, greatly simplified.

The Respiratory System

The respiratory system originates with the oral and two nasal openings. These paths join to become one main tube, the *trachea*. The trachea divides into the two main bronchial tubes which are again divided into branches. By the time the small air sac, the *alveoli* is reached, the original trachea has divided into more than a million separate paths which terminate in over 300 million alveoli. These small air sacs have an average diameter of around .010 inches and very thin membrane walls that allow gas to be exchanged freely with the small pulmonary capillaries of the blood circulatory system.

The blood that arrives at the pulmonary capillaries is oxygen-depleted and rich in carbon dioxide. The alveoli gas has a higher oxygen concentration and a low level of carbon dioxide. The difference in concentrations causes the gas to exchange, resupplying the blood with oxygen and removing most of the carbon dioxide. Respiration brings fresh atmospheric gas into the alveoli and removes the carbon dioxide waste.

Gas is moved in and out of the lungs by two sets of muscles. The prime mover of the respiratory system is the dome-shaped muscle separating the abdomen from the chest cavity, the *diaphragm*. When inhaling, the central part of the diaphragm contracts downward, enlarging the chest cavity or *thorax*. Simultaneously, the muscles of the thorax cause the ribs to distend slightly, also increasing the volume.

This larger volume reduces the pressure in the air tracks and alveoli, causing fresh atmospheric air to rush in through the nose and mouth to supply enough molecules for pressure equalization. Exhalation is a passive process, relaxation of the diaphragm and thorax muscles reduces the volume and causes exhalation. The initial part of the exhalation normally consists of gas from the air track system that has not reached the alveoli. The last part of the exhalation is alveolar gas with a high carbon dioxide content and low oxygen concentration.

The elasticity of the thorax, caused by the diaphragm and rib muscles, is the controlling factor in the free-diving process. When the diver leaves the surface, his air track is closed and his lungs undergo increased pressure and compression. The diaphragm and thorax muscles allow the volume to decrease so that the air pressure within the lungs is always equal to the surrounding water pressure. So long as the maximum point of elasticity is not passed, no differential pressure will develop and no discomfort will be felt by the diver.

The thorax muscles are not capable of expelling all air from the lungs. In a trained diver with normal physiological characteristics, the volume of air that remains after a forced exhalation is about 20%. Examinations of experienced divers show that they do develop a large lung capacity by exercise. Residual volume, or the remaining volume, differs little between divers and non-divers.

The adult repeats the breathing process ten to fourteen times per minute in normal activity. The respiratory system is capable of a much higher rate, but is seldom called upon to reach maximum effort. During each breath the tidal volume, or the amount of air entering and leaving the system, is about 7 liters per minute. At maximum capacity the tidal volume can reach as high as 200 liters per minute, but even during strenuous activity the volume seldom reaches much higher than 100 liters per minute. Of the total volume, about one-third of the inspired gas actually enters the alveoli.

Hyperventilation

The diffusion of gas from the alveoli into the pulmonary capillaries through the porous membrane of the air sac operates in the same manner as gas over an open container of liquid. Placed over the surface of a liquid, a gas dissolves in the liquid in proportion to the pressure of the gas and the nature of the liquid. When an equal number of gas molecules are simultaneously entering and leaving the liquid, the system is said to be in equilibrium. If the pressure of the gas above the liquid is raised, more gas molecules will be striking the surface and more entering into solution with the liquid. At the respiratory/circulatory interface in the alveoli, the blood arrives with a low number of dissolved oxygen molecules and a high level of carbon dioxide molecules. The mixture on the other side of the membrane has different concentrations and the gas exchanges through the membrane in an attempt to reach equilibrium.

Gas Diffusion

An interesting study of gas exchange in the *ama* divers was conducted by Drs. Suk Ki Hong and Hermann Rahn several years ago. According to their data, a diver resting out of the water showed an alveolar gas mixture of 14.3% oxygen, 5.2% carbon dioxide and 80.5% nitrogen. By the process of hyperventilation, the composition of the mixture was changed to 16.7% oxygen, 4% carbon dioxide and 79.3% nitrogen.

Hyperventilation is a breathing technique that employs maximum exhalation of several successive breaths. This forced exhalation expels additional air from the alveoli causing increased ventilation of the air sacs with fresh air; the result is a lowering of carbon dioxide and a rise in oxygen content in the blood.

On a free dive to a depth of 40 feet, the lung volume was compressed to slightly less than half its normal surface volume so that air within the lungs had twice the density as that at the surface. The increased pressure caused more oxygen to pass through the diffusion membrane in an attempt to equalize. The oxygen level within the lungs fell to about 11.1%. Simultaneously, a less desirable action look place. The carbon dioxide level within the lungs also fell—from 4% to 3.2%. The increasing pressure caused the normal flow of carbon dioxide from the blood into the alveoli to reverse and carbon dioxide actually flowed from the lungs into the blood.

On ascent, the lungs expanded to their normal volume with a concurrent drop of internal pressure and gas density. The excess carbon dioxide that went into the blood was quickly released and diffused into the lungs. The decreasing density of the oxygen had a more marked effect. It was found during these tests that oxygen content actually falls to a level which does not permit any further equalizing exchange from the lungs into the blood. Some investigators claim that at this point, the process actually reverses and the blood loses oxygen to the lungs. It was deduced that this may have precipitated deaths which occurred in deep sport skin diving. However, if care is exercised in allowing the respiratory/circulatory system to equalize after each dive, there does not seem to be any cumulative effect.

The Law of Partial Pressures

One of the important physical aspects of gases that will be repeatedly considered in discussing diving physiology, techniques and technology is the *law of partial pressures*. Formidable as it might sound, it is the simplest of all the gas laws. To use an analogy, a group of weights placed upon a scale all exert pressure on the scale pan independent of the other weights present. *Translating this simple concept into the action of gas molecules in a container, the part of the pressure which is due to one type of gas molecule, called the partial pressure of that species, would be the same no matter what other gases are present.*

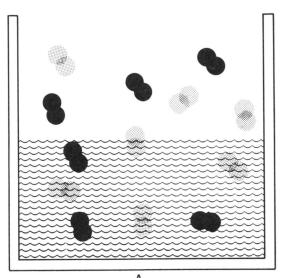

A

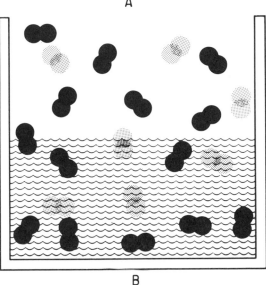

B

Figure 1-13. The gas in A has reached a point of equilibrium with both black and gray molecules. In B the pressure of the gray molecules was raised, forcing more molecules into the liquid.

Going back to the fluid in a container with gas over it, this time the gas will be defined as a mixture of oxygen and nitrogen. If the gas mixture and liquid have reached equilibrium and more oxygen molecules are injected into the gas, or the oxygen partial pressure is effectively increased, more oxygen molecules will be forced into the liquid solution (Figure 1-13). This action will not affect the pressure exerted on the gas by the nitrogen nor

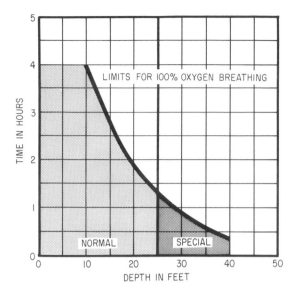

Figure 1-14. This graph, based on the work of E. Lanphier, shows exposure limits for breathing 100% oxygen. The lightly shaded area is considered normal operating region and is generally safe. The dark area is used only on special opera-

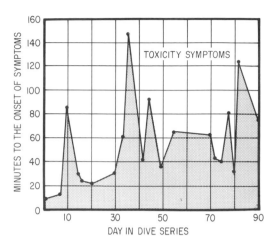

tions. Figure 1-15. Divers have wide variations in tolerance of elevated oxygen pressure. This graph shows the results when the same diver is exposed to pure oxygen at a pressure equivalent to 70 feet in water over a period of 90 days. The time to onset of toxicity symptoms (lip twitching) on each 20-dive series is represented by a point.

will it affect the amount of nitrogen in the liquid solution.

Gas Tolerance

The body, which is not unlike the container, has a tolerance limit to each gas that is inspired. Even oxygen can become toxic under conditions of highly elevated partial pressures. There is a specific partial pressure range within which the system can sustain life. Below that range, sufficient oxygen is not present to support the life functions. Above that range, high oxygen levels cause chemical toxicity, tissue damage and death. Complicating the problem of a critical pressure range is the wide variance in individual tolerance. The point of toxicity is a function of this tolerance which varies from day to day and with the length of exposure, with the ergometric level during exposure and the environment within which the exposure takes place.

The rule of thumb generally applied in diving is that the diver should never inspire a breathing mixture which contains a partial pressure of oxygen in excess of 2 atmospheres absolute (2 AtA). Breathing pure oxygen at the surface is equivalent to oxygen at a partial pressure of 1 atmosphere absolute; breathing pure oxygen at a depth of 33 feet is equivalent to breathing a gas mixture of oxygen at 2 AtA. This level can be exceeded for short periods of time with no apparent damage, but statistically the risk of oxygen toxicity is extremely high.

Dr. Edward H. Lanphier has developed an oxygen limit curve for man (Figure 1-14) which illustrates the safe diving limit using pure oxygen as a breathing medium. Earlier experiments by K. W. Donald in 1947 testing the variation in oxygen tolerance of a single human under repeated conditions show wide differences. The experiments, conducted over a 90-day period with the same diver, consisted of exposing the subject to an oxygen pressure of 3.12 AtA until the onset of oxygen toxicity symptoms. In twenty separate runs, the length of exposure required varied from as little as 9 minutes to as long as 150 minutes. Figure 1-15 is a plot of this time on each of these exposures versus time to the appearance of symptoms. The wide variance in the sensitivity of a single man to Oxygen High Pressure (OHP) is obvious. Even under the "safe" limit of 2 AtA, there is some evidence that long exposures to OHP cause significant pulmonary distress. Other experiments indicate that under an OHP of 0.5 to 1.0 AtA the vital capacity of the lungs is reduced.

So long as a diver is breathing air and circulation is sufficient to provide a constant flow, the problem of Low Oxygen Partial Pressure (LOP) is not normally encountered. However, breathing exotic gas mixtures where the constituents are synthetically or mechanically mixed, the possibility exists that the diver may receive an oxygen level insufficient to support the body. Some deviation from the normal surface concentration of 0.20 AtA can be tolerated, but not much. Breathing a mixture of 15% oxygen, or 0.15 AtA, causes drowsiness and loss of mental acuity in many persons. A drop in oxygen partial pressure to 0.10 AtA, or 10% concentration, causes the oxygen tension in the blood to decrease to a point where deterioration of life functions begins. In normal short-period diving operations, a lower limit of 0.18 AtA has been established. Regardless of depth, therefore, the range of oxygen partial pressures that are considered generally acceptable for breathing falls between 0.18 and 1.8 AtA. For equipment design, these figures are usually set at 0.20 and 1.4 AtA respectively. Pressure and breathing mixtures are tested in simulators like the one in Figure 1-16.

Carbon dioxide, normally present in the lungs at a partial pressure of 0.04 AtA, can cause dramatic reactions in the body if the level rises to 0.10 AtA. Initially, it acts upon control functions in the brain and increases the respiration rate. In some individuals, this first reaction is not marked and if the high partial pressure is being caused by a condition that is cumulative, even higher carbon dioxide levels can be reached without the awareness of the victim. At a level of 0.15 AtA, the body reacts by spasms and unconsciousness. The primary cause of high carbon dioxide levels is lack of lung ventilation. The mechanical cause can be either faulty equipment, diving procedures, or the density of the breathing medium.

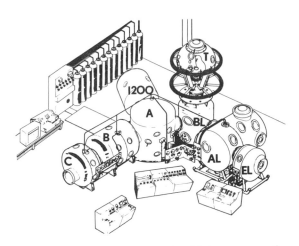

Figure 1-16. The most complex high pressure diving simulator is owned by the French diving company, COMEX. Physiological effects of pressure and breathing mixtures can be tested down to 3,600 feet.

Gas Density

As the depth of diving increases, carbon dioxide retention, or improper exchange, becomes a formidable problem. Part of the trouble is the increasing density of the gas itself. At 692 feet, for instance, a breathing medium would be twenty-five times its surface density and proportionally more energy is required to move it in and out of the lungs. Aside from the energy requirement, gas movement becomes more turbulent and ventilation with its concurrent carbon dioxide elimination is then considerably less effective. Experiments conducted with liquid breathing tend to focus on this aspect of the problem.

Water Breathing

The possibility of man being a total reentrant mammal has fascinated physiologists for many years. Oxygen is present in normal sea water and if the differential partial pressure of the oxygen dissolved in the water is greater than the oxygen tension in the blood, a gas exchange across the alveoli membrane takes place. Although filling the lungs with water is commonly called "drowning," lungs can actually operate when filled with liquid. One of the early experimenters in this area was Dr.

Johannes A. Klystra. Using an oxygen-saturated isotonic salt solution similar to blood plasma, Dr. Klystra immersed mice under pressure. "Breathing" this oxygen solution, the mice did manage to survive for a limited period—some as long as 18 hours.

The experiments were then conducted on other animals more closely approximating human size. In all cases, the animals did survive for a period of time breathing water and many recovered completely from the exposure. In the final analysis, the animals that did not survive apparently died for two reasons—exhaustion from moving the dense breathing medium in and out of the lungs and insufficient carbon dioxide transfer in water, which is 6,000 times less than the transfer in air.

The process of actually flushing the lungs with a saline solution to remove pathological secretions is not uncommon today and is called lung lavage. This procedure was extended to water breathing using a volunteer professional diver, Francis Falejczk. Air in one lung was replaced with an oxygenated saline solution at near body temperature. Five hundred milliliters of the solution were breathed in and out of the lungs for a short period of time. According to Dr. Klystra, the solution was then removed and the subject showed no ill effects. Experiments in water breathing are continuing and may eventually lead to a breathing system that is not affected by Boyle's Law.

Inert Gases Underwater

Water breathing is a technology of the future; in 1971, the diving medium consists of oxygen and an "inert" carrier. The purpose of the carrier is to deliver the oxygen to the lungs, carry away excess carbon dioxide and provide the molecular volume necessary to maintain the required gas pressure. Although gases such as nitrogen, helium, argon and krypton are theoretically inert at a pressure of one atmosphere, their effect on the body when present in the system at elevated partial pressures is anything but inert. The most common inert gas in diving is nitrogen. When breathed in atmospheric air, its partial pressure is approximately 0.80 AtA. As diving depth increases, nitrogen pressure also increases so that at a depth of 100 feet the partial

pressure has reached 3.2 AtA. At this point, although many investigators say earlier, the nitrogen begins to act upon the body as a narcotic.

The degree of narcosis is much more pronounced in inexperienced divers, but even in the experienced diver some reaction is felt. As the nitrogen partial pressure increases with further depth, so does the narcotic reaction. At depths over 200 feet, very few divers have the performance ability, either motor or mental, that they possess at the surface. It has only been in recent years that hypotheses regarding the mechanics of inert gas narcosis have come close to substantiation.

In the early 1800s, narcosis was considered a primary effect of pressure. As late as 1960, the mechanics of narcosis were still unknown. Several theories, including pointing the accusing finger at carbon dioxide, were under consideration. Today, with a better, but by no means complete understanding, there is a general acceptance of the fact that action of the inert gases on man is chemically similar to anesthesia, with one exception—as the partial pressure of the inert gas is reduced, recovery from the symptoms of anesthesia are immediate.

The narcotic effects of high partial pressure of nitrogen have been well known for years, but only since 1968 have the effects of "inert" helium been publicized. The reason is simple and obvious. Helium does not seem to have any detrimental effects until depths near 1,000 feet are reached. There have been very few excursions to this depth, even in chambers. It has been observed on several occasions that at lesser depths, helium as the carrier gas causes uncontrollable tremors in the diver. This, with some other symptoms, was classified as High Pressure Nervous Syndrome (HPNS). In chamber tests conducted by COMEX of France, depths equivalent to nearly 1,200 feet were reached. On one dive to a simulated depth of 1,180 feet performed in June, 1968, the symptoms of HPNS were clearly delineated. A portion of the log follows:

Physalie II   *June 11, 1968*

Divers: R. W. Brauer, R. Veyrunes
Time of dive including descent: 120 minutes

Time spent below 300 meters (descent and ascent): 77 minutes

Total time of ascent: 114 hours, 31 minutes

Breathing mixture on the sea bottom:
O_2 - 1.8%, N_2 - 4%, He - 94.2%

Ascent under PiO_2 of 0.5 AtA from 152 meters

Air from 18.2 meters: Pure O_2, from 9 to 0 meters

Speed of descent: 6 to 2 meters/minute

Average speed of ascent: 5.2 centimeters/minutes

There were several similarities between this dive and previous ones, however:

1. The trembling fits, the fasciculations, the difficulties in movement coordination were more apparent if not more precocious in R.W.B. (confirmed by electromyogram).
2. The sleepiness, more apparent with R.V. occurred around 340 meters This phenomenon had never been so visible before reaching 335 meters.
3. This sleepiness showed the characteristics of disappearing during action, especially during mental tests, and reappeared as soon as the subject's attention slackened, thus reminding Paul Valery's definition: "To sleep is to be uninterested . . . ."
4. The E.E.G. (at last carried out on both subjects) revealed much more than the clinical aspect. Alterations appeared quite soon (220 meters). The disorders increased with the depth, persisted at the beginning of the ascent and only disappeared round about 230 meters.
5. From the psycho-emotional viewpoint, no anxiety or over-excitement, but around 300 meters a progressive aboulia. Beyond 340 meters the impression that it would be dangerous to carry on but no effort to ask by sign (or even less in writing) to stop.

It should be noted that during almost the whole descent (until 320 meters) both divers breathed pure heliox without nitrogen through a mask.

According to Brauer, the absence of nitrogen could have encouraged neuro-muscular disorders, but breathing a mixture with 12% nitrogen brought no change. This does not mean that nitrogen is useless, for as Brauer notes, ". . . in our experiments with animals, such mixtures in no way reduce the trembling fits, but have notable effects on the convulsive phenomenology!!" The log continues:

During ascent and as usual around 175 meters, R.W.B. showed a slight vestibular syndrome, well reduced with Cyclizine and Atropine, but which lasted quite long.

Under the threat of a few short "pains," we had to slightly modify the ascent profile and lengthen it towards the end.

R.V. and R.W.B., when out, felt a mild and very acceptable tiredness.

They knew that they would need less than a week to recover a good condition and that is why they accepted to start again ten days later.

Decompression

Until now, the time element has played no part in our discussion of gas equilibrium. It is, however, one of the more important parameters in applying the physics of gas equilibrium to the physiology of the body under pressure. Gas diffuses through the alveolar membrane rapidly. Once it has been transferred into the blood, it is carried throughout the body tissue where it is absorbed. The quantity of gas taken up by the tissue is a function of the pressure difference, or gradient, the type of tissue, and time.

Each type of tissue has a characteristic absorption rate according to its structure. Tissue material that absorbs gas quickly will likewise release the gas rapidly. When lowering the pressure to which a diver is exposed, as during ascent, extreme care must be exercised to assure that the gas is released from the tissue at such a rate that bubbles are not formed. The pressure gradient that precipitates bubble formation depends upon the physical characteristics of the gas and the tissue, and therefore diving practice and techniques must be adjusted to conform to the peculiarity of the breathing gas constituents. The shaded area in Figure 1-17 shows the safe "no decompression" limit.

Except in extremely rapid ascent, or explosive decompression, the oxygen content of the breathing mixture does not enter into consideration in this problem. Oxygen, within its partial pressure limitations, is carried to body tissue where it is chemically depleted by the metabolic process. The inert or carrier gas, however, enters and leaves the system in the same form, and is therefore the critical element. The process of relieving the partial pressure at a rate that does not permit the carrier gas to bubble in the tissue is called *decompression.*

Decompression techniques date back to the middle 1800s when use of underwater caissons in setting bridge footings became popular. Decompression had actually been used before, but it was at this time that the medical effects of pressure on the human body came under the scrutiny of such men as Paul Bert. Bert made major contributions to the study that are still accepted as the basis for current research. Continuation of active research for over 150 years finds us, in 1971, still with many questions unanswered, even with regard to the basic physiology and pathology involved. One discrete solution applicable to all phases of diving has never been found, but great advancements in knowledge and techniques have been made in the last 10 years, such as the deck decompression complex, shown in Figure 1-18, which can decompress eight divers simultaneously.

The core of the decompression problem is the solubility of the different inert gases in body tissue. Nitrogen and helium are absorbed by the same tissue at different rates—and are released at different rates.

Tissue Saturation

Body tissue is the main storage reservoir for excess gas absorbed by the body due to elevated

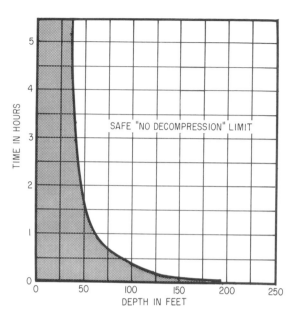

Figure 1-17. The depth to which a diver can descend and the time he can spend at that depth without going through the process of decompression is specifically limited. The shaded area of the graph shows the safe "no decompression" limit. Time is measured from the moment the diver leaves the surface until he starts his ascent.

partial pressure. The blood, which accounts for less than 7% of the total body weight, acts as the conduit. According to Capt. Robert Workman, M.C. USN, of the National Naval Medical Center, Bethesda, Maryland, if the body were composed of a uniform tissue, equally and effectively reached by the circulatory system, 1/26th of the excess nitrogen would be taken up by the tissue in each circulation cycle. In twenty cycles, or 10 minutes, the body would have absorbed over 50% of the maximum nitrogen. The time necessary for nitrogen equilibration at any given pressure would therefore be a simple calculation. Determination of body tissue nitrogen tension would be equally simple. Unfortunately, body tissue is not a homogeneous substance equally supplied by the circulatory system; rather, it is a heterogenous system that contains areas reached by only the sparsest blood flow. These tissues are extremely slow to absorb or give up gas.

Figure 1-18. Facilities for decompression from deep saturation dives are massive and expensive. This deck decompression complex (DDC) for the U.S. Navy's MARK I Deep Diving System can handle eight divers simultaneously. (Photo: FMC.)

Regardless of specific time, all tissues, fast or slow, become saturated or desaturated with inert gas according to the same mathematical principles. Figure 1-19 shows a plot of this curve for any body tissue at a constant pressure. The graph shows that a tissue requiring 6 hours to equilibrate to the inert gas partial pressure would reach 50% of its saturation level in less than 1½ hours. Similarly, upon removing the excess partial pressure, the tissue would give up 50% of the absorbed inert gas in less than one-fourth of the saturation time. Considering the slowest tissue, the body as a whole follows the same curves.

When nitrogen is the inert gas, saturation time is 24 hours. If helium is the inert gas, the time scale is about 12 hours. Figure 1-19 does not, of course, represent a decompression table because it assumes instantaneous application and removal of pressure. In practice this would be disastrous. It does, however, provide a valuable tool in computing the absorbed inert gas at any given depth.

Body tissue can develop positive differential pressure between the absorbed inert gas and the blood or adjacent tissue without the formation of bubbles. In this condition, it is said to be supersaturated for that depth. How great this internal pressure differential may be is one of the controversial aspects of decompression.

Decompression Schedules

The earliest decompression schedules were developed by J. S. Haldane for the Royal Navy in 1908. Based upon the supersaturation hypothesis, they called for the diver to ascend in steps, always keeping the ratio of the tissue partial pressure to ambient pressure no greater than 2:1.

Although the tables worked well for moderate exposures to depths of 165 feet, they proved inadequate for more extensive or deeper exposures. Other investigators in the United States and Great Britain, using different supersaturation methodology, developed subsequent tables for air decompression that are still in use today. Most of the later work was performed empirically, using statistical occurrence of decompression sickness symptoms as the control.

Repetitive Diving Tables

The general use of Self-Contained Underwater Breathing Apparatus in the military brought about the need for new decompression tables and procedures. In standard diving gear, the normal procedure was for the diver to work underwater for a specific time and then go through the decompression process. With Self-Contained Underwater

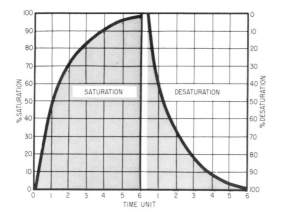

Figure 1-19. This graph shows saturation and desaturation curves for body tissue. Time unit is a function of the tissue considered and the inert gas in the breathing mixture.

Breathing Apparatus, the diver's time was limited by the amount of air he could carry in his equipment. For any major underwater task, a series of SCUBA dives was required instead of one continuous effort. The existing tables could not adequately meet these needs without the diver spending excessive time in decompression procedures. To solve this problem, the Repetitive Diving Tables were developed. With this method, the residual nitrogen in the diver's system is calculated. The time he spends on the surface between dives during which his tissues are releasing nitrogen is also calculated. On the next and all subsequent dives, the residual nitrogen is accurately taken into consideration, resulting in a considerable savings of decompression time.

The use of helium instead of nitrogen as the carrier gas in diving equipment has also caused revamping of decompression tables and procedures. Because helium enters and leaves the tissues at a higher rate, the calculations used for nitrogen decompression were found to be invalid. In the helium decompression technique, tissues that give up gas rapidly become the controlling parameter. Comparison of decompression times required for safe return to the surface after dives using helium or nitrogen as the inert gas in the breathing mixture shows little variation.

Saturation Diving

The need for prolonged exposure at deep depths has created a new diving technology called *saturation diving.* Referring back to Figure 1-19, it is obvious that for any given depth, a level of dissolved gas is reached in a specific time, after which there is only negligible increase regardless of the length of continued exposure. Once the tissue has reached saturation, decompression time is independent of additional time spent at that pressure. In saturation diving projects carried out, such as Sealab I and II, continuous ascent decompression was employed. On the Sealab II project, conducted at a depth of 204 feet, only one minor case of bends resulted, with a decompression rate of 10 minutes per foot. On subsequent Sealab III trials, conducted at a pressure equivalent to 450 feet in the ocean, the steady decompression rate was increased to 15 minutes per foot.

New Decompression Techniques

Through the use of new decompression techniques, decompression of divers is gradually becoming safer and requiring less time. Pure oxygen used during the shallow depth decompression stages is now used on all decompression schedules from deep depths. A switch in breathing mediums increases the rate of desaturation of the slower tissues by lowering the partial pressure of the inert gas in the blood and increasing the rate of transfer. Several diving techniques have been developed that actually switch breathing mediums during the working portion of the dive. Dr. Albert Buhlmann and Hannes Keller used multiple mixed gas mixtures throughout their deep diving experiments that eventually permitted Keller to pass the 1,000-foot depth mark.

There is apparently an increasing tolerance to high partial pressure of gas in the human system. An analysis of the occurrence of bends in compressed air workers has shown that regular exposure to compression and decompression reduces the incidence of bends. This increasing adaptation to pressurization has been attributed to the destruction of gas bubble nuclei in the system. It has been hypothesized that these gas nuclei actually precipitate the formation of bubbles in the same manner as particulate matter in the air causes the formation of water droplets.

A statistically higher occurrence of bends in older compressed air workers, especially those over 50 years of age, was also observed. This has been attributed to a lack of physical tone in the older workers. Several years ago, it was universally assumed that obesity also encouraged the bends. Recent work, however, indicates that fatty tissue is associated with decompression sickness only indirectly—the relationship of excessive fat tissue to poor physical condition appears to be the direct link.

Compression

Compared to the process of decompression, the actual compression of a diver has resulted in only minor problems. With the advent of deep diving, however, some aspects of compression have assumed vital importance. For using air decompression tables, the rate of descent to any depth had been fixed and calculated at 75 feet per minute. Very few divers were able to exceed this rate because free air spaces such as the middle ear and the sinus cavities required at least that much time to equalize with the ambient pressure. On deeper dives, however, it became obvious that the rate of descent was a critical factor in the diver's performance on the bottom.

One of the great advantages of slow descent is the minimization of HPNS. By using slow descent rates, particularly beyond 600 feet, symptoms attenuate and occur at deeper depths than with rapid descent.

Although the rate of diffusion of inert gas into the tissues, regardless of the tension, never reaches the "bubble" formation stage during descent, diffusion into tissue such as cartilage can cause specific discomfort. On rapid decompression experiments to relatively deep depths, divers complain of pain in the joints—appropriately named "no joint juice." The effect is transient and disappears after several hours without apparent tissue damage.

Figure 1-20. A U.S. Navy scientific diver passes through an opening in the ice at the Arctic Pool facility of the Naval Electronics Laboratory Center for an underwater inspection. Heat loss in the 0°C water limits immersion time to a few minutes. (Photo: U.S. Navy, NELC.)

Immersion

In the past few years, the question has been raised in several quarters about the major contributory causes of "drowning." Although the classical definition, "to die by suffocating in water or other liquid," is still accepted, some investigators such as Dr. W. R. Keatinge of Great Britain, feel that in cold water, immersion may not only play a role as a contributory cause, but in many cases may be the actual cause of death. The diver in Figure 1-20 will be limited to only a few minutes of immersion time because of the extremely cold water.

In diving, retention of body heat is a major problem. Massive heat loss is known to reduce motor capability and mental acuity, and to have a detrimental effect upon the transfer of gas in the pulmonary system. The advent of mixed gas diving moved the problem from one of chronic discomfort to an acute question of survival.

The body is a sensitive thermodynamic machine that operates with an internal *core temperature* of approximately 36°C. If that temperature drops more than a few degrees, the effects are externally obvious. At a temperature of 34°C amnesia is common. At 32°C unconsciousness usually results.

Heat Conduction

Humans lose body heat in proportion to their metabolic rates and the amount of insulating subcutaneous fat present. In air, the body, with little or no external insulation, can remain comfortable and thermally stable for long periods at an ambient temperature around 22.2°C. Water, however, conducts heat away from the body twenty-five times faster than air. Heat conduction from the immersed body is so rapid that the heat loss rate is actually controlled by how fast the blood can deliver heat from the core to the skin. For comfort and thermal equilibrium while immersed, the temperature gradient between the water and the body core temperature cannot exceed 0.5°C. Considering that the temperature differential between body core and ambient water is seldom less than 15°C, the extent of the problem is obvious.

Many animals have specialized organs for the production of body heat or *thermogenesis*. Man has no apparent thermal-generating organs and relies on involuntary muscular contraction to increase heat production. During the Sealab II experiments, physiologists reported that after short exposure to thermal stress, a rise in core temperature and a concomitant drop in temperature of the extremities could be observed. The temperature change was attributed to retarded blood flow at the surface and was considered a primary cause of "paradoxical shivering." This phenomenon manifests itself as uncontrollable shivering with a simultaneous feeling of warmth.

Voluntary muscular action such as exercise during cold water exposure does not aid in the prevention or postponement of hypothermia—it does in fact, hasten it. Muscular action produces additional heat, but the resulting increase of blood flow to the surface of the skin actually enhances heat transfer. In tests conducted by Keatinge, conclusive results were obtained showing that with a group of immersed subjects having similar physiological structures, exercise hastened heat loss and incidence of hypothermia.

For thermal stress exposures of short duration during air diving, the solution is simply to insulate the skin surface from the ambient water to reduce heat conduction from skin to water. The degree of insulation required depends upon the temperature gradient and the length of exposure. When the diver is breathing mixed gas that consists of oxygen and a light molecular weight gas such as helium, the protection process becomes more complicated.

When breathing air, a small but measurable quantity of heat is lost through the respiratory system, mainly in warming inspired air. Compared to heat loss due to water conduction, it accounts for less than 15% and except in extreme environmental conditions is not considered a major contributor to the onset of hypothermia. When a helium/oxygen mixture is the environment or the breathing medium, the situation changes drastically. Helium conducts heat at a rate six times that of air and subsequently, the lungs become an important center for heat transfer.

On deep-diving operations where the breathing medium consists of a mixture containing 90% or more helium, insulation of the body to maintain thermal balance is no longer sufficient. External heat must be provided. Even in undersea habitats with helium environments, the ambient temperature must be maintained at a temperature of about 30°C to keep off-duty divers comfortable.

There is some adaptation to thermal stress in divers who are repeatedly exposed to cold water over a long period of time. Most saturation diving experiments report that the divers show an increasing resistance to hypothermia and a simultaneous rise in basal metabolism after periodic stressing. Similar observations were made about the effects of cold on the *ama* divers of Korea and Japan. It was found that these divers, who use no protective clothing except a white cotton coverall, were able to withstand repeated immersion in 10°C water.

Investigation of their basal metabolism showed that during the winter months when water temperatures were low, the basal rate increased by 25%. During the summer months with warmer water, their rate was the same as for non-divers.

Speaking in the Water

Communication plays a vital role in man's underwater performance. We have discussed some of the problems involved in underwater hearing, concluding that sound perception deteriorates because of a combination of physical and physiological mismatches. The problems associated with speaking underwater are somewhat more complicated and fit into three general categories—those associated with underwater life-support equipment, those associated with the physics of the water and breathing gases, and finally, the response of the vocal system to these phenomena.

On all of the evaluations of living in the sea projects, there has been concurrence on the difficulties in adapting to the undersea environment in one area—communication. Recent tests conducted by the Communication Sciences Laboratory of the University of Florida point out the magnitude of the problem at deeper depths. Figure 1-21 shows the percentage of intelligibility on spoken communication at depths down to 850 feet. According to this data, intelligibility across a wide section of listeners and talkers degrades at the rate of 50% each 200 feet, compared to normal level in the atmosphere. Speaking with the mouth occluded with water produces no intelligible results. Using diving equipment where the voice-producing tract is in contact only with the breathing medium does not solve the problem except at shallow and very moderate depths.

Speech Distortion

With air, the increased damping of the pressure causes major changes in the sound-producing process that result in distortions. Characteristic changes are an increase of pitch and loss of some of the delicate phonemes produced by mouth and tongue action. As pressure increases, the laryngeal

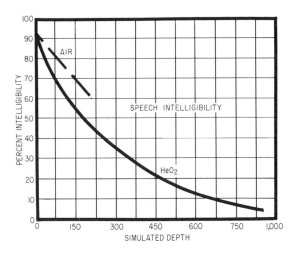

Figure 1-21. Correction of speech distortions caused by pressure and gas density requires sophisticated electronics and a special lexicon. This graph shows the results of intelligibility tests run with a "helium speech unscrambler" at the U.S. Navy Experimental Diving Unit. (Photo: University of Florida, CSL.)

walls become more transparent to the sound, and speech output is further degraded by the loss of energy through the tissue. At the deepest practical limits for air diving, distortions reach a point where the use of a special lexicon is necessary for reliable communication.

Even at shallow depths, underwater speech with a light inert gas has major defects which worsen in proportion to the increasing pressure. The problem is complex because there are so many parameters involved in distortion. The sounds that we customarily hear and interpret are produced in the voice system by several components—resonant chambers, vibrating cords and air movement through articulated cavities. When the gas medium is altered to a point where the density and speed of sound in that medium differ from the norm, part of the sounds produced are radically changed—others are not. The result is a confused sound output with complex waveforms.

In the case of helium the basic, purely mechanical output of the vibrating cords remains unchanged. The sounds produced in the resonant cavities dependent upon the speed of sound in the medium are shifted upward. At the same time, distortions resulting from the pressure are integrated. The speech output, generally called helium or Donald Duck speech because of its high-pitched nasal quality, is not understandable at pressures beyond those equivalent to a depth of 450 feet.

Considerable effort has gone into reconstructing helium speech, but to date the improvement by electronic processing has not been completely successful. A certain amount of adaptation has been reported by aquanauts under elevated helium pressures. Improvements in both speaking and listening can be accomplished and are a function of experience and length of exposure. A combination of experience, a special lexicon and electronic conversion will be necessary to bring intelligibility levels up to normal.

2. A Brief History of Diving

Man has been intrigued with the idea of diving since he first got his feet wet. That his ancestors started in the ocean may account for this urge to see and be part of the underwater world. Sir Allister Hardy, a man of letters and the sea, even hypothesized further, suggesting that the near-shore environment may have been instrumental in changing man into an erect biped. He theorizes that man may have been forced to forage in the shallow water for existence and that during these food gathering expeditions man's posture became erect and he learned to swim. Hardy points to some possible evolutionary mutations that are still present today, such as the manner in which body hair is arranged in patterns suggesting fluid flow, the subcutaneous, insulating layer of fat common to humans and reentrants such as seals. Perhaps it is all true and somewhere in the long chain is a genetic call to the sea.

Historically, the apparent call to the sea has been motivated by more immediate reasons, i.e., curiosity, food, profit or power. Some of the causes are, of course, interrelated, but seldom has anyone attempted to dive beneath the sea without at least one of them.

The history of diving dates back to the earliest remnants of identifiable archaic cultures, and a diving history buff can find archaic references in almost any era that will titilate his cultural sensitivity. How many of these anecdotes are actually true is open to conjecture, but the fact that the references exist gives some credence to ancient diving. Ancient "divers" were undoubtedly swimmers, probably with the capability of traveling only very short distances underwater. At most, they were equipped with breathing tubes comparable to our present day snorkels. Even so equipped, the possibility of their swimming long distances "underwater" without detection is remote. There are spotty references to diving in very early manuscripts, but only in a comparative, never definitive manner. The oldest reference to the function of divers is in connection with warfare; Homer refers to divers in the *Illiad,* but no description of their activities has been found. According to Thucydides, divers were used in the Trojan wars, but again, these were primarily swimmers operating under the veil of darkness. Diving was taking place at this time—free diving not unlike that which the aboriginal sponge divers of the South Pacific are still engaged in today.

Diving as we know it today has very little basis in antiquity. To overcome the physiological limitations imposed by evolution, it is necessary to provide mechanical substitutes in lieu of eonic adaptation. The span of man's recorded history in geologic time is miniscule—a span so short that if man is to regain entrance to the sea, it will necessarily be by his invention and knowledge.

From the standpoint of diving technology, there were two awakenings in antiquity that started man along the path to eventual conquest of the sea. The first was the invention of the snorkel or breathing tube; the second was the diving bell. These assume great importance not on the basis of their technology, but on their lateral thinking. In analysis, they are the seeds for the two diving methods in use today. With the snorkel, regardless of its primitiveness, we find man with the knowledge that he can sustain life underwater if he maintains contact with his life-giving atmosphere. In the concept of the diving bell, we find our struggling Homo sapien with the awareness that his life-sustaining atmosphere is in reality a portable medium. At the risk of oversimplification, these two ideas are the basis for the existence of man underwater.

As the twentieth century fibrilates its way into the closing one-third of its existence, we tend to pride ourselves on the advancements and sophistication of our millenium of technology. In the world of industry, particularly the aerospace field, we have accustomed ourselves to the appearance of the future in the form of artists' sketches. The pace with which some of these sketches are turned into reality has almost led us to accept as feasible an idea that exists as a sketch. This is certainly sophisticated, but based upon their technology, our predecessors were certainly as sophisticated.

Early Diving Apparatus

Unburdened by a knowledge of the physics of gases and water and not restricted by an awareness of the physiologic limitations of the human body, the early inventors put forth a wide variety of artists' sketches depicting apparatus which did indeed solve the problem. They certainly cannot be criticized for inventiveness—only for their test procedures. None of the schemes could have worked. The strange aspect of this inventiveness was its proliferation over several centuries. Evidently no one actually tried the apparatus and what we have are artists' sketches to the second order.

Even Leonardo da Vinci fell into the trap and made sketches for underwater breathing equipment. For some reason, Leonardo never made the invention public even though he had been commissioned for the design. And just as well; a diver could never have provided the lung power necessary to suck air through the apparatus. His designs for self-contained breathing apparatus were equally impractical, but no less imaginative.

Diving Bells

The design and development of diving bells faired a little better than that of diving apparatus, perhaps because of their inherent simplicity and the fact that the initial concept of an inverted cup open at the bottom did not fight the pressure, but just let it happen. Diving bells make their first recorded appearance during the fourth century B.C. Aristotle refers to letting down an inverted kettle filled with air to sponge divers so that they could increase their stay on the bottom. About the same time, a legend sprang up about Alexander the Great using a glass diving bell to inspect Macedonian boom fortifications at the Island of Tyre. Unfortunately, there are no surviving drawings of either of these bells. Those that have been found in manuscripts were drawn hundreds of years later and their accuracy is certainly questionable.

Diving bell technology seemed to lay dormant for nearly 2,000 years. Not until the middle of the sixteenth century do we see their appearance again in written history. What use of the technique had been made during the intervening period, if any, is obscure. During the sixteenth century, diving bells began to appear all over Europe. The major improvements seem to have been greater size and more dependable construction. Successful use of these crude devices in salvaging lost cargoes is widely recorded.

Figure 2-1. The "walking diving bell" was invented and tested by Franz Kessler around 1615. The bell has a wooden framework covered with leather. Weights used to sink it to the bottom were released by the diver to surface. Figure 2-2 shows a two-man diving bell, designed and built by

COMEX of France in the late 1960s, which represents the apogee of diving bell technology started by Dr. Edmund Halley in the eighteenth century. This bell is capable of taking two men to 1,000 feet at either atmospheric or ambient pressure. (Photo: COMEX.)

It was about 1615 that diving apparatus was conceived. A German, Franz Kessler, designed, built and successfully tested a one-man diving bell (Figure 2-1). The mechanism consisted of leather barrel built over a wooden framework and was open at the bottom. The barrel was short enough for the diver's feet to extend below, and was equipped with glass view ports for visibility. We find our first mobile diver on the bottom with a self-contained air supply. The bell was held down in the water by a weight that could be released for surfacing—and what a ride that must have been to the surface!

The idea of replenishing the air supply in diving bells was first made a near practicability by Dr. Edmund Halley about 1720. Halley's scheme was to lower casks of fresh air to the bell. While providing an additional breathing supply for the divers, it also permitted the bell to reach greater depths than ever before. It was only a short tech-

nological step from lowering casks of air to pumping air down to the bell from the surface—a step that awaited the design of suitable pumps. By 1800 this step was completed and the history of diving as we know it today begins. Figure 2-2 shows a modern two-man diving bell.

At best, diving in the early 1800s was a "do it at your own risk" situation. Knowledge of the pressure effects on the body was for all practical purposes non-existent, and some of the hypotheses offered to explain the frailty of the human body underwater make humorous reading today. But, for practitioners of this mysterious and dangerous art, nineteenth-century diving techniques were anything but humorous. The major diving function at this time was salvaging the remains of sunken vessels in shallow water or in harbors. Diving served the additional function of recovering and removing other submerged navigation hazards.

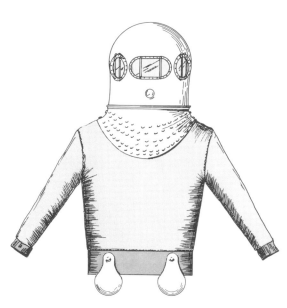

Figure 2-3. This open diving dress, designed and built by Augustus Siebe about 1820, was the direct ancestor of our deep sea dress. Air was pumped from the surface to the helmet and exhausted at the bottom of the tunic. The same design is used today in remote areas of the world.

Figure 2-4 shows modern diving dress built by Advanced Diving Equipment Company. This dress uses a fiberglass helmet and buoyancy control that allows the diver to work in any position, even upside down, without danger of flooding. (Photo: Author.)

Diving System Development

Considering that diving had been taking place for nearly 2,000 years, the advances made between 1800 and 1850 are remarkable. Forerunners of diving equipment still in use today were invented during that period.

Although there had been several artists' sketches of earlier helmet diving systems, it was Augustus Siebe who first built and tested the recognizable ancestor of the deep-sea diving helmet of the twentieth century. He designed the first workable version of the open diving dress (Figure 2-3). The system consisted of a brass helmet, with several viewing ports, attached to a tunic. Air was pumped into the helmet and exhausted through the bottom of the tunic. The system worked well at moderate depths except for one major hazard—in any position but vertical, the system would flood. Siebe's open dress was used successfully by divers with a good sense of equilibrium for more than twenty years. An American, L. Norcross, modified Siebe's diving system by adding an ex-

haust vent in the helmet, eliminating the waist vents and the system's inherent attitude sensitivity.

By 1837, Siebe had modified his diving suit to include all Norcross' improvements and went a step further. He included the first controllable exhaust valve. This permitted the diver to control the rate of exhaust and therefore to change his buoyancy by controlling the amount of air in the diving dress. Figure 2-4 shows a diver wearing a modern suit with this feature. As more was learned about man underwater, the system was continually improved, but the present day deep tethered diving is still based upon the same simple principles.

A diver, free from any attachment to the atmosphere and carrying his own life-support system, is depicted in an Assyrian bas-relief dating back to the first millenium B.C. sucking air from an animal bladder. The underwater swimmer seems completely at home. This quaint idea was suggested as a diving system for centuries to come. It was not, however, until 1825 that practical SCUBA was actually designed. An Englishman,

William James, proposed a self-contained air supply, worn around the waist in the form of a donut-shaped container. The air supply was connected to a leather helmet by a single hose with a flow control valve; expended air was to be exhausted through another valve at the crown of the helmet. Although the James suit (Figure 2-5) never became a working reality, a similar design by an American, Charles Condert, was actually used for shallow diving. The design was, however, somewhat advanced for the manufacturing capability of the times and Condert lost his life on a shallow dive because of equipment failure.

The diving apparatus described by Jules Verne in *Twenty Thousand Leagues Under the Sea,* was the first actual genesis of modern SCUBA. When Verne wrote his fantastic odyssey in 1870, it was received with much the same attitude as interplanetary travel novels were received in the middle of the twentieth century—an interesting diversion but hardly couched in fact. In retrospect, however, we find that Jules Verne's references to diving techniques were postulated from at least an acquaintance with the current state of the diving art. His description of the apparatus is particularly interesting. Verne describes the standard apparatus of the day as follows:

> "... but under these conditions the man is not at liberty; he is attached to the pump which sends him air through an India-rubber pipe, and if we were obliged to be thus held to the 'Nautilus,' we could not go far."

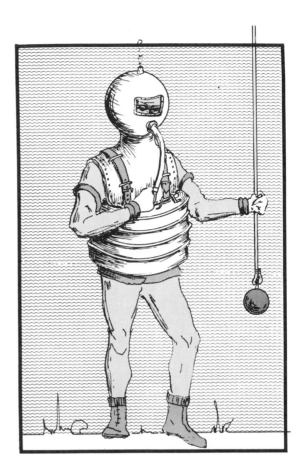

Figure 2-5 (upper right). Self-contained underwater breathing apparatus (SCUBA) was conceived as far back as 1825. This design by William James never became a reality, but a similar design was tested and used by an American, Charles Condert.

Figure 2-6 (lower right) shows the U.S. Navy's MARK VIII underwater breathing apparatus, worn by two Sealab aquanauts on training exercises at Anacapa Island, California. This device is like the Rouquayrol system described by Jules Verne. It can be used as SCUBA or from a central gas supply. (Photo: U.S. Navy, by Hasha.)

Then he proceeds to describe the SCUBA apparatus available to the crew.

> "It is to use the Rouquayrol apparatus . . . which I have brought to perfection for my own use, and which will allow you to risk yourself under these new physiological conditions, without any organ whatever suffering. It consists of a reservoir of thick iron plates, in which I store air under a pressure of fifty atmospheres. This reservoir is fixed on the back by means of braces, like a soldier's knapsack. Its upper part forms a box in which air is kept by means of a bellows, and therefore cannot escape unless at its normal tension. In the Rouquayrol apparatus such as we use, two India-rubber pipes leave the box and join a sort of tent which holds the nose and mouth; one is to introduce fresh air, the other to let out the foul, and the tongue closes one or the other, according to the wants of the respirator."

Although Verne wrote this passage about 5 years after the actual invention of the first demand-type SCUBA, the Rouquayrol-Denayrouze apparatus, he evidently did not really believe that man could be exposed to the pressure of depth with merely a pressure equalizing breathing apparatus, because he continues:

> "But I, encountering great pressure at the bottom of the sea, was obliged to shut my head, like that of a diver, in a ball of copper; and it is to this ball of copper that the two pipes, the inspirator and the expirator, open."

This passage is an excellent clue to the state of consciousness of the educated man at that time. He was willing to accept the fact that man could bring his life-support medium with him to depth, but in all events, he must be protected from the tremendous pressure involved. It is certainly not fair to castigate man for this ignorance because, today in 1971, if the average man were to be canvassed on the street and asked the purpose of using a diving helmet and suit, the concensus would undoubtedly be that they protect the body from the terrible pressure. Vern's later description of dressing the diver out was probably built upon firsthand observation plus some extrapolation, so it is worthy of note in addition to being interesting.

> "At the Captain's call two of the ships crew came to help us to dress in these heavy, impervious clothes, made of India rubber without seam, and constructed expressly to resist considerable pressure. One would have thought it a suit of armour, both supple and resisting. This formed the trousers and waistcoat. The trousers were finished off with thick boots, weighted with heavy, leaden soles. The texture of the waistcoat was held together by bands of copper, which crossed the chest, protecting it from the great pressure of the water, and leaving the lungs free to act; the sleeves ended in gloves, which in no way restrained the movement of the hands. There was a vast difference between these consummate apparatuses and the old cork breast-plates, jackets, and other contrivances in vogue during the eighteenth century."

The Rouquayrol diving system was undoubtedly the first demand breathing system to gain any degree of popularity. The "bellows" which Verne described were actually the diaphragm portion of a regulator that supplied air to the diver through a crude oral—nasal mask. The apparatus was incorporated into many standard tethered diving systems as an emergency or "bail-out" air supply. The small metal cylinder was worn on the back and served as a reservoir. The air supply hose from the surface could be disconnected and the diver could operate autonomously until his air supply was exhausted. It is similar to the modern design in Figure 2-6. The limited quantity of air that could be carried in the relatively low pressure cylinder made it impractical for general usage.

Aside from military applications—and there was considerable resistance to the development of diving as an adjunct to naval operations—little was accomplished in diving technology in the early 1900s. Firms like Siebe-Gorman Company of England continued development, but only on a limited scale. The development of the submarine as an effective weapon, however, did focus the needs for underwater breathing apparatus.

Interest in submarine escape schemes built around simple SCUBA equipment led to the development of the first dependable "rebreather" systems. This equipment was based upon the principle of removing the carbon dioxide waste from the diver's exhalation and regenerating the air by adding oxygen. By 1940, we find standard deep sea diving gear having progressed to a point of standard operational and generally safe equipment. With the work of Haldane and with an increasing knowledge of underwater physiology, divers could now safely descend to reasonable depths and expect to return to the surface with no physical penalty. SCUBA, on the other hand, was in its infancy, and there was no really dependable equipment available for either civilian or military operations. The technology and the need was there, but a catalyst was required.

The catalyst appeared in June, 1943, when Jacques Cousteau and Emile Gagnon first dove the Aqualung in the Mediterranean. Aside from some minor mechanical problems, the equipment was dependable, relatively inexpensive and, above all, simple to operate. Cousteau and Gagnan continued their testing and development in secret until the end of World War II. Then began what might be called an underwater population explosion. The Aqualung was, of course, not a mechanically conceptual invention. It brought demand SCUBA into the context of twentieth-century technology. Equally important, the prolific and imaginative writing of Cousteau brought diving into focus with such startling beauty that men in the sea became a concept to which everyone could relate.

Advances Since 1950

Considered from the vantage point of 1970, the history of man underwater divides naturally at the year 1950. For some 2,500 years, man had been attempting to penetrate the air/sea interface with the statistical probability of returning to the surface, as Jules Verne aptly put it, ". . . without any organ whatever suffering." At that time, Siebe's diving system had gone through continual development so that it reached a point of maximum performance for the configuration. Mixed gas equipment and basic helium/oxygen techniques had been tried and the Aqualung had gone through several rapid generations of technical improvement. Underwater physiology, although relatively gross, had produced a level of knowledge which was to open the way for critical study of the human body underwater. It was time for the diving community to broaden—for diving to cease being an adventurer's art and become a broad technical effort to really make man a sea creature.

The decade from 1950 through 1960 was the era of diving proliferation. Scientific diving became a new underwater entity, sport diving became an international pastime, military diving developed man as an undersea weapon system, and commercial diving took on the look of a profit-making adventure as resource exploitation of the continental shelf became a reality. The decade also marked an important change in the attitude toward underwater equipment development. Prior to 1950, the controlling force in diving technology resided in two specific entities—the navies of Great Britain and the United States. The world-wide professional diving community, outside of the military, numbered no more than eight hundred.

Research and development is an expensive undertaking, one that could hardly be supported by an industry with a total employment of eight hundred and a questionable financial prognosis. Consequently, underwater equipment development was federally funded, with the proviso that only activities shown to support "operational needs" could receive financial aid. This rather ambiguous phrase was, and still is, construed in many military circles to mean "after we experience some sort of catastrophe that points up a sore need, we will begin development of a system to prevent it happening again."

By the end of the 1950s, there was a dramatic shift in diving research and development—profit-oriented industries were pressing the professional diving business for more capability and simultaneously threatening it with promises and attempts to substitute unmanned robots for divers; scientific divers were pressing for more and better equipment and instrumentation; the ranks of sport

Figure 2-7. The last two decades represent an underwater population explosion. By 1960, sport divers had grown from a handful of addicted free divers to 50,000 addicted SCUBA divers. In 1975 sport divers will number over 4,000,000. This group has just returned from a 200-foot dive in the Bahamas. (Photo: International Underwater Explorers Society, by Dave Woodward.)

divers had grown from a handful of addicted free divers to a group of 50,000 addicted SCUBA divers (Figure 2-7). The operational needs of the diving community could no longer be satisfied by military research. Diving was at the threshold of becoming a business.

The last decade has been one of technical progress for the diving business. Old depth barriers, arbitrarily set on the basis of incomplete information, were broken. Saturation diving and living in the sea changed from fantasy to reality and commercial diving won the race with mechanical robots for the continental shelf. Prototype vehicles like the one in Figure 2-8 are being tested for use in the 1970s. Considered in the light of the past 2,500 years, the events of the 1960s are enough to stun the imagination.

Breaking the 1,000-foot Barrier

In the twentieth century, we are accustomed to having feats of great accomplishment performed by men supported by pyramiding technical teams. There are few areas in which a single man can conceive, muster support, and carry out a program that represents a breakthrough in technology. It did happen in diving on December 3, 1962. This was the day man penetrated beyond the 1,000-foot depth barrier in the open sea. The place was the Pacific Ocean, just off Catalina Island. The man was Hannes Keller (Figure 2-9), a young Swiss mathematician. There was such bad publicity and so much tragedy associated with the dive that the feat itself was obscured. But in 1971 it stands as the benchmark for deep diving.

Figure 2-8 (left). Nearly 2,500 years after Alexander the Great was said to have used a glass diving bell, engineers at the Maikai range on the Island of Hawaii test a plexiglass submersible. This bell is a prototype of a vehicle that will travel from Hawaii to California during the 1970s. (Photo: Maikai Range.) In Figure 2-9 (above), Hannes Keller (arrow) prepares to break the 1,000-foot barrier. The dive was a technical success but adverse publicity and a double tragedy obscured the feat for many years.

Figure 2-10. The diving bell Atlantis transported Keller and Small to 1,015 feet. An external gas manifold developed a massive leak, causing last minute changes in dive plans and the death of Peter Small. (Photo: Author.)

Some years ago at the University of Zurich, a physician/physiologist, Dr. Albert Bühlmann, developed a theory of using multiple mixtures of inert gases for descent to great depths with minimum decompression. Keller, who had been conducting his own diving experiments in Lake Lucerne, Switzerland, felt that Bühlmann's theory together with his mathematics and knowledge of diving technology would provide a combination that could push man through the imaginary depth boundary of 600 feet where he had been stymied for many years. Bühlmann agreed and together they developed a sequential mixed gas breathing medium scheme and decompression tables to

accompany it. After several successful demonstrations to 700 feet, they were prepared to prove their system in the open ocean to a depth of 1,000 feet, but there were no takers.

Confident that he had the solution, Keller began organizing the experiment with his own funds and credit. His first financial breakthrough was the sale of the publication rights of the story to several newspapers and the *Saturday Evening Post.* By summer of 1962, he had the loan of the *Eureka,* a vessel from Shell Oil Company, a borrowed diving bell, donated gas, borrowed underwater television cameras, his own diving equipment and a promise of financial aid from the United States Office of Naval Research. In late November, everything was in readiness and the test dives began.

Because the actual diving techniques and gas mixtures that Keller planned to use were secret, a thin veil of mystery began to cloak the entire project. And, since the news rights had been sold, access to the site was restricted. Local news media were anything but gracious. Mechanical problems plagued the tests and the other member of the proposed diving team, Peter Small, had several mild "hits" of the bends on shallower test dives. Pressed by financial and equipment loan commitments, the project began to take on an aura of desperate urgency. If these external pressures could have been relieved, it is doubtful that the expedient actions—actions that later resulted in tragedy—would have been taken.

The dive plan was simple; Keller and Small would descend in the bell (Figure 2-10) to a depth of 1,000 feet. Here they would exit with Self-Contained Underwater Breathing Apparatus, swim for approximately 5 minutes, plant the American and Swiss flags, reenter the bell and start for the surface. If all went well, decompression using Bühlmann's method, would be short.

There was no voice communication between the bell and the surface. The only communication link was to be two underwater television systems loaned by Hydro Products of San Diego. One camera was set up outside one of the bell's portholes to monitor the divers and the internal pressure gages during descent and ascent. The second camera and light system was set up under the bell to monitor the divers after they had exited at 1,000 feet. Two stand-by safety divers were selected to assist in rigging the bell and to perform hookup of surface air when the bell reached shallow depths on its ascent. One of the divers, Dick Andersen, proved to be a competent safety man and in fact was instrumental in bringing Keller back safely to the surface. The other, Christopher Whittaker, proved to be less experienced and paid for it with his life.

On the morning of December 3, with the *Eureka* using her position-holding system, the bell started its descent. Keller and Small were wearing constant volume diving suits to protect them from the 40°F water at the bottom. Viewed on the TV screen, the descent was normal. When the divers reached 1,000 feet, they found that a leak had caused one of the external gas manifolds to empty and they were forced to make a change in the operational plans—Keller would exit alone, Small would remain inside. Keller exited through the lower hatch in the bell and swam to the bottom.

From this point on, exactly what happened is confused, but review of the videotapes after the dive showed Keller swimming down with the two flags, becoming entangled with the American flag, and then swimming up to the bell and entering the hatch. Both Small and Keller were visible on the camera that was monitoring the inside of the bell. When questioned later, Keller said that because of low gas pressure, he had exhausted his oxygen supply in the SCUBA equipment and had returned to the bell "weak and dizzy." Both divers lost consciousness and the bell was hauled up to a depth of 180 feet. Keller temporarily regained consciousness at this time and was trying to render assistance to Small.

At this depth, the surface dive station reported that they could not maintain pressure within the bell. Since the divers could not survive if the pressure were lowered any further, the standby divers were called to correct the problem. It had been planned for the divers to descend to the bell to make connections at a shallower depth. On first inspection, the divers could not find the cause for the lack of pressure integrity. Whittaker surfaced and reported. When he arrived at the surface, his condition indicated that he should not make another dive and was advised not to. The surface was still unable to pressurize the bell so Whittaker returned to Andersen.

Andersen found that one of the divers' swim fins had lodged in the hatch seal and was preventing it from closing. After the correction was made, Andersen signaled for Whittaker to go to the surface and report that the bell could be pressurized. Christopher Whittaker was seen for the last time over the closed-circuit TV as he started for the surface.

Andersen could not understand the delay and proceeded to the surface to find the reason– it was then that the surface crew knew Whittaker was missing. An immediate search and recovery effort was started, but to no avail. Whittaker's body was never recovered.

When the chamber was brought aboard the *Eureka,* Keller and Small were unconscious again. Both divers regained consciousness shortly; Keller was to completely recover within a few hours and Small to succumb to lung damage caused by embolism.

Stripped of all the bad publicity, secrecy and improper support, the experiment was still a success in reaching a scientific goal that proved man could penetrate the ocean to the 1,000-foot mark and safely return. For some time after the dive, various explanations were offered for Keller's survival, the most popular being that he was a physiological freak and not subject to normal reactions in the presence of elevated partial pressures. This, of course, has proven to be incorrect and other experimenters have repeatedly reached greater depths in simulated chamber dives and will reach that depth in the ocean in 1971—but Keller was there first.

Underwater Living Experiments

Each segment of technological advancement has a proponent who emerges as undisputed motivator. Without question Capt. George F. Bond, M.C. USN, was the prime mover and researcher in extending man's underwater capability from "bounce" visits to the sea floor to semi-permanent residency. Beginning with laboratory experiments on animals, Bond and his co-workers proved that mammals could undergo complete saturation with inert gas and then, after suitable decompression, return to the surface environment with no apparent physiological penalty. He had placed volunteer divers under saturation for periods of up to six days and brought them back with no bends symptoms. The laboratory experiments continued and, throughout the world, plans were laid for man to attempt habitation of the ocean floor. True, many of these plans were modest in scope, but living in the open sea is quite different than living in simulated conditions of the laboratory test chamber.

The contingencies and external emotional stress generated by separation from the normal source of a life-giving oxygen are incomprehensible to anyone who has not been in that situation. It is not unlike a space traveler orbiting the earth—to bail out of the vehicle and to deaccelerate into the atmosphere would be terminal. To the aquanaut living on the ocean floor, rising to the surface in an emergency procedure is no less dangerous even though the physics involved are different.

When Edwin A. Link started the Man-in-the-Sea project in September, 1962, a new era of underwater technology was initiated. It is true that in comparison to more recent ambitious projects such as Sealab III, Tektite I, and UWL-Helgoland, Man-in-the-Sea I is like a 10-second flight compared to our modern jets. It is nevertheless important because it showed for the first time that man actually could live under the sea. He has to bring his own food, he has to bring his own air, and his own heat, but he can survive.

Since Link's Man-in-the-Sea program in 1962, there have been approximately 35 laboratory experiments conducted throughout the world to prove that man could live in the sea. A detailed report of these would take volumes. For example, the Sealab II project final report was 425 pages long. The equipment and techniques used in these projects are described in the technology section of this book. The following summary includes most of the major underwater laboratory endeavors in the past decade.

Man-in-the-Sea

The well known explorer and inventor of the Link Trainer, Edwin A. Link, organized and directed the first underwater living experiment, Man-in-the-Sea, in September, 1962. Compared to later projects, it was a modest undertaking, but all the objectives were accomplished. A young diver, Robert Stenuit, was lowered to a depth of 193 feet in the Mediterranean off Villefranche. His "habitat" for the 24-hour dive was a small two-man submersible decompression chamber.

During the experiment, Stenuit breathed a 3% oxygen, 97% helium mixture supplied from a surface support vessel. The dive itself was a complete success. Stenuit was saturated and brought through decompression with no apparent ill effects. The first step had been taken in changing Homo sapien to Homo aquaticus!

Conshelf I (Precontinent I), France

Shortly after the Man-in-the-Sea experiment by Edwin Link, Jacques-Yves Cousteau started the Precontinent I project. Although the depth of water at the first site was only 10 meters, or 33 feet, two divers, Alber Falco and Claude Westy, remained underwater for a full week for the first time. Link's experiments showed that man could live for 4 days at nearly 200 feet; Cousteau's experiment extended the time at a lesser depth. The station was equipped with two one-man chambers so that underwater decompression could be carried out in emergencies. The divers who had been breathing air during the experiment were decompressed for a period of only 2 hours before proceeding to the surface.

Conshelf II (Precontinent II), France

Encouraged by the success of Conshelf I, Cousteau's group undertook a much more ambitious program, Conshelf II, in June, 1963. Although it was neither the deepest nor longest underwater living experiment, it stands as the most imaginative and creative underwater enterprise ever conducted.

Unlike the other experiments, Conshelf II was actually an underwater colony consisting of a main habitat embodying the most comfortable and pleasing accommodations that have ever, or probably will ever, be placed upon the ocean floor. Shaped like a sea creature, it was nicknamed "Starfish House" because its basic configuration was four separate chambers radiating from one central section. Sleeping eight comfortably, the habitat contained all living, laboratory and maintenance capabilities. In addition to the main habitat, there was an underwater hanger for the two-man diving saucer that permitted complete maintenance or repair without surfacing. During the underwater experiment, the saucer was used for dives through the deep scattering layer down to depths of 900 feet.

One goal of the program was for two aquanauts to dive to a smaller deep habitat, or Deep House, and occupy it for 7 days. During this period it was planned that they would make excursion dives down to depths of over 300 feet. The plan was that the Deep House would be established at a depth of 90 feet.

The site selected for the experiment was in the Red Sea off the coast of Sudan. The Starfish House was established in a depth of 33 feet at a location known as Roman Reef. The Deep House was set up in 90 feet of water further down the slope, but as with most underwater work, not without incident. The actual ledge planned to hold the Deep House proved too small for the base. After modification, the house was lowered into position. During the installation, with several men inside, the house broke loose and started tumbling down the steep incline. One of the men escaped from the hatch, the other was trapped inside. Fortunately, the Deep House struck another ledge at 140 feet and stopped. If it had continued down, there would certainly have been loss of life. From that point on, the entire mission ran as smoothly as a fantasy film, with some minor inconveniences. The divers, who spent seven days at 90 feet did make forays down to 300 feet. In fact, their deepest dive from the Deep House, without decompression, was to a depth of 363 feet. During their work stay they did suffer from the environment in a way that few other aquanauts have or probably ever will.

The project had been scheduled for the early part of the year but, as in any undersea venture, one frustrating delay after another caused the schedule to be moved back so that it was summer when they were actually occupying the habitat. In winter, the thermocline, or interface of cold and hot water, would have been above 90 feet and the Deep House surrounded by cool water. In summer the inside of the habitat was not unlike a sauna bath with 100% humidity. Most other underwater living experiments suffer the opposite problem.

Another unique feature of this experiment was that the aquanauts were not selected on their physical merits or diving experience. The team was selected from the support vessel *Calypso's* regular crew on the basis of their ability to contribute to the community and their adaptability to the task's to be accomplished. A marine biologist, Dr. Vaussuere, supervised the mission's scientific experiments. The team was also unique in that for the last several days of submersion it contained the first woman aquanaut, Mme Cousteau.

Sealab I, USA

After almost superhuman efforts, indescribable political machinations and the aid of everything but an explosion at the Pentagon, Capt. George F. Bond, USN, was able to initiate the Sealab I project. As an outgrowth of the genesis project, Sealab I was almost entirely a low-budget operation. The habitat, built from the scrap and pontoons of a minesweeper at the U.S. Mine Defense Laboratory in Panama City, Florida, proved to be satisfactory. Sealab I was conducted at the Argus tower near Bermuda in 190 feet of water.

With only minimal logistic support and using a cadre of dedicated divers, the Sealab I project was successfully carried out under Bond's supervision, permitting four aquanauts to live in the habitat for 11 days. Life support and energy were provided from the surface ship. During the operation a breathing mixture of 4% oxygen, 17% nitrogen and 79% helium was used as a standard breathing mixture.

Sealab II, USA

Encouraged by the success of Sealab I, a full-scale scientific expedition with adequate funds was now initiated by the U.S. Navy. The habitat (Figure 2-11) was placed in 205 feet of water in Scripps Canyon off the coast of California. Three teams of aquanauts spent approximately 15 days

Figure 2-11. Aquanauts make final adjustments to the Sealab II habitat as it begins its descent to the bottom off the California coast. (Photo: U.S. Navy.)

each in the sea-floor habitat. Energy and logistic support were provided from a ship. Breathing gas, a 4% oxygen, 9% nitrogen, and 87% helium premix, was self-contained in storage bottles mounted externally on the laboratory. Water temperature in the canyon was 48°F (8.89°C). Visibility varied from 5 to 30 feet. During the experiment, the excursion dives were made from the habitat to depths of 300 feet without decompression and without incident. A prodigious quantity of physiological, oceanographic and ocean engineering data was gathered during the project and Sealab II remains to date the largest deep underwater laboratory project brought to a successful completion.

The objectives of the program outlined by the U.S. Navy's Deep Submergence Project Office were as follows:

1. Determination of man's general ability to do useful work at a depth of 200 feet in a realistic ocean environment under saturated diving conditions.
2. Determination of physiological changes in man as a result of extended diving.
3. Measurement of performance to determine work degradation or improvement, as compared to surface-diver operations, and as a function of dive time.
4. Determination of stressful conditions and their effects on the group interactions of the aquanauts.

The degree of success of each planned work function varied considerably because of the limited availability of diving time. In general, it can be stated that man's ability to do work under saturated diving conditions at a depth of 205 feet was more than amply proved. Actually, the diversity of tests, while providing a considerable overall project enhancement, tended to limit the measurement of man's capability in each specific field of interest. The determination of man's general ability to do useful work at 205 feet in a realistic ocean environment under saturated diving conditions implies a wide range of work involvement. Such a diversity of work was planned and undertaken in the Sealab Program. The various work assignments were as follows:

All three teams:
    Touch-sensitivity tests
    Arithmetical tests
    Aquasonic intelligibility tests
    Light and form visibility tests
    Stationary target array identification
    Contrast and resolution studies
    Visual acuity tests
    Auditory range studies
    Sound localization studies
    Water clarity meter correlation
    Strength testing
    Triangle assembly tests
    Two-hand coordination tests
    Group assembly tests
    Hookah evaluation
    Time-lapse photography
    Heated wet-suit evaluation
    Plankton studies
    Underwater weather station assembly,
       calibration and inspection
    Fish rake census
    Sediment coring
    Fish cage placement and stocking
Teams 1 and 2 only:
    Sand movement studies
    Portable EEG, EKG recorder evaluation
Teams 2 and 3 only:
    Excursion diving
Team 1 only:
    Fish migration studies
    Fish gas bladder studies
Team 2 only:
    Porpoise evaluation
Team 3 only:
    Foam-in-salvage evaluation
    Salvage tools and equipment evaluation
    Geological airlift evaluation
    Geological corer evaluation

The investigation of physiological changes in man as a result of the extended saturation diving of Sealab II resulted in a conviction that:

1. No significant short-time physiological changes occur which result in deterioration of the aquanauts' physical condition.
2. Acclimation to stressful temperature changes of Sealab habitat living (85°F or 29.4°C) and ocean-floor swimming (47°F to 54°F or 8°C to 12°C) occurs, with the result that aquanauts can perform better and for longer periods of time in the surrounding ocean waters.

The work performance measurement to determine work degradation or improvement as compared to surface-diver operations and as a function of dive duration resulted in the following:

1. The results of life-and-pull strength tests showed a decrease in exertable strength between dry land and Sealab.
2. The individual triangle assembly (manual dexterity) tests revealed a 37% decrease in performance between dry land and Sealab.
3. The two-hand coordination test showed a 17% decrement in performance in Sealab.
4. The three-dimensional group assembly task took twice as long in Sealab as on dry land.
5. No decrement was found between predive and Sealab mental arithmetic tests.

Determination of stressful conditions and their effects on the aquanauts' group interactions leave little doubt that the Sealab environment was stressful. The conditions which contributed to this stressful environment included the following:

1. The water was cold and visibility poor.
2. The work schedule, requiring long hours of preparation, was very often interrupted, delayed or revised.
3. Communications were difficult because of helium-speech distortion.
4. Sleep was often disrupted for most men by the long hours of work, high humidity, poor air circulation, and physical complaints of headaches, minor ear infections, and skin rashes.

However, in spite of the stressful aspects of the situation, the motivation and morale of the men were extremely high. Group cohesiveness, measured by the divers' choices of their own team members, increased for each of the three teams from pre- to post-experiment measures. Despite a general feeling of accomplishment, many men were dissatisfied with the amount of work they personally accomplished.

Man-in-the-Sea II, USA

Immediately after the success of the initial Man-in-the-Sea project, Edwin Link began planning a more ambitious program to bring man to deeper depths. Unlike the first dive, this attempt was to include a complete habitat on the bottom and two divers. Robert Stenuit, who had been the diver on the 200-foot, 24-hour Man-in-the-Sea I experiment, was selected as team leader. Jon Lindberg was selected as the second team member, bringing a wealth of deep-diving experience to the attempt.

The small habitat, an inflatable rubber sausage-shaped container, was set on the ocean floor at Great Stirrup Cay in the Bahama Islands in 432 feet of water. The SPID (Submersible Portable Inflatable Dwelling) was connected to the support vessel by an umbilical that provided constant monitoring of the habitat environment and the aquanauts over closed-circuit TV. Breathing mixture for the experiment was 3.6% oxygen, 5.6% nitrogen, and 90.8% helium. The submersible decompression chamber used on the first project served as an elevator to deliver the divers to the SPID and to provide a means of escape in the event of habitat failure. When the divers arrived at the habitat, they found that water had flooded the air purifying system and without it the SPID was uninhabitable because of carbon dioxide buildup. Repairs were made on site by Lindberg and the team stayed two days and two nights at the bottom. Completed in July, 1964, the project stood for many years as the deepest open-sea saturation diving programs.

Kitjesch, Russia

Using a converted railway tank car as a pressure hull, the Russians made their first public attempt at an underwater laboratory project.

Placed on the bottom off the Crimean Coast at a depth of 45 feet during the summer of 1965, the habitat was designed to sustain four aquanauts. Life-support gas and energy were provided from the land. As with their space activities, the Russians left a considerable amount of information out of their report so that operational details and the planned program are unknown.

Conshelf III (Precontinent III), USA

Using a spherical habitat resembling artists' concepts of what ocean-bottom cities would look like, Cousteau again conducted a successful underwater living project. Six aquanauts spent three weeks in the laboratory located on the bottom in 190 feet of water, and during the experiment made untethered excursions down to 300 feet. The habitat obtained gas and power from the surface for a portion of the experiment and the autonomous capability of the system was successfully tested. Gas mixtures with oxygen varying from 1.9 to 2.3%, and 1% nitrogen with the remainder being helium, were used during the experiment.

Permon II, Czechoslovakia

This project was plagued with problems from its inception. Habitat handling problems, poor engineering and an equally poor choice of timing and location added to the miseries. The first trial, carried out off the Yugoslav Coast at Split, was a comedy of errors. Heavy seas tossed the magnetite-filled ballast barrels all over the bottom, the restraining hawsers broke, and the ballast barrels disappeared. A second attempt to put the Permon II on the bottom was no more successful. This time, when the habitat broke loose, it floated into a busy shipping lane and was almost lost by collision. The experiment was aborted at this point and the Czechoslovaks took their habitat and went looking for a better operating location.

Ikhtiandr 66, Russia

In August, 1966, the Russians again placed a laboratory in the Black Sea off the Crimean Coast.

This lab was established in 33 feet of water and air from a compressor station on land was used as life support. The single chamber laboratory supported a team of two aquanauts for one week. No further information on this experiment has been released.

Sadko 1, Russia

Again in the Black Sea, the Russians placed a laboratory at a depth of 120 feet. The lab, actually a 5-foot diameter sphere, was equipped with only minimal creature comforts and was evidently designed for short stays. Although the laboratory is purported to have been used for 1 month at a depth of 25 meters, the only information available is that two aquanauts used the habitat for 6 hours. Logistic support was reported to be from the land on the 6-hour dive and from a ship on the 1-month dive.

Caribe I, Cuba

Somewhere near the end of 1966, the Cubans placed a habitat off the north coast of their island at a depth of approximately 45 feet. Information about the experiment is scarce, but reports indicate that the habitat was approximately 9 feet long, 5 feet in diameter, and was fitted out in a rather primitive manner. Caribe I was supposedly a preliminary experiment for a larger undertaking, Caribe II, scheduled for 1967. According to the reports, Caribe I supported two aquanauts for 3 days with logistics from the surface ship. No further information has been released on the Caribe series experiments.

Permon III, Czechoslovakia

Still unconvinced that the middle of winter was not the best time to perform underwater experiments, the Czechoslovaks tried their luck at underwater living once again in March, 1967. This time the Permon laboratory was placed in 30 feet of water under ice in a stone quarry. Low temperatures and heavy surface winds complicated the logistic support but the laboratory was finally situated on the bottom. Gas was supplied from an integrated tank storage system. Two aquanauts lived for 4 days on the bottom.

Medusa I, Poland

By the summer of 1967, it seemed to be a question of national pride to have conducted an underwater living experiment. Medusa I, placed on the bottom of Lake Clodono in July, 1967, at a depth of 24 meters, was evidently a precursor of the more ambitious experiments, Medusa II and Medusa III. Using logistic support from the land, two aquanauts lived for 3 days in the small habitat with surface-supplied life support and gas of 37% oxygen and 63% nitrogen.

Hebros I, Bulgaria

The same month as the Medusa experiment was conducted, the Bulgars placed a slightly larger habitat in the Bay of Warna and conducted their first underwater living experiment. There is very little information about the operation with the exception that the habitat was shaped like the Russian Kitjesch and was approximately 16 feet long by 6 feet in diameter. It is reported that the habitat was designed for two aquanauts but no further details on logistic support or results of the experiment have been made public.

Oktopus, Russia

In the same month, July, 1967, the Russians conducted an underwater living experiment. Using an inflatable hemispherical habitat with air life support from a land station, three aquanauts are reported to have lived on the bottom, at 33 feet, for several weeks.

Ikhtiandr 67, Russia

Using the same habitat of the 1966 experiment, the Russians conducted an underwater living experiment with five aquanauts during August, 1967. There is no information available about this experiment except that the habitat was a three-chamber configuration and that five aquanauts were employed in the experiment.

Sadko 2, Russia

Using a double sphere habitat (each sphere the same as that used in Sadko 1), the Russians established an underwater living experiment in the Black Sea sometime during the summer of 1967. The habitat was placed on the bottom in 75 feet of water and excursions were made from the habitat down to depths of 190 feet. Two aquanauts lived in the station for 6 days. No report is available about the breathing gas used, with the exception that it was self-contained in an externally-mounted tank storage bank.

Kockelbockel, Netherlands

Using a small habitat, approximately 15 feet high and 7 feet in diameter, the Dutch conducted a short-term underwater living experiment with two teams of four divers each. The habitat was autonomous and used air for the diving depth of approximately 50 feet.

UWL-Adelaide, Australia

With the exception of a newspaper account in September, 1967, no data has been released on this underwater experiment by the Australians. According to the report, the habitat included sleeping and living quarters and an acoustic laboratory for the measurement of ambient and fish-generated noises. Apparently the logistic support was from a surface barge and the experiment was conducted over a period of several weeks.

Romania LS I, Rumania

Again, only a minimum amount of information is available about this underwater experiment. It is purported to have been conducted in 1968 in Bicaz Lake. The habitat, intended to accommodate a crew of two aquanauts, was approximately 18 feet long and 7 feet in diameter.

Karnola, Czechoslovakia

The third underwater living experiment by the Czechoslovaks has had little information released about the activities. The habitat was said to have been designed to maintain a crew of five at a depth of approximately 50 feet.

Tschernomor, Russia

Once again in the Black Sea at the Bay of Belendshik, the Russians conducted their most ambitious underwater living experiment during July, 1968. At a depth of 45 feet, the habitat provided living quarters for thirty aquanauts in five teams of six each. Creature comforts in the habitat were minimal. One of the interesting techniques used during the experiment was the change-over of diver teams at the surface. The laboratory was brought from the bottom to the surface after each team completed its task. Decompression of about 10 hours was required. An ambitious program of physical oceanography and marine biology was carried out during the stay on the bottom.

Medusa II, Poland

The second Polish underwater living experiment was conducted in a small habitat designed to support three aquanauts for 14 days. Logistics were supplied from a surface ship for the habitat situated in about 95 feet of water in the Baltic Sea.

Robinsub I, Italy

The Italians' first underwater living experiment could hardly be called a well-organized scientific endeavor. In fact, it was more analgous to an underwater promotional sporting event. Using an inflatable plastic tent, one diver lived at a depth of 33 feet for approximately 48 hours.

Hebros II, Bulgaria

Using a considerably larger and more sophisticated habitat, the Bulgars conducted their second underwater living experiment in 1968. The project is reported to have supported two aquanauts for 10 days at a depth of about 95 feet. Logistic support was from a surface craft but no details are available as to the breathing gas or the scientific results of the activity.

Sprut, Russia

The exact date of the experiment has not been released and few technical details are available with the exception of a brief description of the habitat. The habitat was an inflatable fabric underwater balloon similar in design to the Oktopus experiment, but larger and considerably more sophisticated. The lab was designed to support two or three aquanauts for periods of 14 days. During the 1968 experiment, the site selected was in 33 feet of water and it is probable that air was used as a breathing gas.

BAH I, Federal Republic of Germany

The experiment was to support two aquanauts for 11 days at a depth of 33 feet, with air and energy provided through an umbilical from the surface vessel. Size, design, equipment and working depth were anything but spectacular, but a start was made. Unfortunately, the project was overshadowed by Dr. Horst Hartmann's death during a reconnaissance dive.

Ikhtiandr 68, Russia

In September, 1968, the Russians conducted another UWL experiment off the Crimean Coast. The habitat was reported to be a "glass chamber" with a water displacement of approximately 15 cubic meters. At a depth of 40 feet, the lab supported several crews for a total of 8 days. Air and energy were supplied from the shore.

Malter I, German Democratic Republic

Although the first German Democratic Republic underwater laboratory was scheduled to be Berlin I, a group of four divers completed a small habitat called the Malter I and conducted their experiment in 25 feet of water at the Malter Dam. The Malter I represents very little in technical advances, but it demonstrates that a group of enthusiastic divers with minimum financial support can actually place a habitat in a reasonable depth of water and economically conduct an underwater living experiment. Without a scientific program associated with this type of activity, however, there is very little purpose in the effort.

Tektite I, USA

In February, 1969, the longest sustained underwater living experiment was conducted off St. John Island in the Virgin Islands, using a team of four scientist-aquanauts. The ideal underwater conditions in the Virgin Islands made the 60 day experiment an unqualified success from the standpoint of both underwater living and acquisition of valuable marine data.

The habitat (Figure 2-12) consisted of two 12-foot diameter chambers, each 18 feet high, connected by a 4-foot diameter funnel. One chamber was used as a laboratory and living quarters, while the other served for observation and an equipment locker. The complex was supported from a surface barge and a breathing mixture of 8% oxygen and 92% nitrogen was used in both the habitat and the SCUBA diving equipment.

In addition to the scientific task pursued by the four aquanauts, an extensive physiological and behavioral observation program was conducted to determine the effects of long-term confinement, the physiological effects of increased partial pressures of nitrogen and the psychological effects of isolation with a small crew in situations requiring maximum cooperation. The overall cost of the 60 day experiment was estimated at $2.5 to $3 million. Because of the extensive physiological and psychological monitoring, an abnormally large support crew was required—eleven observers monitored the aquanauts at all times. Figure 2-13 shows the monitoring station.

Sealab III, USA

The most ambitious and certainly the most expensive underwater living project ever attempted by man was the Sealab III experiment (Figure 2-14). Cost figures ranging from $10 million to $36 million (depending upon the method of accounting) have been reported for this project which was scheduled to support five teams of twelve aquanauts at a depth of 600 feet.

The underwater habitat (the refurbished Sea-lab II habitat) was supplied with gas and power from a specially redesigned surface vessel. Back-up logistics were also provided from a base at San Clemente Island in the event that bad weather forced the surface support ship to leave the site. The complex consisted of a complete deck decompression system, personnel transfer capsules (PTC) for transportation of divers, and the sea-floor habitat. The habitat had provisions for an autonomous operation and complete facilities for supporting twelve to fifteen men simultaneously.

The project was indefinitely postponed because of the death of aquanaut Berry L. Cannon. Although differing in minor aspects, testimony of the principals and participants at the

Figure 2-12 (above). Cutaway drawing shows the Tektite habitat used by the U.S. Department of Interiror aquanauts for their 60-day underwater living experiment. The four-man scientific diving team brought saturation diving into the realm of the civilian scientific diving community. (Photo: Ocean Industry, Feb. 1969.) Figure 2.13 (right). Behavioral monitor's post. Four TV monitors display one habitat room. Two monitors are for external v/w cameras. (Photo: U.S. Navy.) Figure 2-14 (lower right). Artist's concept shows the U.S. Navy's Sealab III experiment. The habitat is shown on the bottom with umbilicals attached to its mother, the USS Elk River. A personnel transfer capsule (PTC) is shown descending to transfer divers from the surface compression complex to the habitat. (Photo: U.S. Navy.)

subsequent board of inquiry was in basic agreement on the circumstances surrounding Cannon's death. The following is a synopsis of testimony given under oath during the inquiry by Sealab III participants.

The Sealab III habitat was rigged and lowered to the ocean floor in 610 feet of water off San Clemente Island. At the pre-lowering inspection, several gas leaks were observed when the habitat was submerged and pressurized to 12 psi. The leaks were considered minor, and according to the on-scene commander, J. M. Tomsky, not sufficient to abort the lowering.

Leaks were seen from the submersible *Deepstar-4000* after the habitat was on the bottom. Comdr. Scott Carpenter, the observer on the *Deepstar-4000,* reported seeing seven streams of gas apparently coming from cable penetrators. Gas loss at this time had increased to 3,000 standard cubic feet (scf) per hour. Sealab Command felt that these leaks could be stopped from inside the habitat and decided upon sending the first diving team to the habitat to open it up and make the necessary on-site repairs.

A four-man diving team consisting of Chief Warrant Officer Robert A. Barth, team leader; Berry L. Cannon; Richard M. Blackburn, and John F. Reaves was placed in the deck decompression chamber (DDC) and started through the compression cycle at a rate of 240 feet per hour. No ill effects were observed during the compression except for "no joint juice" symptoms at the 180 to 200 feet depth mark.

After reaching transfer pressure, the diving team locked into the PTC and started on their descent. In the helium atmosphere, under high pressure, all the divers complained of extreme cold and uncontrollable trembling. No provision was made in the PTC to control ambient temperature. When the PTC reached 500 feet, Barth and Cannon were dressed out in the MARK IX underwater breathing apparatus—a tethered life-support system operating on an umbilical from the PTC—and standard wet suits. They swam to the habitat and began the "unbottoning" procedure. Before they could complete the task, Cannon returned to the PTC. According to Blackburn, the stand-by diver, Cannon was dazed, incoherent and weak. Not long afterward, Barth returned to the PTC in the same condition—in his own words, "dizzy, weak, about to pass out and in need of assistance."

The MARK IX system, even when operated on emergency by-pass, was unable to satisfy the respiratory needs of the divers under working conditions in 50°F water.

Earlier testimony by a project officer from the Experimental Diving Unit (EDU) at Washington, D. C., pointed out that the MARK IX systems used by the aquanauts had not been tested in simulated Sealab III conditions because the Navy had no facility to perform such tests. Testimony from Sealab participants indicated the Mark IX had not been previously used in training dives deeper than 60 feet. Late modifications, designed to prevent breathing gas cutoff, were not tested at EDU to ascertain any breathing restrictions they might cause under heavy work conditions.

The diving team was brought to the surface and transferred to the DDC. No ill effects were apparent in the team after they had been warmed and fed.

Leaks in the Sealab habitat continued to worsen. Now gas leaked at rates as high as 15,000 scf/hour. Available gas supplies would permit operation for only several more hours; when the gas supply was gone, the habitat would flood and probably be lost. Gas loss was further increased by gas hosing up through the main electrical cable, causing the sheathing to burst in several places. During the unbuttoning procedure, ballast tanks were flooded so that the habitat's gross weight now exceeded the lifting capacity of the support ship, *Elk River.*

Sealab III Command decided to attempt another entry and to have a diver team repair the habitat. A second team of divers started compression in the other DDC about 30 minutes after Team 1 started but at a compression rate of 40 feet per hour.

Although they had been awake and working for about 20 hours, Team 1 was again given the diving task. Hot water hoses were rigged to the outside of the PTC. It was planned that after egress from the PTC at the bottom, Barth and Cannon would connect the hoses to hot water diving suits locked into them for this second trip.

The divers experienced the same cold during their return to the bottom, even with blankets and foul weather gear they brought along. At bottom, Barth and Cannon were dressed out in the hot water suits and MARK IX gear. A standard skin-diving face mask and bite-type mouthpiece were used because, according to testimony, the commercial clamshell full face mask would not provide a satisfactory seal when used with the positive-pressure MARK IX system.

Barth wore coveralls and a wet suit hood under his hot water suit; Cannon wore nothing under his suit. When the divers were out of the PTC and ready to hook up their hot water, they found that the water from the hoses was as cold as the ambient water. They left the PTC without the hot water supply.

The divers were expected to open the access hatch into the habitat at the diving station. According to Comdr. Scott Carpenter, this hatch has a dead weight of 150 pounds and must be pried open with a crowbar to break the seal. Barth started this process, working slowly because of the life-support system deficiency.

After several minutes, he looked for Cannon and saw him lying on his back, convulsing, mouthpiece free-flowing above him and mask partially off. When Barth got to him, he tried to force the emergency demand regulator into Cannon's clenched mouth, but couldn't. He started dragging him back to the PTC about 40 to 50 feet away. It was not possible to summon help because the MARK IX has no communication system. The umbilicals became fouled, causing Barth again to over-breathe the system. He was forced to leave Cannon about 30 feet from the PTC. He arrived at the PTC in a near unconscious state. In his own words, "If it had been 20 feet further, I wouldn't have made it." Part of the problem had been observed on the surface over closed-circuit television, and the stand-by divers in the PTC had been alerted to render assistance. Blackburn was dressed out when Barth returned to the PTC.

Cannon was brought aboard the PTC by Blackburn, and emergency mouth-to-mouth resuscitation and external heart massage procedures were followed during the ascent. Cannon was presumed dead during the ascent and no medical officer was locked in when the PTC surfaced and was mated to the DDC.

Results of an autopsy on Cannon at the San Diego County coroner's office report that death was caused by cardiorespiratory failure attributable to carbon dioxide poisoning. During testimony, Dr. F. J. Luibel, pathologist, stated that this opinion was predicated on a toxologist's report of tests on traceal air performed many hours after the accident, and that if based on the pathology findings alone, no causation could have been determined.

Subsequent testimony by Lt. Comdr. Lawrence Raymond, an expert on thermal stress, brought out that the divers' lack of protection from cold, both in the PTC and the water, could have been a prime cause of the accident. According to Dr. Raymond, the divers had suffered severe thermal stress on the first dive and he felt that their core temperatures had not returned to normal at the beginning of the second dive. Combined with their lack of rest, breathing cold gas with a higher thermal conductivity, and carbon dioxide build-up at low body temperatures could have initiated cardiac arrhythmia through the body's adrenalin mechanism.

No physiological tests had been performed prior to the accident to determine if men could survive in 50°F (10°C) water with minimal thermal protection, performing moderately heavy work, after several hours of chilling and breathing cold helium oxygen gas from the Mark IX underwater breathing apparatus.

On-site inspection of the four MARK IX rigs used from the PTC at the time of the accident revealed that one of the baralyme canisters (carbon dioxide absorbent) in the MARK IX units was empty. But no record of diver equipment combinations was kept. It is therefore impossible to determine if this was the equipment used by Cannon on the second dive.

Testimony by Capt. Robert Workman, head of the Environmental Stress Division of the Naval Medical Research Institute, Bethesda, Maryland, indicated that even at the surface, a diver performing moderate work while breathing by means of a MARK IX life-support system would be in serious trouble in less than a minute if the baralyme

carbon dioxide absorbent were not present. He also testified that the breathing mixtures the divers were using could not have resulted in oxygen toxicity. Even when the divers operated the by-pass on the MARK IX equipment and breathed the oxygen-rich gas (7.0% oxygen) from the emergency backpack, the maximum obtainable partial pressure converted to atmospheric equivalents was 1.4 atmospheres.

The findings of the hearing were inconclusive. It was the opinion of the board that Cannon used the equipment with the faulty scrubber unit, but proof of this was never established. Regardless of the actual cause, the accident was a serious blow to military diving research—one that will probably take a decade from which to recover. All planning for future projects of this scope was shelved, and during 1969 funds for Sealabs suddenly became unavailable.

UWL Helgoland, West Germany

The last major underwater living experiment of the decade was conducted by the West German Government during July, 1969. Although the experiment was conducted at a modest depth of 60 feet, it introduced new concepts that will set the standard for future living experiments. The site of the project was in the North Sea, approximately 1½ miles east of Helgoland, Germany, an area subject to violent and unpredictable weather. To avoid dependence upon surface support ships, a completely self-contained underwater laboratory complex was designed and built by Drägerwerk of Lubeck, Germany.

The habitat was a cylinder 30 feet in length and 8 feet in diameter, mounted on four adjustable legs. The weight in water, including two ballast pontoons, was approximately 60 tons. Unlike most other habitats, Helgoland was designed to remain underwater for several years and the ballast pontoons could be flooded or "blown" to facilitate movement of the complex to other sites.

The habitat was connected by an umbilical cable to a large surface buoy containing electrical power supply, oxygen make-up supply, radio and TV transmitting and receiving equipment. The 10-foot diameter buoy was capable of providing the submerged aquanauts with life support and power for 2-week periods without resupply. An external supply depot was established on the ocean floor adjacent to the habitat for storing extra food and underwater equipment that could not be stored within the habitat.

The atmosphere within the habitat was normal air—carbon dioxide was continually removed by an absorbent blower system and the oxygen was maintained at 20% by an automatic metering system which controlled the supply from the surface buoy. Gas levels were constantly monitored at a shore control station over a telemetry link with its transmitter in the surface buoy.

One of the unique emergency features of the system was a one-man "escape chamber." In the event of illness or injury, an aquanaut could be placed into the capsule, sealed, and sent to the surface. Designed for pickup by helicopter, the chamber contained an integral life-support system to sustain the diver under pressure until he could be brought to the emergency shore station. The shore system consisted of a standard decompression facility for treatment under pressure.

The initial Helgoland experiment was primarily conducted to determine the practicability of protracted scientific sea-floor investigations in the hostile environment of the North Sea. Its moderate depth precluded gathering any spectacular physiological data and the short time frame would not permit acquisition of isolation behavioral data not already collected. The basic system was designed for use at depths to 330 feet and system components for deeper operations, such as personnel transfer capsules, were included to permit future deep operations.

The segment of the program successfully performed in 1969 consisted of a 22-day operation test of the system and concept and a limited medical and marine biological research schedule. The medical investigations consisted of basal studies and ergometric capability evaluations under a varying controlled diet. Behavioral studies were also conducted to determine the reaction of aquanauts under the influence of relative isolation in extremely poor environmental conditions. The experiment revealed no measurable changes in the divers' physical well being and no one had any meaningful adverse emotional responses to the ordeal. The marine biological program, which was closely correlated with the ergometric diver study consisted of the following major tasks:

1. Measurement of water temperature gradients near and in the sea floor with thermistor probes
2. Measurement of current velocity and direction near the sea floor and correlation of current to turbidity and light level
3. Collection and quantifying plankton concentration as a function of tide action
4. Collection of sediments and cores on a regular basis
5. Microbiological and microfaunistic research
6. Bottom sediment transport studies

The last team of aquanauts surfaced on August 19, 1969, but complete data on the experiment will not be available until 1971. From both an operational and equipment performance standpoint, the project was successful. One interesting comment by Dragerwerk at the conclusion indicates that it was educational—"The total expenses for performance of investigations at the sea ground are greater than expected."

From the standpoint of underwater capability, it is obvious that the period of 1950 to 1970 was the opening of the ocean to man. The basic technology for man to live in the sea was either acquired or merged with his capability and turned into reality. We see a decade of underwater living experiments culminating 2,500 years of struggling attempts to mate the physiology of man with the physics of the underwater environment. The translation of experimental technology into working reality does not, however, occur overnight. During 1969, saturation diving became an accepted technique in commercial diving, reaching a peak during that summer when more than six saturation diving projects were operating simultaneously in the offshore oil leases in the Gulf of Mexico. But many basic questions of physiology remain unanswered today. Much of the equipment required to assure safe and continued underwater operations still remains to be developed.

Table 2-1 is a summary of underwater living experiments.

Table 2-1
Sustained Underwater Living Experiment in the Open Sea

Project	First Trial	Country	Crew (Number, Time)	Depth	Respiration Gas
Man-in-the-Sea I	1962 (Sept.) Med. France	USA	1 person 1-4 days	61 m.	3% O_2; 97% He
Conshelf I	1962 (Sept.) Med. France	France	2 persons 1 week	10 m.	Air
Conshelf II	1963 (June) Red Sea	France	5 persons 29-31 days (1 week)	11 m. (27 m.)	Air (5% O_2; 20% N_2; 75% He)
Man-in-the-Sea II	1964 (June/ July) Bahamas	USA	2 persons 49 hours	132 m.	4% O_2; 5% N_2; 91% He
Sealab I	1964 (July) Bermuda	USA	4 persons 11 days	59 m.	4% O_2; 17% N_2; 79% He
Sealab II	1965 (Aug.) Pacific, Cal.	USA	28 persons (3 teams, 10 days each)	60 m.	4% O_2; 9% N_2; 87% He
Kitjesch	1965 (summer) Crimean coast	USSR	4 persons	15 m.	
Conshelf III	1965 (Oct.) Med. France	France	6 persons 3 weeks	100 m.	1.9-2.3% O_2; 1% N_2
Permon II	1966 (July) discontinued Yug. coast	CSSR	2 persons	30 m. planned	
Ikhtiandr 66	1966 (Aug.) Black Sea Crimean coast	USSR	1-2 persons 7 days	11 m.	Air
Sadko 1	1966 Black Sea Caucas. coast	USSR	2 persons 6 hours (1 month at 25 m. Oct. 1966)	40 m.	
Caribe I	1966 (end of year)	Cuba	2 persons 3 days	15 m.	
Permon III	1967 (March) Sea at Bruntal	CSSR	2 persons 4 days	10 m.	
Medusa I	1967 (July) Lake Klodno	Poland	2 persons 3 days	24 m.	37% O_2; 63% N_2
Hebros I	1967 (July) Bay of Warna	Bulgaria	2 persons	10 m.	

(Table 2-1 continues on p. 52.)

Table 2-1 *(continued)*

Project	First Trial	Country	Crew (Number, Time)	Depth	Respiration Gas
Oktopus	1967 (July) Black Sea Crimean coast	USSR	3 persons (several weeks?)	10 m.	Air
Ikhtiandr 67	1967 (Aug.)	USSR	5 persons		
Sadko 2	1967 (summer) Black Sea Caucas. coast	USSR	2 persons 6 days	25 m. (50-60 m.)	
Kockelbockel	1967 Sloterplas	Netherlands	2-4 persons short period	15 m.	Air
UWL-Adelaide	1967-1968	Australia	several persons		
Romania LS I	1968 ? Bicaz Lake	Roumania	2 persons		
Karnola	1968 ?	CSSR	5 persons	8-15 m.	
Tschernomor	1968 (June) Crimean coast	USSR	5 x 6 persons total 1 month	14 m. (poss. 30 m.)	
Medusa II	1968 (July) Baltic	Poland	3 persons 14 days	30 m.	
Robinsub I	1968 (July) Ustica Island	Italy	1 person	10 m.	Air
Hebros II	1968 Cape Maslennos	Bulgaria	2 persons 10 days?	30 m.	
Sprut	1968 Crimean coast	USSR	2-3 persons 14 days	10 m.	
BAH I	1968 (Sept.) Baltic	Federal Republic of Germany	2 persons 11 days	10 m.	Air
Ikhtiandr 68	1968 (Sept.) Crimean coast	USSR	several crews total 8 days	12 m.	
Malter I	1968 (Nov./ Dec.) Malter Dam	German Democratic Republic	2 persons 2 days	8 m.	Air
Tektite I	1969 (Feb.)	USA	4 persons	12.7 m.	8% O_2 ; 92% N_2
Sealab III	1969 (Feb.) discontinued San Clemente, Cal.	USA	5 x 12 persons	183 m.	2% O_2 ; 6% N_2 ; 92% He
UWL-Helgoland	1969 (July) North Sea	Federal Republic of Germany	4 persons 10 days each	23 m.	Air

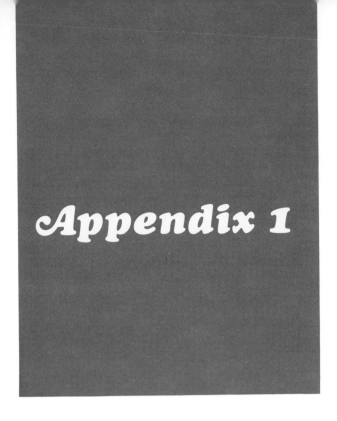

Appendix 1

Adapting to the Underwater Environment

Seeing Underwater—Adaptive Devices

Although magnification provided by refraction of light passing through the glass interface of a diving mask or helmet can assist in near point underwater work, the loss of visual field accompanying magnification often creates hazards for the diver. Restoring this peripheral vision requires either a modification of diving equipment or the addition of a new adaptive device. In the past several years, different approaches to accomplish this have been explored. The greatest success has come with two methods: use of spherically-shaped face plates in helmets, and the development of Skindiver Contact Air Lenses (SCAL).

Face Plates

The use of a spherical face plate has been successfully accomplished in both the injection-molded "bubble" diving helmet developed by General Electric Corporation in conjunction with their closed circuit mixed gas SCUBA, and in the PIEL

Model EP100 Helmet, developed by Societe Industrielle Des Etablissments Piel of France. The problem in successfully using this technique is centering the spherical plate, which must remain constant with respect to the eyes. The helmet design must include geometry allowing a 180° field of view without interference, which in most configurations on the market is an impossibility. The PIEL helmet is designed much like a motorcycle crash helmet and the spherical face plate fits in the same manner as a visor.

Adaptation to near-normal underwater vision, even by experienced divers, has proven to be rather slow. Having developed adaptive reflexes for a different visual perception underwater, the diver still needs from 5 to 6 hours of diving to coordinate his visual-motor system actions with unmagnified vision.

Skindiver Contact Air Lenses

Although development of a normalizing face mask for free-swimming divers has been investi-

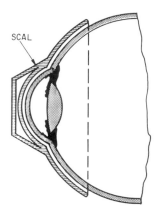

SCAL

Figure AI-1. Skindiver contact air lens (SCAL) fits over the corneal protrusion and part of the sclera. The flat frontal surface can include corrective prescription.

gated and some experiments conducted under government research contracts, no practical solution has been forthcoming. One area of investigation that does offer possibilities for a general solution to the problem is the Skindiver Contact Air Lens. Some highly successful work has been performed in this area by Dr. Allan H. Grant of Wheaton, Maryland. Under a research contract from the U.S. Navy, Dr. Grant developed a scleral contact lens that enables the diver to see clearly underwater without the use of face mask, goggles, or helmet. Using the lenses the wearer can see clearly above water and, theoretically, if the water surface bisects the pupil, the wearer would see clearly in both media simultaneously. With these lenses, underwater vision is limited only by water conditions and illumination.

Contact lenses are not new to the diver; many divers wear contact lenses with their diving equipment to correct refractive errors. The scleral lens, however, is quite different from the normal contact lens, which covers only the cornea of the eye. Scleral lenses are based upon underwater physics and eye physiology.

When the human eye is immersed in water, approximately 75% of its refracting power is lost because the refractive index of the cornea and water are about the same. Loss of the cornea/air interface results in the loss of the 43 diopters of verging power normally present. The visual effect is the same as gross hyperopia, or farsightedness. With this type of refractive error, light rays normally bent by the cornea to permit point focus by the lens onto the retina pass through the eye at an angle causing them to focus well beyond the retinal surface. The result is a blurred image on the retina—an image that cannot be focused by action of the internal eye lens.

Figure AI-1 shows a SCAL fitted to the eye. The lens is essentially a double-walled contact lens, with a discrete air or vacuum space between the two lens layers. The inner layer is a scleral contact lens into which the diver's refractive correction may be incorporated if necessary. The outer layer, called the cap, consists of an optically flat, circular wall from which extends an integrally formed conical flange mounted upon the outer surface of the scleral layer. The air space between layers is dehumidified to prevent condensation from temperature gradients.

When the diver is submerged with the scleral lens in place, he effectively has a miniature face mask over each eye. The magnification experienced with a standard face mask is still present, but the loss of visual field caused by the equipment is absent. In the average face mask, visual field is reduced by approximately 45%; with SCAL, the reduction is only 10%.

Tests were conducted using Navy divers to determine the effectiveness of SCALs compared to a standard face mask. Divers were tested for visual acuity and visual field in a wet test chamber at simulated depths from 10 to 300 feet. Figure AI-2 shows the results of the visual field tests on twelve subjects. Visual acuity tests revealed that vision was unaltered from the surface to all test depths in subjects wearing SCALs. Following chamber tests, the divers wore the contact lenses during underwater operations for periods varying from several minutes to several hours.

The only significant difference between openwater operations and the results obtained in the fresh water chamber was that all divers felt uncomfortable stinging or burning. The degree of discomfort varied with individual sensitivities. Subsequent opthalmoscopic tests revealed no objective evidence of corneal changes associated with the irrita-

Air

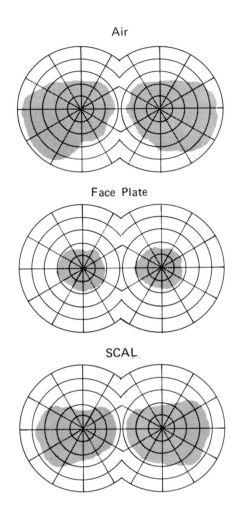

Face Plate

SCAL

Figure AI-2. Mean visual acuity of 12 subjects is compared between air, underwater with face mask, and underwater with SCAL.

tion. Several "wetting" solutions were tried in an attempt to minimize irritation. The most viscous solution was found to delay the onset of discomfort. Objectively, properly fitted lenses can be worn for periods of up to 8 hours in air. Extensive underwater use of scleral lenses is now only awaiting the availability of a wetting solution that will be relatively impervious to sea water and provide thin film protection against irritation.

Object Recognition at Depth

Considerable controversy exists among divers regarding the most easily recognizable colors un-

derwater. A large quantity of data is available concerning visibility and recognition of objects at shallow depths. In Sealab II experiments conducted at a depth of 205 feet, quantified results were obtained indicating that much of the data extrapolated to deeper depths from shallow water experiments are not applicable.

Six divers were used in the Underwater Detection and Recognition experiment and a total of twenty observations were made on each target. The targets selected were a black circle, white square, yellow triangle, and a white cross. Table AI-1 shows the results.

It is apparent that the black disc was detected and recognized from a further distance than either the white or yellow targets, even though the black disc was smaller. All divers on the project provided subjective correlation to the visual test data. They stated that the white Sealab habitat was far more visible than the reddish-orange personnel transfer capsule. Some estimates of the difference in visibility between the white and orange targets were as high as three times.

Effects of Pressure

Squeeze

Being mostly liquid, the human body is relatively unaffected by elevated ambient pressures within limits and under conditions of equilibration. However, when a differential pressure is created within a tissue structure or by failure to equalize enclosed free air spaces to ambient pressure, the tissues over which the ΔP is developed are subjected to squeeze and possible damage.

Table AI-1
Underwater Detection and Recognition

Parameter	Feet From Target			
	Black Circle	White Square	Yellow Triangle	White Cross
Detection	24.4	18.3	16.7	16.5
Recognition	20.0	14.2	13.5	13.4

Most squeezes encountered in diving are relatively minor and can be reduced or avoided by proper diving procedures. Squeezes of ear, sinus, face mask and suit are well documented in most standard diving procedure manuals. Two forms of squeeze, both poorly documented until recently, are generally fatal—body squeeze and thoracic squeeze. Body squeeze occurs when the entire body is forced into a diving helmet and has been rare in recorded diving history. Because of improvements in equipment dating back to Gunner Stilson in the early 1900s, few cases have been recorded in the last 50 years.

Thoracic squeeze is also a relatively uncommon accident, but is a possibility in all present-day forms of diving. Late in 1969, the U.S. Navy Submarine Medical Center released an exceptionally well documented report on a fatality involving thoracic squeeze. Excerpts from the case report by Lt. M. B. Strauss and LCDR P. Wright are quoted here.

Case Report:

R. C. (28-year-old Lt., USN), an experienced underwater demolition team diver, was engaged in recreational diving at Subic Bay, Republic of the Phillipines, with another UDT diver. They were performing deep breath-hold dives in water which has a maximum depth of eighty feet. They dove alternately so that each could observe the other.

After a series of dives without incident, the companion diver first observed R.C. "floating" face up in a semiflexed position at a depth of forty feet, the estimated limit of visibility from the surface. R. C. appeared to be ascending slowly and was immediately brought to the surface by his partner. Total time of submersion was less than three minutes. Upon surfacing, R. C. was unconscious, apneic, and bleeding frothy bright red blood from the mouth. Five mouth-to-mouth respiratory exchanges were performed and the victim was carried to a small craft anchored fifty yards away.

Examination by a physician with special training in diving medicine, within ten minutes of the accident, revealed a deeply cyanotic, unconscious patient with frothy red hemoptysis, irregular respirations, and intermittent generalized clonic seizures. Conservative care, including postural drainage, gentle restraint, and maintenance of an oral airway was instituted while the patient was promptly transferred to a recompression facility. Soon after arriving at the recompression facility, approximately thirty minutes after the accident, R. C. improved substantially while breathing air at ambient pressure. It was elected not to pressurize the chamber. During this interval the patient gradually regained consciousness and became coherent in speech. His blood pressure and pulse rate were normal. The peripheral cyanosis reflected in lips and nail beds had cleared. Periodically, the patient would cough up and expectorate small amounts of clotted dark red blood. Intermittent breathing of one hundred percent oxygen was poorly tolerated, for it provoked episodes of coughing. Resonant rales were heard throughout the chest but were more accentuated in the right hemithorax. Ten minutes later, the patient's pulse rate gradually increased to 120 per minute. Breathing 100% oxygen, which was better tolerated at that time, lowered the rate to 90 per minute. The patient, while intermittently breathing oxygen, was transferred by ambulance to the naval hospital.

During the latter portion of the ambulance trip, R.C. complained of increasing difficulty in breathing and said he was "feeling bad again." At hospital admission, approximately one hour after the accident, peripheral cyanosis and tachycardia were again present. The blood pressure was 110/70. One hundred percent oxygen was administered with positive pressure ventilation via a naso-tracheal tube. Ringer's lactate was infused intravenously. Moderate amounts of bright red frothy blood were aspirated from the naso-tracheal tube. Auscultation revealed loud resonant rales in all lung fields. The patient's condition continued to decline despite these measures and hypotension gradually developed. Eight hundred milliliters of plasma and 1250 ml. of Ringer's lactate solution were transfused over a period of 90 minutes but failed to counteract the tachycardia and declining blood pressure. The patient gradually lost consciousness.

Cyanosis continued to be unaffected by positive pressure ventilation with oxygen. Approximately three hours after the initial insult, cardiac arrest occurred. The arrest was refractory to all resuscitative measures and the patient expired.

Post Mortem Examination

Necropsy examination revealed no evidence of external injury or suggestion of bites or stings by venomous sea animals. Pertinent findings were confined to the lungs which weighed approximately three and one-half times their normal weight (right lung: 1250 grams; left lung: 1150 grams). On gross examination, the lungs were dark red and hemorrhagic throughout. Only slight crepitance was present. Cut sections revealed a consolidated purplish parenchyma which exuded large amounts of bloody fluid. Microscopic examination disclosed congested alveolar capillaries and multifocal areas of interstitial and intra-alveolar hemorrhage throughout the lungs. These findings were consistent with a diffuse bilateral vascular injury with intravascular congestion, interstitial edema, diffuse disruption of small vessels, and intra-alveolar hemorrhage. These findings were not consistent with the post mortem picture of drowning.

Discussion

The clinical findings, necropsy examination, and pathophysiology in this case fit the syndrome thought to result from a thoracic squeeze. This disorder is an unusual medical problem associated with diving in that it requires a considerable degree of expertise before conditions can be achieved to precipitate it. The breath-hold diver must have sufficient breath-holding capabilities to remain submerged for over a minute as well as great facility in rapidly clearing his ears and sinuses with descent. Theoretically, thoracic squeeze occurs when the air comprising the total lung capacity (after maximal inspiration on the surface) is compressed by the increasing ambient pressures while descending to an amount less than the residual lung volume. Such pressure volume relationships would respond in accordance with Boyle's Law, namely $P_1 \times V_1 = P_2 \times V_2$. Since few divers know their lung capacities and carefully observed world record setting breath-hold dives have far exceeded the so-called total lung capacity/residual volume threshold, calculation of the threshold, in general, is an exercise in theory only. It has been estimated that in the "average"-sized man, this limit would occur at 80 to 100 feet.

Historically, this condition was noted long before the self-contained breathing apparatus (SCUBA) was invented. In 1823, S. N. Smith stated, "...if the depth be considerable, the water on the breast and organs is so great that it occasions the eyes to become bloodshot and produces spitting of blood, and if the practice is persisted in, it most likely proves fatal." One may think of the thoracic squeeze injury as the converse of the air embolism disorder. In the latter disorder, lung tissue is contused and rent by overexpansion of lung tissue due to failure to exhale sufficiently while ascending. In thoracic squeeze, the alveoli have been compressed to their smallest non-collapsible volume (i.e., residual volume). As the pressure gradient increases, the "low pressure" alveolar capillaries become distended. Transudation, diapedesis, and finally rupture with hemorrhage into the interstitium and alveoli occur. Theoretically, if a large gradient is created rapidly enough, bronchioles and blood vessels could rupture simultaneously and air embolism could result. However, since one is dealing with a compressed volume of air rather than an expanding volume of air as in embolism, intravascular air emboli is much less likely to manifest itself. Production of explosive gradients, e.g., rupture of a submarine hull while "at depth" could cause a crushed chest, rupture of the diaphragm, and/or displacement of the viscera into the thorax.

The presented case raises four significant questions. First, how does one explain the appearance of a condition most likely to be a thoracic squeeze, occurring at a depth well below the most conservative estimates of the victim's breath-hold diving threshold? Several explanations are apparent. His total lung volume could have been substantially reduced by failure to inhale fully before descending, expenditures of large amounts of air while descending because of difficulty clearing ears, sinuses, or face mask, or accidentally losing tidal volume due to coughing. A fifty percent reduction in the total lung capacity at the surface would reduce the depth threshold by one-half. While Linaweaver states, "considerable injury can occur in the absence of pain," Schaefer and Dougherty, in a personal communication, have noted in their observations that pain associated with thoracic squeeze appears before serious signs and symptoms are manifested. The explanation of shallow water blackout and loss of tidal volume while on the bottom is not adequate, for the unconscious patient could never passively exhale to a point greater than the residual volume of air in the lungs. If this happened after the diver successfully reached the bottom, the pressure gradient would become essentially zero and no lung injury should occur. On the other hand, if the patient lost consciousness, for example, while ascending, passively lost his tidal volume and then began to descend due to negative buoyancy, conditions to precipitate a thoracic squeeze are present. Judgment as to which explanation best accounts for the course of events that occurred in the case reported herein, is reserved. Since the victim was apparently ascending when rescued, loss of consciousness, passive loss of tidal volume, and descent to a point below the new depth threshold seems the less likely explanation.

The second question that arises is how does one account for the initial unconsciousness, hemoptysis, and clonic seizures?   Clonic seizures are often associated with hypoxia, but usually the hypoxic insult is greater than three minutes duration. The possibility of traumatic rupture of air conducting structures as well as blood vessels and resultant cerebral air emolism is entertained. This was the rationale underlying the immediate transfer of the victim to a recompression facility. However, the spontaneous improvement and absence of air emboli in the cerebral vasculature on sectioning the brain, makes this supposition difficult to prove.

Third, how can the initial transient improve while breathing air and receiving conservative treatment thereafter followed by a continuous gradual deterioration to a hypoxic death be explained? This course appears to reflect a progressive, alveolo-capillary diffusion block. After recovery from the "shock" of the initial insult, ventilation was adequate to supply the body's oxygen requirements and was characterized by the remarkable improvement. Subsequent decline occurred when the alveolo-capillary block became great enough to interfere with adequate respiratory exchange. When this became manifest a "vicious cycle" resulted. The deficient ventilation caused further hypoxia to the already damaged alveolar capillaries. The next step in the chain of events is pulmonary edema and diapedesis from the alveolar capillaries. These two sequelae cause further hypoxic insult. The deterioration after seeming improvement is not unlike the thirty to sixty minute latency periods noted before rapid fluid losses appear in the severely burned patient or the swelling response manifests itself in the newly traumatized joint.

Finally, how can one rule out a case of drowning secondary to losing consciousness in the water? The short duration of submersion, immediate onset of hemoptysis and clonic convulsions is not typical of the drowning victim. The rapid recovery initially is consistent with near drowning, but one would expect sustained improvement (barring pneumotinic complications) until normal, rather than the deterioration observed in this case. Gross and microscopic necropsy findings of diffuse bilateral pulmonary congestion, interstitial edema and interstitial and intra-alveolar hemorrhage suggest a thoracic squeeze injury. Pulmonary edema is the predominant finding in the near drowning victim.

It is interesting to note that post mortem intracardiac electrolyte studies revealed a mild hemoconcentration in the left side of the heart as compared to the right side. This is thought to be consistent with drowning.

Appropriate treatment for this condition would include cardiovascular support as indicated and supportive oxygen ventilation. Pneumonitis should be anticipated and treated appropriately. Hyperbaric oxygen could be beneficial in counteracting, to a limited degree, the alveolo-capillary diffusion block. The extent of alveolo-capillary block should be monitored by arterial gas studies. In theory, a heart-lung machine could provide adequate gas transfer to and from the blood, but is virtually impractical.

Summary

An unusual diving accident characterized by pulmonary congestion and edema; interstitial and intra-alveolar hemorrhage is described. The clinical course was marked by transient improvement and incomplete recovery; then followed shortly thereafter by progression to a fatal outcome. The clinical and necropsy findings are discussed in relationship to the phenomenon of thoracic squeeze.

Addendum

In light of the previous discussion, it is interesting to compare the breath-hold diving abilities of humans and the "diving" mammals. Descents to 3,300 feet have been recorded in the sperm whale while harbor seals, which have approximately the same lung capacities as humans, have dived to 825 feet. The record human breath-hold dive is two hundred and forty feet.

Several adaptations have been observed in the diving mammal which make the extraordinary depth excursions possible. First, the lungs of seals and whales can collapse to a point where they become atelactic and yet not separate from the chest wall. Compensation for the decreased lung volume is made possible by the marked compressibility of their thoracic cages. This would ameliorate the intrathoracic pressure gradients between the chest wall, alveoli, and pulmonary fasculature which would otherwise result from descending. Another factor which may be important is the shunting of blood from the extremities to the central fasculature. This has been well documented in the oxygen conserving (diving) reflex. Seals are noted to have tortuous vena cavae and the diaphragm is thought to act like a sphincter on these vessels while diving. The blood from the dilated vena cavae could, in theory, compensate for the void produced by the collapsed lungs. Since blood is noncompressible and would directly transmit the hydrostatic pressure while descending, the pressure gradients which precipitate thoracic squeeze would not be produced.

These adaptations have not been observed in the human breath-hold diver. However, certain adjustments have been noted. First, a significant increment in the tidal lung capacity and a relative decrease in the residual lung volume is observed after repeated breath-hold diving. The greater the changes, the greater the depth threshold for thoracic squeeze. Large discrepancies between the calculated depth threshold based on the total lung capacity, residual volume ratio for world record breath-hold divers R. A. Croft and Jacques Mayol are apparent. Schaefer, et al., have demonstrated a shift in blood volume to the thorax with descents in water. If this blood displaced a corresponding amount of air in the lungs, a degree of protection from thoracic squeeze would be realized and, in part, accounts for the discrepancy between Croft's record descent to 240 feet and his predicted thoracic squeeze depth threshold of 197 feet. The factor of chest wall compressibility has not been measured in the human, but may afford a measure of protection analogous to that predicted in the diving manual.

Bone Damage

Considerable effort has gone into determining what, if any, effects compression and decompression have on the human body. Because of the close supervision of Workman's Compensation authorities, pressure work performed in tunnel construction has received excellent documentation. Recent experience shows that repeated compression and decompression, even to moderate depth equivalents, can result in structural damage to the body. This damage manifests itself as *aseptic bone necrosis,* (ABN) which appears under radiographic examination as areas of increased bone density in the head, neck, and the long bones in the arm and leg. When a joint such as the thigh or shoulder is involved, aseptic bone necrosis can be crippling.

There have been few documented cases of ABN among American divers, but several have been found in Europe. The disparity between the incidence of ABN among divers and tunnel workers has been attributed to the difference in decompression schedules used for pressure operations. For example, on-decompression time used after an 8-hour shift at 40 psi in construction of the East River Tunnels in New York was only 48 minutes. The same exposure would require a 98-minute decompression on the Standard U.S. Navy (1962) Air Decompression Tables.

A comprehensive study of bone lesions attributable to decompression problems was conducted in England under the sponsorship of the Medical Research Council of Great Britain. Their principal findings were

1. Of 241 men who worked at a pressure of approximately 2 AtA on the Clyde River Tunnel project, 19% or 47 men exhibited one or more bone lesions under radiography. Of these, 10% of the men had lesions that were juxta-articular and potentially disabling.
2. The lesions were apparently a function of the number of times the worker was decompressed and the higher incidence was with men who had greatest susceptibility to the bends.

Later test programs in the United States indicate that in addition to a relationship to incidence of bends, there is a direct correlation with the patient's age. In the ABN cases detected, the average age of the men was over 45.

Although the etiology of aseptic bone necrosis is relatively obscure today, physiologists feel that the primary cause is due to cell damage in the fatty bone marrow resulting from differential pressure. According to Capt. Albert A. Benke, Jr., M.D., M.C. USN (Ret.), the bone structure, from the standpoint of body systems, is the major deterrent to long exposures of man to elevated air pressure.

Psychomotor Capability

Under pressure and conditions where no deterioration of either physical or mental capability by inert gas narcosis is obvious, a marked decrease in the normal (surface) psychomotor capability of divers has been observed. Although the degree of reported deterioration varies in the data obtained by different investigators, the consensus is that some deterioration is observed at all depths and levels of problem difficulty. Under stress conditions at deep working depths, deterioration may be as great as 50%.

The U.S. Navy's Office of Naval Research has supported a number of behavioral underwater studies through the Physiology-Psychology Branch. One of the more complete analyses of psychomotor deterioration was performed during Sealab II. A series of tests was designed to measure strength, manual dexterity and spacial relationships, as well as group coordination, with a complex spacial problem. The strength tests indicated a low order difference between underwater data and pre-dive baseline studies, with mean deviations ranging from 4 to 15%. The manual dexterity and spacial relationship testing, however, revealed gross deviations between surface and underwater capability. The results should be considered in the design of underwater equipment and operational planning.

The individual assembly tests measured manual dexterity and the ability to form spacial re-

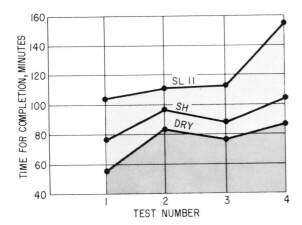

Figure AI-3. The time required to complete tasks increases with water depth and level of thinking required. This chart shows results of four separate tests of increasing complexity. DRY = baseline land study; SH = same test in shallow water; SL II = same test at Sealab II site.

lationships. The test required the diver to assemble three 1-foot lengths of steel into a triangle by joining the corners of the plates with nuts, washers and bolts. A combination of bolt sizes and hole symmetries were selected to provide four forms of test. While it was possible to assemble the symmetrical plates in any combination, only one combination provided exact fit. The test was selected as representative of underwater work operations requiring assembly, adjustment and manipulation of small equipment. The test was performed using a platform mounted on the end of the habitat shark cage that enabled the diver to work while standing on the bottom.

Figure AI-3 shows the data obtained during baseline studies, in shallow water and at the Sealab site. The specific test configuration, that is, the size of the bolts provided and specification of more exact construction requirements details, had a marked effect on performance deterioration. The data show a 37% increase in the time required to complete the tasks underwater over the mean time during surface baseline tests. With more detailed specifications for construction, the difference in means between underwater and surface times increased to nearly double.

The *group assembly task* was devised to observe the manner in which a group of four men planned and carried out a task requiring perception of complex spacial relationships and cooperative assembly of components. The work consisted of cooperative assembly of a three-dimensional structure consisting of short lengths of ½-inch pipe and connectors. A drawing showing the final assembly was provided to the divers and they were required to work out the operational construction details. The plan of attack was worked out in the habitat and the actual work took place at the platform attached to the shark cage on the outside of the habitat. Due to scheduling difficulties, the test was actually performed only once during the program, but meager as these data are, they demonstrate the difference in surface/underwater capabilities.

Only one group assembly test was conducted outside the habitat. The team discussed their construction method and practiced the task immediately before leaving the habitat. The best dry land assembly time recorded during baseline testing was 6 minutes. The underwater assembly task took 12 minutes, 20 seconds.

Decompression

The basis for present day theory on both decompression and decompression sickness dates back to 1907 and J. S. Haldane. His original work has undergone considerable modification and extension, but few current publications on the subject of decompression do not reference his methodology. Haldane was concerned only with what we would consider dives to relatively moderate depths (165 feet) and of short duration. His findings were that body tissue could be supersaturated with inert gas and not form bubbles if pressure difference (P) between the inert gas in the tissue and the ambient pressure did not exceed a 2:1 ratio. Based on his experiments, a man would become saturated for a specific ambient air pressure in approximately 5 hours. The half-time, or time increment to reach a 50% saturation level, would be 75 minutes.

Later work in the United States showed that although these tables were safe at moderate water

Table AI-2
Atmospheric Nitrogen Content of Tissues and Organs

UNIT	TISSUE-ORGAN GROUP				
	I	II	III	IV	V
	Blood Brain Heart Kidney	Muscle Skin Spinal Cord Nerves	Bone Cellular (fat-free)	Bone Marrow (fat-rich)	Adipose Tissue
WEIGHT g	15,000	37,000	3,500	1,500	9,500
WATER CONTENT g	12,000	30,000	2,000	240	2,000
FAT CONTENT g	350	100	–	1,200	7,000
NITROGEN ml	126	275	18	63	368
BLOOD PERFUSION ml/min	4,000	1,200	80	50	375
NITROGEN TRANSPORT (1st minute)	40	12	0.8	0.5	3.75
N, ELIMINATION Half-time minutes	1.9	16	16	85	69

depths, they did not hold up for deep diving. Experimentation indicated that rather than saturation being accomplished in 5 hours, it took closer to 24 hours and that in this light, tissue with half-times up to 240 minutes must be considered. The half-times considered in calculation of decompression schedules and the safe ratios that can be approached are a function of the depth of dive. The controlling half-time tissue will change with depth of exposure and method of decompression.

In 1951, O. E. Van Der Aue, in calculating schedules for decompression from air dives with the use of oxygen, developed the following ratios:

20	3:54:1
40	3:54:1
75	2:94:1
120	2:60:1

These ratios were used to calculate the U.S. Navy Standard Decompression Tables.

According to Behnke, the atmospheric nitrogen content of various tissues and organs in relation to blood perfusion and nitrogen elimination when breathing oxygen at rest is as shown in Table AI-2. Calculated on the basis of a lean man having no more than 10% body fat, these half-time values extend up to 85 minutes for bone marrow and are usable for moderately deep air diving schedules.

Calculation of Decompression Schedules with the Haldane or P Method. As reported by Workman, the Haldane method is based on the assumption that the time-rate of change (dP/dt) of the inert gas tissue tension (P) is proportional to the difference between the partial pressure of the inspired inert gas (P_I) and the instantaneous value of P.

Thus

$$dP/dt = K (P_I - P) \qquad (AI-1)$$

If P_I is constant from time zero (t=0), and if P_0 is the inert gas tension at t = 0, then the inert gas tissue tension (P) at time t can be determined as a function of P_I, P_0 and t. Separating the variables and inserting limits,

$$\int_{P_0}^{P} [dp/(P_I - P)] = \int_{0}^{t} K \, dt$$

The solution of this equation is

$$ln (P_I - P) \Big|_{P_0}^{P} = -Kt \qquad \text{(AI-2)}$$

or

$$P_I - P = (P_I - P_0) \, e^{-Kt}$$

Subtraction of P_0 from both sides and rearrangement of terms lead to the desired expression

$$P = P_0 + (P_I - P_0) \, (-e^{-Kt}) \qquad \text{(AI-3)}$$

During decompression, P_I is reduced in steps at intervals intended to effect rapid but safe elimination of the inert gas absorbed in the tissues (P). The time for $(P_I - P)$ to decrease to one half its value immediately after a reduction of P_I is called the half-time ($t_{1/2}$). This time is characteristic of the particular tissue and inert gas involved, but is independent of the actual values of P_I and P within the range of pressures for which equation AI-1 is valid. Then from equation AI-2

$$ln (P_I - Pt_{1/2})/(P_I - P_0) = ln \, \tfrac{1}{2} = -Kt_{1/2}$$

and

$$K = ln \, 2/t_{1/2} = 0.693/t_{1/2} \qquad \text{(AI-4)}$$

The exponent in equation AI-3 can then be expressed as

$$-(0.693t/t_{1/2})$$

The acceptance of equation AI-1 requires the assumption that inert gas is uniformly distributed throughout the tissue represented by $t_{1/2}$ at all times. Two methods are currently used to determine the critical supersaturation value. The supersaturation ratio is the sum of dissolved gas tension (P_i) to ambient or hydrostatic pressure (H_p):

$$\sum_{i=0}^{n} \Sigma \, P_i/H_p$$

The supersaturation gradient (ΔP) is the difference between the sum of the dissolved gas tissue tension (P_i) and the hydrostatic pressure:

$$\Delta P = \sum_{i=0}^{n} \Sigma \, P_i - H_p$$

where n is the number of gases with significant partial pressures within the tissue ($t_{1/2}$). P may vary as a function of the rate constant (K) and with water depth (D). That is

$$\Delta P = f \, (D \, AtA, K)$$
$$\Delta P = P_s \, tgD$$

where g = rate of change of P as function of depth, P_s = P at surface (1 AtA).

Isobaric (Oxygen Window) Decompression. The use of oxygen in the final stages of decompression appreciably shortens shallow water stops. The principle involved in using a single gas without an inert carrier is based on the manner in which the human body uses oxygen. Unlike inert gas, oxygen transferred from the capillaries to tissue is used and a space, or partial pressure vacancy, exists.

When air is inspired at atmospheric pressure, the partial pressure of oxygen in the arterial blood is about 100 mm of Hg. After the blood circulates, the oxygen level in the venous capillaries is down to 40 mm of Hg. The difference, or *oxygen window,* of 60 mm Hg is available for the transport of inert gas. When the oxygen level in the arterial blood is elevated to 300 mm Hg by enriching the oxygen in the inspired gas, the oxygen level in the venous capillaries does not increase proportionally, but remains nearly the same, 50 mm Hg. The oxygen window then is widened to 250 mm Hg.

If pure oxygen is used in the last stages of decompression, the oxygen window is large enough to eliminate inert gas six times faster than when breathing air. With air breathing, nitrogen is eliminated at approximately 1 pound each 30 minutes; with isobaric oxygen breathing the rate is 1 pound each 5 minutes.

Decompression Tables

The American decompression tables in use today, regardless of whether they are for single or repetitive dives, on air or heliox, are based on the ΔP principle.

Single and Repetitive Dive Tables

The purpose of single dive tables is to permit the diver to perform his underwater task and return to the surface with a safe inert gas tissue tension. While on the surface, his system continues to eliminate inert gas. If a second dive is required, repetitive tables are used that give consideration to the level of inert gas partial pressure in his system at the beginning of the second dive. Repetitive dive tables have been developed for both air and mixed gas diving. The mixed gas diving tables, developed at the U.S. Navy Experimental Diving Unit, were released in mid - 1970.

Air Tables

Three basic air decompression tables are in use today. The U.S. Navy Standard Air Decompression Table, the U.S. Navy Air Decompression Table for Exceptional Exposures, and the U.S. Navy Standard Repetitive Air Dive Tables. The use of these tables (AI-3 through AI-7) are discussed in the following paragraphs.

The Standard Air Decompression Table, AI-3, provides a safe decompression schedule for dives between 40 and 190 feet. Incidence of bends using these tables is 0.69%. The table is straightforward and requires little explanation. It is, however, an "equal or next highest" computation. For example, if a dive were made to 62 feet for 75 minutes, the user would read down the *Depth* column to 62 feet, or the next highest number which in this case would be 70 feet.

The same procedure is applicable to time. Reading across to *Bottom Time,* time increments are in 10-minute intervals and the user would go to the next highest, or 80 minutes. No interpolation is permitted between depth or time intervals. Bottom time on all U.S. Navy tables is computed from the time the diver starts his descent from the surface until the time he starts his ascent. The standard ascent rate on air tables is set as a maximum of 60 feet per minute to allow short half-time tissues to desaturate. Maximum descent rate is limited to 75 feet per minute.

The column *Repet. Group* is included for use in repetitive dives. The letter found under this column for a specific exposure is one of sixteen that indicate the level of residual inert gas tension in the tissues at the conclusion of the dive.

U.S. Navy Standard Air Decompression Table for Exceptional and Extreme Exposures, Table AI-4, was designed for emergency use and is not generally used for standard operations. The table takes into consideration the longer half-time tissues, 160-minute and 240-minute, which become the major controlling tissues on long, deep exposures. These schedules were tested at 140 feet for 6 hours and at 300 feet for 1 hour. No repetitive dive group number is provided since a dive requiring decompression on this schedule would be classified as an emergency and no repetitive dives should be made.

U.S. Navy Standard Repetitive Air Dive Tables, AI-5, 6, and 7, were created to make SCUBA equipment use feasible for tasks that could not be accomplished in a single dive. The tables were calculated on the concept that after completion of a dive to any depth the diver's tissue will contain an excess of nitrogen for a period of 24 hours. Further, if the nitrogen tension present after a specific exposure were known, the rate at which the inert gas tissue tension would decrease at the surface would be predictable. Using these factors, and taking into consideration the amount of time spent at the surface between dives, the residual nitrogen in the system could be equi-

(Text continues on p. 71.)

Table AI-3 (USN Table 1-10)
U.S. Navy Standard Air Decompression Table

Depth (feet)	Bottom time (min)	Time to first stop (min:sec)	Decompression stops (feet)					Total ascent (min:sec)	Repetitive group
			50	40	30	20	10		
40	200	---						0	0:40	(*)
	210	0:30						2	2:40	N
	230	0:30						7	7:40	N
	250	0:30						11	11:40	O
	270	0:30						15	15:40	O
	300	0:30						19	19:40	Z
50	100	---						0	0:50	(*)
	110	0:40						3	3:50	L
	120	0:40						5	5:50	M
	140	0:40						10	10:50	M
	160	0:40						21	21:50	N
	180	0:40						29	29:50	O
	200	0:40						35	35:50	O
	220	0:40						40	40:50	Z
	240	0:40						47	47:50	Z
60	60	---						0	1:00	(*)
	70	0:50						2	3:00	K
	80	0:50						7	8:00	L
	100	0:50						14	15:00	M
	120	0:50						26	27:00	N
	140	0:50						39	40:00	O
	160	0:50						48	49:00	Z
	180	0:50						56	57:00	Z
	200	0:40					1	69	71:00	Z
70	50	---						0	1:10	(*)
	60	1:00						8	9:10	K
	70	1:00						14	15:10	L
	80	1:00						18	19:10	M
	90	1:00						23	24:10	N
	100	1:00						33	34:10	N
	110	0:50				2		41	44:10	O
	120	0:50				4		47	52:10	O
	130	0:50				6		52	59:10	O
	140	0:50				8		56	65:10	Z
	150	0:50				9		61	71:10	Z
	160	0:50				13		72	86:10	Z
	170	0:50				19		79	99:10	Z
80	40	---						0	1:20	(*)
	50	1:10						10	11:20	K
	60	1:10						17	18:20	L
	70	1:10						23	24:20	M
	80	1:00					2	31	34:20	N
	90	1:00					7	39	47:20	N
	100	1:00					11	46	58:20	O
	110	1:00					13	53	67:20	O
	120	1:00					17	56	74:20	Z
	130	1:00					19	63	83:20	Z
	140	1:00					26	69	96:20	Z
	150	1:00					32	77	110:20	Z

(Table AI-3 continues on p. 66.)

Table AI-3 *(continued)*

Depth (feet)	Bottom time (min)	Time to first stop (min:sec)	Decompression stops (feet)					Total ascent (min:sec)	Repetitive group
			50	40	30	20	10		
90	30	---------	---------	---------	---------	---------	0	1:30	(*)
	40	1:20	---------	---------	---------	---------	7	8:30	J
	50	1:20	---------	---------	---------	---------	18	19:30	L
	60	1:20	---------	---------	---------	---------	25	26:30	M
	70	1:10	---------	---------	---------	7	30	38:30	N
	80	1:10	---------	---------	---------	13	40	54:30	N
	90	1:10	---------	---------	---------	18	48	67:30	O
	100	1:10	---------	---------	---------	21	54	76:30	Z
	110	1:10	---------	---------	---------	24	61	86:30	Z
	120	1:10	---------	---------	---------	32	68	101:30	Z
	130	1:00	---------	---------	5	36	74	116:30	Z
100	25	---------	---------	---------	---------	---------	0	1:40	(*)
	30	1:30	---------	---------	---------	---------	3	4:40	I
	40	1:30	---------	---------	---------	---------	15	16:40	K
	50	1:20	---------	---------	---------	2	24	27:40	L
	60	1:20	---------	---------	---------	9	28	38:40	N
	70	1:20	---------	---------	---------	17	39	57:40	O
	80	1:20	---------	---------	---------	23	48	72:40	O
	90	1:10	---------	---------	3	23	57	84:40	Z
	100	1:10	---------	---------	7	23	66	97:40	Z
	110	1:10	---------	---------	10	34	72	117:40	Z
	120	1:10	---------	---------	12	41	78	132:40	Z
110	20	---------	---------	---------	---------	---------	0	1:50	(*)
	25	1:40	---------	---------	---------	---------	3	4:50	H
	30	1:40	---------	---------	---------	---------	7	8:50	J
	40	1:30	---------	---------	---------	2	21	24:50	L
	50	1:30	---------	---------	---------	8	26	35:50	M
	60	1:30	---------	---------	---------	18	36	55:50	N
	70	1:20	---------	---------	1	23	48	73:50	O
	80	1:20	---------	---------	7	23	57	88:50	Z
	90	1:20	---------	---------	12	30	64	107:50	Z
	100	1:20	---------	---------	15	37	72	125:50	Z
120	15	---------	---------	---------	---------	---------	0	2:00	(*)
	20	1:50	---------	---------	---------	---------	2	4:00	H
	25	1:50	---------	---------	---------	---------	6	8:00	I
	30	1:50	---------	---------	---------	---------	14	16:00	J
	40	1:40	---------	---------	---------	5	25	32:00	L
	50	1:40	---------	---------	---------	15	31	48:00	N
	60	1:30	---------	---------	2	22	45	71:00	O
	70	1:30	---------	---------	9	23	55	89:00	O
	80	1:30	---------	---------	15	27	63	107:00	Z
	90	1:30	---------	---------	19	37	74	132:00	Z
	100	1:30	---------	---------	23	45	80	150:00	Z
130	10	---------	---------	---------	---------	---------	0	2:10	(*)
	15	2:00	---------	---------	---------	---------	1	3:10	F
	20	2:00	---------	---------	---------	---------	4	6:10	H
	25	2:00	---------	---------	---------	---------	10	12:10	J
	30	1:50	---------	---------	---------	3	18	23:10	M

Table AI-3 *(continued)*

Depth (feet)	Bottom time (min)	Time to first stop (min:sec)	Decompression stops (feet) 50	40	30	20	10	Total ascent (min:sec)	Repetitive group
	40	1:50				10	25	37:10	N
	50	1:40			3	21	37	63:10	O
	60	1:40			9	23	52	86:10	Z
	70	1:40			16	24	61	103:10	Z
	80	1:30		3	19	35	72	131:10	Z
	90	1:30		8	19	45	80	154:10	Z
140	10						0	2:20	(*)
	15	2:10					2	4:20	G
	20	2:10					6	8:20	I
	25	2:00				2	14	18:20	J
	30	2:00				5	21	28:20	K
	40	1:50			2	16	26	46:20	N
	50	1:50			6	24	44	76:20	O
	60	1:50			16	23	56	97:20	Z
	70	1:40		4	19	32	68	125:20	Z
	80	1:40		10	23	41	79	155:20	Z
150	5						0	2:30	C
	10	2:20					1	3:30	E
	15	2:20					3	5:30	G
	20	2:10				2	7	11:30	H
	25	2:10				4	17	23:30	K
	30	2:10				8	24	34:30	L
	40	2:00			5	19	33	59:30	N
	50	2:00			12	23	51	88:30	O
	60	1:50		3	19	26	62	112:30	Z
	70	1:50		11	19	39	75	146:30	Z
	80	1:40	1	17	19	50	84	173:30	Z
160	5						0	2:40	D
	10	2:30					1	3:40	F
	15	2:20				1	4	7:40	H
	20	2:20				3	11	16:40	J
	25	2:20				7	20	29:40	K
	30	2:10			2	11	25	40:40	M
	40	2:10			7	23	39	71:40	N
	50	2:00		2	16	23	55	98:40	Z
	60	2:00		9	19	33	69	132:40	Z
	70	1:50	1	17	22	44	80	166:40	Z
170	5						0	2:50	D
	10	2:40					2	4:50	F
	15	2:30				2	5	9:50	H
	20	2:30				4	15	21:50	J
	25	2:20			2	7	23	34:50	L
	30	2:20			4	13	26	45:50	M
	40	2:10		1	10	23	45	81:50	O
	50	2:10		5	18	23	61	109:50	Z
	60	2:00	2	15	22	37	74	152:50	Z
	70	2:00	8	17	19	51	86	183:50	Z

(Table AI-3 continues on p. 68.)

Table AI-3 *(continued)*

Depth (feet)	Bottom time (min)	Time to first stop (min:sec)	Decompression stops (feet)					Total ascent (min:sec)	Repetitive group
			50	40	30	20	10		
180_ _ _ _ _ _ _ _ _ _ _	5	_ _ _ _ _ _ _ _ _ _	_ _ _ _ _ _ _	_ _ _ _ _ _ _	_ _ _ _ _ _ _	_ _ _ _ _ _ _		0	3:00	D
	10	2:50	_ _ _ _ _ _ _	_ _ _ _ _ _ _	_ _ _ _ _ _ _	_ _ _ _ _ _ _		3	6:00	F
	15	2:40	_ _ _ _ _ _ _	_ _ _ _ _ _ _	_ _ _ _ _ _ _		3	6	12:00	I
	20	2:30	_ _ _ _ _ _ _	_ _ _ _ _ _ _	1	5		17	26:00	K
	25	2:30	_ _ _ _ _ _ _	_ _ _ _ _ _ _	3	10		24	40:00	L
	30	2:30	_ _ _ _ _ _ _	_ _ _ _ _ _ _	6	17		27	53:00	N
	40	2:20	_ _ _ _ _ _ _	3	14	23		50	93:00	O
	50	2:10	2	9	19	30		65	128:00	Z
	60	2:10	5	16	19	44		81	168:00	Z
190_ _ _ _ _ _ _ _ _ _	5	_ _ _ _ _ _ _ _ _ _	_ _ _ _ _ _ _	_ _ _ _ _ _ _	_ _ _ _ _ _ _	_ _ _ _ _ _ _		0	3:10	D
	10	2:50	_ _ _ _ _ _ _	_ _ _ _ _ _ _	_ _ _ _ _ _ _	1		3	7:10	G
	15	2:50	_ _ _ _ _ _ _	_ _ _ _ _ _ _	_ _ _ _ _ _ _	4		7	14:10	I
	20	2:40	_ _ _ _ _ _ _	_ _ _ _ _ _ _	2	6		20	31:10	K
	25	2:40	_ _ _ _ _ _ _	_ _ _ _ _ _ _	5	11		25	44:10	M
	30	2:30	_ _ _ _ _ _ _	1	8	19		32	63:10	N
	40	2:30	_ _ _ _ _ _ _	8	14	23		55	103:10	O
	50	2:20	4	13	22	33		72	147:10	Z
	60	2:20	10	17	19	50		84	183:10	Z

*See table AI-5 for repetitive groups in no-decompression dives.

Table AI-4 (USN Table 1-14)
U.S. Navy Standard Air Decompression Table for Exceptional Exposures

Depth (ft)	Bottom time (min)	Time to first stop (min:sec)	Decompression stops (feet)													Total ascent time (min:sec)
			130	120	110	100	90	80	70	60	50	40	30	20	10	
40_ _ _ _	360	0:30													23	23:40
	480	0:30													41	41:40
	720	0:30													69	69:40
60_ _ _ _	240	0:40												2	79	82:00
	360	0:40												20	119	140:00
	480	0:40												44	148	193:00
	720	0:40												78	187	266:00
80_ _ _ _	180	1:00												35	85	121:20
	240	0:50											6	52	120	179:20
	360	0:50											29	90	160	280:20
	480	0:50											59	107	187	354:20
	720	0:40										17	108	142	187	455:20

Table AI-4 *(continued)*

Depth (ft)	Bottom time (min)	Time to first stop (min:sec)	Decompression stops (feet)													Total ascent time (min:sec)
			130	120	110	100	90	80	70	60	50	40	30	20	10	
100___	180	1:00	----	----	----	----	----	----	----	----	----	1	29	53	118	202:40
	240	1:00	----	----	----	----	----	----	----	----	----	14	42	84	142	283:40
	360	0:50	----	----	----	----	----	----	----	----	2	42	73	111	187	416:40
	480	0:50	----	----	----	----	----	----	----	----	21	61	91	142	187	503:40
	720	0:50	----	----	----	----	----	----	----	----	55	106	122	142	187	613:40
120___	120	1:20	----	----	----	----	----	----	----	----	----	10	19	47	98	176:00
	180	1:10	----	----	----	----	----	----	----	----	5	27	37	76	137	284:00
	240	1:10	----	----	----	----	----	----	----	----	23	35	60	97	179	396:00
	360	1:00	----	----	----	----	----	----	----	18	45	64	93	142	187	551:00
	480	0:50	----	----	----	----	----	----	3	41	64	93	122	142	187	654:00
	720	0:50	----	----	----	----	----	----	32	74	100	114	122	142	187	773:00
140___	90	1:30	----	----	----	----	----	----	----	----	2	14	18	42	88	166:20
	120	1:30	----	----	----	----	----	----	----	----	12	14	36	56	120	240:20
	180	1:20	----	----	----	----	----	----	----	10	26	32	54	94	168	386:20
	240	1:10	----	----	----	----	----	----	8	28	34	50	78	124	187	511:20
	360	1:00	----	----	----	----	----	9	32	42	64	84	122	142	187	684:20
	480	1:00	----	----	----	----	----	31	44	59	100	114	122	142	187	801:20
	720	0:50	----	----	----	----	16	56	88	97	100	114	122	142	187	924:20
170___	90	1:50	----	----	----	----	----	----	12	12	14	34	52	120		246:50
	120	1:30	----	----	----	----	2	10	12	18	32	42	82	156		356:50
	180	1:20	----	----	----	4	10	22	28	34	50	78	120	187		535:50
	240	1:20	----	----	----	18	24	30	42	50	70	116	142	187		681:50
	360	1:10	----	----	22	34	40	52	60	98	114	122	142	187		873:50
	480	1:00	----	14	40	42	56	91	97	100	114	122	142	187		1007:50
200___	5	3:10	----	----	----	----	----	----	----	----	----	----	----	----	1	4:20
	10	3:00	----	----	----	----	----	----	----	----	----	----	----	1	4	8:20
	15	2:50	----	----	----	----	----	----	----	----	----	----	1	4	10	18:20
	20	2:50	----	----	----	----	----	----	----	----	----	----	3	7	27	40:20
	25	2:50	----	----	----	----	----	----	----	----	----	----	7	14	25	49:20
	30	2:40	----	----	----	----	----	----	----	----	----	2	9	22	37	73:20
	40	2:30	----	----	----	----	----	----	----	----	2	8	17	23	59	112:20
	50	2:30	----	----	----	----	----	----	----	----	6	16	22	39	75	161:20
	60	2:20	----	----	----	----	----	----	----	2	13	17	24	51	89	199:20
	90	1:50	----	----	----	----	1	10	10	12	12	30	38	74	134	324:20
	120	1:40	----	----	----	6	10	10	10	24	28	40	64	98	180	473:20
	180	1:20	----	1	10	10	18	24	24	42	48	70	106	142	187	685:20
	240	1:20	----	6	20	24	24	36	42	54	68	114	122	142	187	842:20
	360	1:10	12	22	36	40	44	56	82	98	100	114	122	142	187	1058:20
210___	5	3:20	----	----	----	----	----	----	----	----	----	----	----	----	1	4:30
	10	3:10	----	----	----	----	----	----	----	----	----	----	----	2	4	9:30
	15	3:00	----	----	----	----	----	----	----	----	----	----	1	5	13	22:30
	20	3:00	----	----	----	----	----	----	----	----	----	----	4	10	23	40:30
	25	2:50	----	----	----	----	----	----	----	----	----	2	7	17	27	56:30
	30	2:50	----	----	----	----	----	----	----	----	----	4	9	24	41	81:30
	40	2:40	----	----	----	----	----	----	----	----	4	9	19	26	63	124:30
	50	2:30	----	----	----	----	----	----	----	1	9	17	19	45	80	174:30

(Table AI-4 continues on p. 70.)

Table AI-4 *(continued)*

Depth (ft)	Bottom time (min)	Time to first stop (min:sec)	130	120	110	100	90	80	70	60	50	40	30	20	10	Total ascent time (min:sec)
220	5	3:30													2	5:40
	10	3:20												2	5	10:40
	15	3:10											2	5	16	26:40
	20	3:00										1	3	11	24	42:40
	25	3:00										3	8	19	33	66:40
	30	2:50									1	7	10	23	47	91:40
	40	2:50									6	12	22	29	68	140:40
	50	2:40								3	12	17	18	51	86	190:40
230	5	3:40													2	5:50
	10	3:20											1	2	6	12:50
	15	3:20											3	6	18	30:50
	20	3:10										2	5	12	26	48:50
	25	3:10										4	8	22	37	74:50
	30	3:00									2	8	12	23	51	99:50
	40	2:50								1	7	15	22	34	74	156:50
	50	2:50								5	14	16	24	51	89	202:50
240	5	3:50													2	6:00
	10	3:30											1	3	6	14:00
	15	3:30											4	6	21	35:00
	20	3:20										3	6	15	25	53:00
	25	3:10									1	4	9	24	40	82:00
	30	3:10									4	8	15	22	56	109:00
	40	3:00								3	7	17	22	39	75	167:00
	50	2:50							1	8	15	16	29	51	94	218:00
250	5	3:50												1	2	7:10
	10	3:40											1	4	7	16:10
	15	3:30										1	4	7	22	38:10
	20	3:30										4	7	17	27	59:10
	25	3:20									2	7	10	24	45	92:10
	30	3:20									6	7	17	23	59	116:10
	40	3:10								5	9	17	19	45	79	178:10
	60	2:40					4	10	10	10	12	22	36	64	126	298:10
	90	2:10		8	10	10	10	10	10	28	28	44	68	98	186	514:10
260	5	4:00												1	2	7:20
	10	3:50											2	4	9	19:20
	15	3:40										2	4	10	22	42:20
	20	3:30									1	4	7	20	31	67:20
	25	3:30									3	8	11	23	50	99:20
	30	3:20								2	6	8	19	26	61	126:20
	40	3:10							1	6	11	16	19	49	84	190:20
270	5	4:10												1	3	8:30
	10	4:00											2	5	11	22:30
	15	3:50										3	4	11	24	46:30
	20	3:40									2	3	9	21	35	74:30
	25	3:30								2	3	8	13	23	53	106:30
	30	3:30								3	6	12	22	27	64	138:30
	40	3:20							5	6	11	17	22	51	88	204:30

Table AI-4 *(continued)*

Depth (ft)	Bottom time (min)	Time to first stop (min:sec)	130	120	110	100	90	80	70	60	50	40	30	20	10	Total ascent time (min:sec)
280___	5	4:20												2	2	8:40
	10	4:00										1	2	5	13	25:40
	15	3:50									1	3	4	11	26	49:40
	20	3:50									3	4	8	23	39	81:40
	25	3:40								2	5	7	16	23	56	113:40
	30	3:30							1	3	7	13	22	30	70	150:40
	40	3:20						1	6	6	13	17	27	51	93	218:40
290___	5	4:30												2	3	9:50
	10	4:10										1	3	5	16	29:50
	15	4:00									1	3	6	12	26	52:50
	20	4:00									3	7	9	23	43	89:50
	25	3:50								3	5	8	17	23	60	120:50
	30	3:40							1	5	6	16	22	36	72	162:50
	40	3:30						3	5	7	15	16	32	51	95	228:50
300___	5	4:40												3	3	11:00
	10	4:20										1	3	6	17	32:00
	15	4:10									2	3	6	15	26	57:00
	20	4:00								2	3	7	10	23	47	97:00
	25	3:50							1	3	6	8	19	26	61	129:00
	30	3:50							2	5	7	17	22	39	75	172:00
	40	3:40						4	6	9	15	17	34	51	90	231:00
	60	3:00		4	10	10	10	10	10	14	28	32	50	90	187	460:00

Extreme exposures—250 and 300 ft

Depth (ft)	Bottom time (min)	Time to first stop (min:sec)	200	190	180	170	160	150	140	130	120	110	100	90	80	70	60	50	40	30	20	10	Total ascent time (min:sec)
250_____	120	1:50							5	10	10	10	10	16	24	24	36	48	64	94	142	187	684:10
	180	1:30					4	8	8	10	22	24	24	32	42	44	60	84	114	122	142	187	931:10
	240	1:30					9	14	21	22	22	40	40	42	56	76	98	100	114	122	142	187	1109:10
300_____	90	2:20					3	8	8	10	10	10	10	16	24	24	34	48	64	90	142	187	693:00
	120	2:00				4	8	8	8	10	14	24	24	24	34	42	58	66	102	122	142	187	890:00
	180	1:40	6	8	8	8	14	20	21	21	28	40	40	48	56	82	98	100	114	122	142	187	1168:00

(Text continued from p. 64.)

librated to a time penalty at the depth of the second dive.

The 120-minute half-time tissue was selected as control and the tissue tension at time of surfacing was subdivided into sixteen groups, each representing a pressure increment of 2 feet of inert gas pressure above zero feet. Each equivalent surface tension was assigned a letter from A through Z with maximum permissible tissue tension carrying the letter Z. These letters, representing the tissue tension, were entered on the Standard Air Decompression Table, AI-3. A "no decompression" table, AI-5, was used to assign inert gas penalties to dives not requiring decompression.

A Surface Interval Credit Table, AI-6, was calculated to credit the diver with inert gas lost during the surface interval between dives.

Table AI-5 (USN Table 1-11)
No-Decompression Limits and Repetitive Group Designation Table
for No-Decompression Air Dives

Depth (feet)	No-decompression limits (min)	Repetitive groups (air dives)														
		A	B	C	D	E	F	G	H	I	J	K	L	M	N	O
10	----------	60	120	210	300	-----	-----	-----	-----	-----	-----	-----	-----	-----	-----	-----
15	----------	35	70	110	160	225	350	-----	-----	-----	-----	-----	-----	-----	-----	-----
20	----------	25	50	75	100	135	180	240	325	-----	-----	-----	-----	-----	-----	-----
25	----------	20	35	55	75	100	125	160	195	245	315	-----	-----	-----	-----	-----
30	----------	15	30	45	60	75	95	120	145	170	205	250	310	-----	-----	-----
35	310	5	15	25	40	50	60	80	100	120	140	160	190	220	270	310
40	200	5	15	25	30	40	50	70	80	100	110	130	150	170	200	-----
50	100	----	10	15	25	30	40	50	60	70	80	90	100	-----	-----	-----
60	60	----	10	15	20	25	30	40	50	55	60	-----	-----	-----	-----	-----
70	50	----	5	10	15	20	30	35	40	45	50	-----	-----	-----	-----	-----
80	40	----	5	10	15	20	25	30	35	40	-----	-----	-----	-----	-----	-----
90	30	----	5	10	12	15	20	25	30	-----	-----	-----	-----	-----	-----	-----
100	25	----	5	7	10	15	20	22	25	-----	-----	-----	-----	-----	-----	-----
110	20	----	-----	5	10	13	15	20	-----	-----	-----	-----	-----	-----	-----	-----
120	15	----	-----	5	10	12	15	-----	-----	-----	-----	-----	-----	-----	-----	-----
130	10	----	-----	5	8	10	-----	-----	-----	-----	-----	-----	-----	-----	-----	-----
140	10	----	-----	5	7	10	-----	-----	-----	-----	-----	-----	-----	-----	-----	-----
150	5	----	-----	5	-----	-----	-----	-----	-----	-----	-----	-----	-----	-----	-----	-----
160	5	----	-----	-----	5	-----	-----	-----	-----	-----	-----	-----	-----	-----	-----	-----
170	5	----	-----	-----	5	-----	-----	-----	-----	-----	-----	-----	-----	-----	-----	-----
180	5	----	-----	-----	5	-----	-----	-----	-----	-----	-----	-----	-----	-----	-----	-----
190	5	----	-----	-----	5	-----	-----	-----	-----	-----	-----	-----	-----	-----	-----	-----

A third table, the Repetitive Dive Timetable, AI-7, was then tabulated to provide the nitrogen penalty in minute equivalents for use on the second dive. Although the tabulations sound rather complex, in practice they are very simple. For example, let us assume that a diver must perform a series of two dives as follows:

Dive #1: Depth to 60 feet, bottom time 50 minutes
Surface Interval Between Dives: 2.0 hours
Dive # 2: Depth 100 feet, bottom time 26 minutes

1. Dive #1 is a "no decompression" dive. To find the inert gas tension use Table AI-5. Enter the table from left to right at 60 feet depth and move right to 50 minutes. At the top of that column is the group letter H.

2. To obtain credit for the 2.0 hour surface interval use the Surface Interval Credit Table, Table AI-6. Entering the table from left to right at H, move right until the next highest surface interval is reached, in this case, 2:23. At the top of that column is the new group letter, E.

3. This letter represents the inert gas level in the tissue at the beginning of the second dive, 2 hours after completion of Dive #1. To convert this letter to minutes of equivalent submersion at the depth of the second dive, use the Repetitive Dive Timetable, AI-7. Enter the table from left

Table AI-6 (USN Table 1-12)
Surface Interval Credit Table for Air Decompression Dives

Z	O	N	M	L	K	J	I	H	G	F	E	D	C	B	A
0:10–0:22	0:23–0:34	0:35–0:48	0:49–1:02	1:03–1:18	1:19–1:36	1:37–1:55	1:56–2:17	2:18–2:42	2:43–3:10	3:11–3:45	3:46–4:29	4:30–5:27	5:28–6:56	6:57–10:05	10:00–12:00*
O	0:10–0:23	0:24–0:36	0:37–0:51	0:52–1:07	1:08–1:24	1:25–1:43	1:44–2:04	2:05–2:29	2:30–2:59	3:00–3:33	3:34–4:17	4:18–5:16	5:17–6:44	6:45–9:54	9:55–12:00*
	N	0:10–0:24	0:25–0:39	0:40–0:54	0:55–1:11	1:12–1:30	1:31–1:53	1:54–2:18	2:19–2:47	2:48–3:22	3:23–4:04	4:05–5:03	5:04–6:32	6:33–9:43	9:44–12:00*
		M	0:10–0:25	0:26–0:42	0:43–0:59	1:00–1:18	1:19–1:39	1:40–2:05	2:06–2:34	2:35–3:08	3:09–3:52	3:53–4:49	4:50–6:18	6:19–9:28	9:29–12:00*
			L	0:10–0:26	0:27–0:45	0:46–1:04	1:05–1:25	1:26–1:49	1:50–2:19	2:20–2:53	2:54–3:36	3:37–4:35	4:36–6:02	6:03–9:12	9:13–12:00*
				K	0:10–0:28	0:29–0:49	0:50–1:11	1:12–1:35	1:36–2:03	2:04–2:38	2:39–3:21	3:22–4:19	4:20–5:48	5:49–8:58	8:59–12:00*
					J	0:10–0:31	0:32–0:54	0:55–1:19	1:20–1:47	1:48–2:20	2:21–3:04	3:05–4:02	4:03–5:40	5:41–8:40	8:41–12:00*
						I	0:10–0:33	0:34–0:59	1:00–1:29	1:30–2:02	2:03–2:44	2:45–3:43	3:44–5:12	5:13–8:21	8:22–12:00*
							H	0:10–0:36	0:37–1:06	1:07–1:41	1:42–2:23	2:24–3:20	3:21–4:49	4:50–7:59	8:00–12:00*
								G	0:10–0:40	0:41–1:15	1:16–1:59	2:00–2:58	2:59–4:25	4:26–7:35	7:36–12:00*
									F	0:10–0:45	0:46–1:29	1:30–2:28	2:29–3:57	3:58–7:05	7:06–12:00*
										E	0:10–0:54	0:55–1:57	1:58–3:22	3:23–6:32	6:33–12:00*
											D	0:10–1:09	1:10–2:38	2:39–5:48	5:49–12:00*
												C	0:10–1:39	1:40–2:49	2:50–12:00*
													B	0:10–2:10	2:11–12:00*
														A	0:10–12:00*

Repetitive group at the beginning of the surface interval from previous dive

*Dives conducted after 12-hour surface interval are not considered repetitive.

to right at group letter E and move right until under the column that lists the depth of the *second dive*, in this case 100 feet, and read the number 18. This means that at the start of the second dive the diver is starting with nitrogen tension in his tissue equivalent to being at a depth of 100 feet for 18 minutes.

4. To find the decompression required for the second dive, 100 feet for 26 minutes, add the 18 minute penalty to the second dive bottom time which gives a total of 44 minutes. On the Standard Air Decompression Table, AI-3, read the decompression for 100 feet with 44 minutes bottom time. The next highest time is 50 minutes and requires two decompression stops.

Helium/Oxygen Decompression Tables

The U.S. Navy Standard Helium/Oxygen Tables were developed on the basis of the Haldane theory, using maximum tissue half-times of 100 minutes and a maximum 1.7 to 1 ratio of tissue helium pressure to the absolute pressure of the decompression stop. Incidence of decompression

Table AI-7 (USN Table 1-13)
Repetitive Dive Timetable for Air Dives

Repetitive groups	Repetitive dive depth (ft) (air dives)															
	40	50	60	70	80	90	100	110	120	130	140	150	160	170	180	190
A	7	6	5	4	4	3	3	3	3	3	2	2	2	2	2	2
B	17	13	11	9	8	7	7	6	6	6	5	5	4	4	4	4
C	25	21	17	15	13	11	10	10	9	8	7	7	6	6	6	6
D	37	29	24	20	18	16	14	13	12	11	10	9	9	8	8	8
E	49	38	30	26	23	20	18	16	15	13	12	12	11	10	10	10
F	61	47	36	31	28	24	22	20	18	16	15	14	13	13	12	11
G	73	56	44	37	32	29	26	24	21	19	18	17	16	15	14	13
H	87	66	52	43	38	33	30	27	25	22	20	19	18	17	16	15
I	101	76	61	50	43	38	34	31	28	25	23	22	20	19	18	17
J	116	87	70	57	48	43	38	34	32	28	26	24	23	22	20	19
K	138	99	79	64	54	47	43	38	35	31	29	27	26	24	22	21
L	161	111	88	72	61	53	48	42	39	35	32	30	28	26	25	24
M	187	124	97	80	68	58	52	47	43	38	35	32	31	29	27	26
N	213	142	107	87	73	64	57	51	46	40	38	35	33	31	29	28
O	241	160	117	96	80	70	62	55	50	44	40	38	36	34	31	30
Z	257	169	122	100	84	73	64	57	52	46	42	40	37	35	32	31

sickness using the schedules has been 0.83% (six cases in 721 dives).

Unlike air decompression tables which deal with a constant percentage of inert gas in the inspired air, helium/oxygen percentages vary with the planned dive depth. Therefore, the tables must be calculated on a helium partial pressure basis. Because there are more variables, the helium/oxygen tables require more calculations for use.

To determine the percentage of oxygen to be used on a dive, both the depth of dive and length of exposure must be considered. The maximum permissible partial pressure for various exposures is obtained from the limit table, AI-8.

When the maximum oxygen partial pressure has been determined based upon exposure, the oxygen percentage for the gas mixture is determined with the following formula:

$$\text{Maximum Oxygen \%} = (L \times 33)/(D + 33)$$

where L = maximum O_2 partial pressure
D = depth of dive

Table AI-8
Maximum Partial Pressure Exposures

Exposure (minutes)	Max. O_2 Partial Pressure
30	2.0 AtA
40	1.9 AtA
60	1.8 AtA
80	1.7 AtA
100	1.6 AtA
120	1.5 AtA
180	1.4 AtA
240	1.3 AtA

The helium decompression tables consist of 36 separate schedules, each representing a discrete helium partial pressure. The 36 schedules cover dives between 60 and 410 feet in increments of 10 feet and provide for exposures of 10 to 240 minutes in 20-minute intervals.

Table AI-9 (USN Table 1-22)
Table of Helium-Oxygen Partial Pressures
(enter this table—select partial pressures)

Depth (feet)	Oxygen percent												
	15	16	17	19	21	23	25	30	35	40	45	50	55
40	64	63	63	61	60	58	57	(*)	(*)	(*)	(*)	(*)	(*)
50	73	72	71	69	68	66	64	60	56	(*)	(*)	(*)	(*)
60	81	80	80	78	76	74	72	67	63	58	54	(*)	(*)
70	90	89	88	86	84	82	80	75	70	64	59	54	(*)
80	99	98	97	94	92	90	88	82	76	71	65	59	54
90	108	106	105	103	100	98	95	89	83	77	71	64	
100	116	115	114	111	108	106	103	96	90	83	76		
110	125	123	122	119	116	113	111	103	96	89	82		
120	134	132	131	127	124	121	118	111	103	95			
130	142	141	139	136	133	129	126	118	110	102			
140	151	149	148	144	141	137	134	125	116				
150	160	158	156	152	149	145	141	132	123				
160	168	166	165	161	157	155	149	139					
170	177	175	173	169	165	161	157	147					
180	186	184	182	177	173	169	165	154					
190	195	192	190	186	181	177	172						
200	203	201	199	194	189	185	180						
210	212	209	207	202	197	192	188						
220	221	218	216	210	205	200	195						
230	229	227	224	219	214	203	203						
240	238	235	233	227	222	216							
250	247	244	241	235	230	224							
260	255	252	250	244	238								
270	264	261	258	252	246								
280	273	270	267	260	254								
290	282	278	275	269									
300	290	287	284	277									
310	299	295	292	285									
320	308	304	301										
330	316	313	309										
340	325	321	318										
350	334	330	326										
360	342	338											
370	351	347											
380	360	356											

*No decompression to 100% oxygen.

The partial pressure table, AI-10, is an "equal or next highest" computation with no interpolation possible. The partial pressure decompression schedules are based upon 100% oxygen being breathed at the 50- and 40-foot decompression stops.

U.S. Navy Helium-Oxygen Repetitive Dive Tables

A table of partial pressures, Table AI-9, is provided to determine which partial pressure schedule should be used. Enter the table from left to right at the dive depth and continue right until under the column headed by the oxygen percentage to be used. No provision is made for change of oxygen percentage during the dive. This table is based upon the formula

$$PP = (D + 33)[1.00 - (O_2 - 0.02)]$$

where D = depth of water
O_2 = decimal equivalent of oxygen percentage
0.02 = assumed loss of oxygen in life-support system

After several years of rigorous field testing and evaluation, the U.S. Navy released its helium-oxygen repetitive dive tables in mid-1970. The table format and the methodology used to determine decompression schedules is similar to the repetitive air dive tables.

The set of tables consists of two standard decompression tables, a no decompression table, surface interval credit table and repetitive dive, or penalty, timetable. One standard dive table provides decompression schedules and repetitive group designators for initial dives using a constant gas mixture through dive and decompression. The alternate standard dive table is used when 100% oxygen decompression techniques are used at the 20 and 30 feet stops.

(Text continues on p. 90.)

Table AI-10 (USN Table 1-23)
Helium-Oxygen Decompression Table

Time of dive	Time to first stop	|180	170	160	150	140	130	120	110	100	90	80	70	60	50	40	Total time
						Feet and minutes										

PARTIAL PRESSURE (60)

Time of dive	Time to first stop	180	170	160	150	140	130	120	110	100	90	80	70	60	50	40	Total time
10	4															0	4
20	4															0	4
30	4															0	4
40	4															0	4
60	4															0	4
80	2															6	8
100	2															7	9
120	2															9	11
140																	
160																	
180																	
200																	
240	2															13	15

PARTIAL PRESSURE (70)

Time of dive	Time to first stop	180	170	160	150	140	130	120	110	100	90	80	70	60	50	40	Total time
10	3															6	9
20	3															7	10
30	3															9	12
40	3															10	13
60	3															15	18
80	3															17	20
100	3															22	25
120	3															25	28
140	3															27	30
160	3															29	32
180	3															31	34
200	3															31	34
220	3															33	36
240	3															33	36

PARTIAL PRESSURE (80)

Time of dive	Time to first stop	180	170	160	150	140	130	120	110	100	90	80	70	60	50	40	Total time
10	3															6	9
20	3															10	13
30	3															13	16
40	3															17	20
60	3															24	27
80	3															32	35
100	3															40	43
120	3															42	45
140	3															45	48
160	3															47	50
180	3															48	51
200	3															48	51
220	3															48	51
240	3															50	53

Table AI-10 *(continued)*

Time of dive	Time to first stop	Feet and minutes															Total time
		180	170	160	150	140	130	120	110	100	90	80	70	60	50	40	

PARTIAL PRESSURE (90)

Time of dive	Time to first stop	180	170	160	150	140	130	120	110	100	90	80	70	60	50	40	Total time
10	3															8	11
20	3															15	18
30	3															18	21
40	3															23	26
60	3															35	38
80	3															45	48
100	3															50	53
120	3															55	58
140	3															58	61
160	3															60	63
180	3															60	63
200	3															62	65
220	3															62	65
240	3															63	66

PARTIAL PRESSURE (100)

Time of dive	Time to first stop	180	170	160	150	140	130	120	110	100	90	80	70	60	50	40	Total time
10	3															10	13
20	3															17	20
30	3															24	27
40	3															31	34
60	3															47	50
80	3															56	59
100	3															63	66
120	3															67	70
140	3															70	73
160	3															72	75
180	3															73	76
200	3															73	76
220	3															73	76
240	3															75	78

PARTIAL PRESSURE (110)

Time of dive	Time to first stop	180	170	160	150	140	130	120	110	100	90	80	70	60	50	40	Total time
10	3															12	15
20	3															21	24
30	3															31	34
40	3															39	42
60	3															56	59
80	3															67	70
100	3															75	78
120	3															78	81
140	3															81	84
160	3															83	86
180	3															84	87
200	3															84	87
220	3															85	88
240	3															86	89

(Table AI-10 continues on p. 78.)

Table AI-10 (continued)

PARTIAL PRESSURE (120)

Time of dive	Time to first stop	180	170	160	150	140	130	120	110	100	90	80	70	60	50	40	Total time
10	3															14	17
20	3															25	28
30	3															36	39
40	3															47	50
60	3															66	69
80	3															77	80
100	3															84	87
120	3															87	90
140	3															90	93
160	3															92	95
180	3															93	96
200	3															93	96
220	3															95	98
240	3															97	100

PARTIAL PRESSURE (130)

Time of dive	Time to first stop	180	170	160	150	140	130	120	110	100	90	80	70	60	50	40	Total time
10	3														0	16	19
20	3														0	29	32
30	3														0	42	45
40	3														0	53	56
60	3														0	73	76
80	3														0	86	89
100	3														0	92	95
120	3														0	96	99
140	3														0	99	102
160	3														10	92	105
180	3														10	93	106
200	3														10	94	107
220	3														10	95	108
240	3														10	96	109

PARTIAL PRESSURE (140)

Time of dive	Time to first stop	180	170	160	150	140	130	120	110	100	90	80	70	60	50	40	Total time
10	3														0	19	22
20	3														0	34	37
30	3														0	49	52
40	3														0	62	65
60	3														0	82	85
80	3														0	94	97
100	3														0	99	102
120	3														10	97	110
140	3														10	98	111
160	3														10	99	112
180	3														12	99	114
200	3														13	99	115
220	3														14	99	116
240	3														15	99	117

Table AI-10 (continued)

PARTIAL PRESSURE (150)

Time of dive	Time to first stop	180	170	160	150	140	130	120	110	100	90	80	70	60	50	40	Total time
10	3													0	10	11	24
20	3													0	10	28	41
30	3													0	10	45	58
40	3													7	10	59	79
60	3													7	10	78	98
80	3													7	10	90	110
100	3													7	10	96	116
120	3													7	11	98	119
140	3													7	13	99	122
160	3													8	15	99	125
180	3													9	15	99	126
200	3													10	16	99	128
220	3													11	16	99	129
240	3													12	16	99	130

PARTIAL PRESSURE (160)

Time of dive	Time to first stop	180	170	160	150	140	130	120	110	100	90	80	70	60	50	40	Total time
10	3												0	0	10	12	25
20	3												0	7	10	33	53
30	3												0	7	10	50	70
40	3												0	7	10	65	85
60	3												0	7	10	84	104
80	3												0	7	10	96	116
100	3												0	7	13	99	122
120	3												0	9	16	99	127
140	3												0	15	16	99	133
160	3												0	18	16	99	136
180	3												0	20	16	99	138
200	3												0	22	16	99	140
220	3												0	23	16	99	141
240	3												7	19	16	99	144

PARTIAL PRESSURE (170)

Time of dive	Time to first stop	180	170	160	150	140	130	120	110	100	90	80	70	60	50	40	Total time
10	3												0	7	10	15	35
20	3												0	7	10	36	56
30	3												0	7	10	55	75
40	3												0	7	10	70	90
60	3												7	6	10	83	109
80	3												7	9	10	98	127
100	3												7	13	14	98	135
120	3												7	17	16	99	142
140	3												8	21	16	99	147
160	3												11	22	16	99	151
180	3												11	23	16	99	152
200	3												12	23	16	99	153
220	3												14	23	16	99	155
240	3												16	23	16	99	157

(Table AI-10 continues on p. 80.)

Table AI-10 (continued)

Time of dive	Time to first stop	Feet and minutes															Total time
		180	170	160	150	140	130	120	110	100	90	80	70	60	50	40	

PARTIAL PRESSURE (180)

Time of dive	Time to first stop	180	170	160	150	140	130	120	110	100	90	80	70	60	50	40	Total time
10	3											0	7	0	10	17	37
20	3											0	7	0	10	41	61
30	3											0	7	1	10	62	83
40	3											0	7	4	10	77	101
60	3											0	7	10	10	92	122
80	3											0	9	14	13	98	137
100	3											7	5	18	15	99	147
120	3											7	9	21	16	99	155
140	3											7	11	22	16	99	158
160	3											7	15	23	16	99	163
180	3											7	17	23	16	99	165
200	3											7	19	23	16	99	167
220	3											7	21	23	16	99	169
240	3											7	23	23	16	99	171

PARTIAL PRESSURE (190)

Time of dive	Time to first stop	180	170	160	150	140	130	120	110	100	90	80	70	60	50	40	Total time
10	4											0	7	0	10	20	41
20	4											0	7	0	10	44	65
30	4											0	7	4	10	67	92
40	4											7	0	8	10	81	110
60	4											7	5	11	10	96	133
80	4											7	9	15	15	99	149
100	4											7	13	19	16	99	158
120	4											7	17	23	16	99	166
140																	
160																	
180																	
200																	
240																	

PARTIAL PRESSURE (200)

Time of dive	Time to first stop	180	170	160	150	140	130	120	110	100	90	80	70	60	50	40	Total time
10	4										0	0	7	0	10	22	43
20	4										0	7	0	2	10	50	73
30	4										0	7	0	7	10	69	97
40	4										0	7	4	9	10	84	118
60	4										0	7	9	13	12	93	138
80	4										7	3	13	18	15	99	159
100	4										7	6	16	21	16	99	169
120	4										7	8	20	23	16	99	177
140																	
160																	
180																	
200																	
240																	

Table AI-10 (continued)

Time of dive	Time to first stop	Feet and minutes															Total time
		180	170	160	150	140	130	120	110	100	90	80	70	60	50	40	

PARTIAL PRESSURE (210)

Time of dive	Time to first stop	180	170	160	150	140	130	120	110	100	90	80	70	60	50	40	Total time
10	4										0	7	0	0	10	25	46
20	4										0	7	0	4	10	53	78
30	4										7	0	3	7	10	74	105
40	4										7	0	7	10	10	86	124
60	4										7	4	10	14	13	98	150
80	4										7	8	14	18	16	99	166
100																	
120																	
140																	
160																	
180																	
200																	
240																	

PARTIAL PRESSURE (220)

Time of dive	Time to first stop	180	170	160	150	140	130	120	110	100	90	80	70	60	50	40	Total time
10	4										0	0	7	0	0	10	28	49
20	4										0	7	0	1	6	10	57	85
30	4										0	7	0	6	7	10	79	113
40	4										0	7	3	9	10	10	90	133
60	4										7	0	9	11	17	13	98	159
80	4										7	3	11	15	20	13	99	172
100																	
120																	
140																	
160																	
180																	
200																	
240																	

PARTIAL PRESSURE (230)

Time of dive	Time to first stop	180	170	160	150	140	130	120	110	100	90	80	70	60	50	40	Total time
10	4										0	0	7	0	2	10	30	53
20	4										0	7	0	3	7	10	61	92
30	4										0	7	2	6	9	10	81	119
40	4										7	0	6	9	11	10	93	140
60	4										7	4	9	12	18	14	99	167
80																	
100																	
120																	
140																	
160																	
180																	
200																	
240																	

(Table AI-10 continues on p. 82.)

Table AI-10 *(continued)*

Time of dive	Time to first stop	180	170	160	150	140	130	120	110	100	90	80	70	60	50	40	Total time

PARTIAL PRESSURE (240)

Time of dive	Time to first stop	180	170	160	150	140	130	120	110	100	90	80	70	60	50	40	Total time
10	4								0	0	7	0	0	3	10	33	57
20	4								0	7	0	1	4	7	10	65	98
30	4								0	7	0	5	7	10	10	85	128
40	4								7	0	3	7	9	13	11	95	149
60	4								7	0	8	10	14	18	15	99	175
80																	
100																	
120																	
140																	
160																	
180																	
200																	
240																	

PARTIAL PRESSURE (250)

Time of dive	Time to first stop	180	170	160	150	140	130	120	110	100	90	80	70	60	50	40	Total time
10	4								0	7	0	0	2	4	10	35	62
20	4								0	7	0	2	5	7	10	68	103
30	4								7	0	2	6	7	10	10	87	133
40	4								7	0	5	8	9	14	12	96	155
60																	
80																	
100																	
120																	
140																	
160																	
180																	
200																	
240																	

PARTIAL PRESSURE (260)

Time of dive	Time to first stop	180	170	160	150	140	130	120	110	100	90	80	70	60	50	40	Total time
10	4								0	7	0	0	2	4	10	37	64
20	4								7	0	0	3	7	7	10	70	108
30	4								7	0	4	6	8	10	10	89	138
40	4								7	2	5	9	9	14	13	96	159
60																	
80																	
100																	
120																	
140																	
160																	
180																	
200																	
240																	

Table AI-10 (continued)

PARTIAL PRESSURE (270)

Time of dive	Time to first stop	180	170	160	150	140	130	120	110	100	90	80	70	60	50	40	Total time
10	4							0	7	0	0	0	4	4	10	40	69
20	4							0	7	0	2	4	6	7	10	74	114
30	4							7	0	2	5	6	9	10	10	92	145
40	4							7	0	3	8	9	10	15	14	96	166
60																	
80																	
100																	
120																	
140																	
160																	
180																	
200																	
240																	

PARTIAL PRESSURE (280)

Time of dive	Time to first stop	180	170	160	150	140	130	120	110	100	90	80	70	60	50	40	Total time
10	4							0	7	0	0	2	3	4	10	42	72
20	4							7	0	0	2	6	6	8	10	78	121
30	4							7	0	3	6	6	9	13	10	93	151
40																	
60																	
80																	
100																	
120																	
140																	
160																	
180																	
200																	
240																	

PARTIAL PRESSURE (290)

Time of dive	Time to first stop	180	170	160	150	140	130	120	110	100	90	80	70	60	50	40	Total time
10	4						0	0	7	0	0	3	3	4	10	46	77
20	4						0	7	0	0	4	6	7	7	10	81	126
30	4						7	0	1	5	5	9	9	12	10	96	158
40																	
60																	
80																	
100																	
120																	
140																	
160																	
180																	
200																	
240																	

(Table AI-10 continues on p. 84.)

Table AI-10 *(continued)*

Time of dive	Time to first stop	Feet and minutes															Total time
		180	170	160	150	140	130	120	110	100	90	80	70	60	50	40	
PARTIAL PRESSURE (190)																	
10																	
20																	
30																	
40																	
60																	
80																	
100																	
120																	
140	4											9	19	23	16	99	170
160	4											11	20	23	16	99	173
180	4											13	21	23	16	99	176
200	4											14	22	23	16	99	178
220	4											15	23	23	16	99	180
240	4											17	23	23	16	99	182
PARTIAL PRESSURE (200)																	
10																	
20																	
30																	
40																	
60																	
80																	
100																	
120																	
140	4										7	11	21	23	16	99	181
160	4										7	15	23	23	16	99	187
180	4										7	17	23	23	16	99	189
200	4										7	18	23	23	16	99	190
220	4										7	20	23	23	16	99	192
240	4										8	20	23	23	16	99	193
PARTIAL PRESSURE (210)																	
10																	
20																	
30																	
40																	
60																	
80																	
100	4										7	12	17	23	16	99	178
120	4										8	15	21	23	16	99	186
140	4										10	17	21	23	16	99	190
160	4										12	17	22	23	16	99	193
180	4										14	18	22	23	16	99	196
200	4										16	18	23	23	16	99	199
220	4										17	19	23	23	16	99	201
240	4										18	20	23	23	16	99	203

Table AI-10 *(continued)*

PARTIAL PRESSURE (220)

Time of dive	Time to first stop	180	170	160	150	140	130	120	110	100	90	80	70	60	50	40	Total time
10																	
20																	
30																	
40																	
60																	
80																	
100	4									7	6	14	19	23	16	99	188
120	4									7	8	18	23	23	16	99	198
140	4									7	11	18	23	23	16	99	201
160	4									7	14	19	23	23	16	99	205
180	4									7	15	20	23	23	16	99	207
200	4									7	16	20	23	23	16	99	208
220	4									8	17	20	23	23	16	99	210
240	4									9	19	20	23	23	16	99	213

PARTIAL PRESSURE (230)

Time of dive	Time to first stop	180	170	160	150	140	130	120	110	100	90	80	70	60	50	40	Total time
10																	
20																	
30																	
40																	
60																	
80	4								0	7	8	12	17	21	16	99	184
100	4								0	7	12	15	20	23	16	99	196
120	4								0	8	14	19	23	23	16	99	206
140	4								0	10	16	20	23	23	16	99	211
160	4								7	6	18	20	23	23	16	99	216
180	4								7	7	19	20	23	23	16	99	218
200	4								7	9	19	20	23	23	16	99	220
220	4								7	11	19	20	23	23	16	99	222
240	4								7	13	19	20	23	23	16	99	224

PARTIAL PRESSURE (240)

Time of dive	Time to first stop	180	170	160	150	140	130	120	110	100	90	80	70	60	50	40	Total time
10																	
20																	
30																	
40																	
60																	
80	4								7	3	10	14	18	23	16	99	194
100	4								7	6	12	17	23	23	16	99	207
120	4								7	7	16	19	23	23	16	99	214
140	4								7	11	16	20	23	23	16	99	219
160	4								7	13	19	20	23	23	16	99	224
180	4								8	15	19	20	23	23	16	99	227
200	4								8	17	19	20	23	23	16	99	229
220	4								9	17	19	20	23	23	16	99	230
240	4								11	17	19	20	23	23	16	99	232

(Table AI-10 continues on p. 86.)

Table AI-10 *(continued)*

PARTIAL PRESSURE (250)

Time of dive	Time to first stop	180	170	160	150	140	130	120	110	100	90	80	70	60	50	40	Total time
10																	
20																	
30																	
40																	
60	4							0	7	4	8	11	14	19	16	99	182
80	4							0	7	7	11	16	18	23	16	99	201
100	4							0	7	10	14	19	23	23	16	99	215
120	4							7	3	12	17	19	23	23	16	99	223
140	4							7	4	15	18	19	23	23	16	99	228
160	4							7	7	16	19	19	23	23	16	99	233
180	4							7	9	17	19	19	23	23	16	99	237
200	4							7	11	17	19	20	23	23	16	99	239
220	4							7	12	17	19	20	23	23	16	99	240
240	4							7	13	17	19	20	23	23	16	99	241

PARTIAL PRESSURE (260)

Time of dive	Time to first stop	180	170	160	150	140	130	120	110	100	90	80	70	60	50	40	Total time
10																	
20																	
30																	
40																	
60	4							7	0	7	9	12	16	21	16	99	191
80	4							7	3	9	13	15	21	23	16	99	210
100	4							7	6	11	14	19	23	23	16	99	222
120	4							7	8	13	19	20	23	23	16	99	232
140	4							7	11	15	19	20	23	23	16	99	237
160	4							8	13	17	19	20	23	23	16	99	242
180	4							9	14	17	19	20	23	23	16	99	244
200	4							10	16	17	19	20	23	23	16	99	247
220	4							11	16	17	19	20	23	23	16	99	248
240	4							13	16	17	19	20	23	23	16	99	250

PARTIAL PRESSURE (270)

Time of dive	Time to first stop	180	170	160	150	140	130	120	110	100	90	80	70	60	50	40	Total time
10																	
20																	
30																	
40																	
60	4						0	7	3	7	10	14	16	21	16	99	197
80	4						0	7	6	10	13	17	23	23	16	99	218
100	4						7	2	9	13	16	20	23	23	16	99	232
120	4						7	4	11	14	19	20	23	23	16	99	240
140	4						7	5	14	15	19	20	23	23	16	99	245
160	4						7	7	15	17	19	20	23	23	16	99	250
180	4						7	9	16	17	19	20	23	23	16	99	253
200	4						7	11	16	17	19	20	23	23	16	99	255
220	4						7	13	16	17	19	20	23	23	16	99	257
240	4						7	15	16	17	19	20	23	23	16	99	259

Table AI-10 *(continued)*

PARTIAL PRESSURE (280)

Time of dive	Time to first stop	180	170	160	150	140	130	120	110	100	90	80	70	60	50	40	Total time
10																	
20																	
30																	
40	4						7	0	2	5	8	8	12	16	13	98	173
60	4						7	0	6	8	10	14	19	23	16	99	206
80	4						7	3	8	11	14	17	23	23	16	99	225
100	4						7	5	11	13	16	20	23	23	16	99	237
120	4						7	8	12	16	19	20	23	23	16	99	247
140	4						7	10	16	17	19	20	23	23	16	99	254
160	4						8	13	16	17	19	20	23	23	16	99	258
180	4						9	14	16	17	19	20	23	23	16	99	260
200	4						10	15	16	17	19	20	23	23	16	99	262
220	4						12	15	16	17	19	20	23	23	16	99	264
240	4						14	15	16	17	19	20	23	23	16	99	266

PARTIAL PRESSURE (290)

Time of dive	Time to first stop	180	170	160	150	140	130	120	110	100	90	80	70	60	50	40	Total time
10																	
20																	
30																	
40	4					0	7	0	4	6	8	9	12	17	15	98	180
60	4					0	7	4	6	8	12	15	18	23	16	99	212
80	4					7	0	7	9	11	15	17	23	23	16	99	231
100	4					7	2	9	11	15	17	20	23	23	16	99	246
120	4					7	4	11	13	16	19	20	23	23	16	99	255
140	4					7	5	13	16	17	19	20	23	23	16	99	262
160	4					7	8	14	16	17	19	20	23	23	16	99	266
180	4					7	10	15	16	17	19	20	23	23	16	99	269
200	4					7	12	15	16	17	19	20	23	23	16	99	271
220	4					7	13	15	16	17	19	20	23	23	16	99	272
240	4					7	14	15	16	17	19	20	23	23	16	99	273

PARTIAL PRESSURE (300)

Time of dive	Time to first stop	180	170	160	150	140	130	120	110	100	90	80	70	60	50	40	Total time
10	5				0	0	0	7	0	0	0	4	3	4	10	49	82
20	5				0	0	7	0	0	2	6	6	6	9	10	83	134
30	5				0	0	7	0	2	5	5	9	9	14	12	94	162
40	5				0	0	7	0	5	7	8	11	13	17	15	98	186
60	5				0	7	0	6	7	9	12	15	20	23	16	99	219
80	5				0	7	2	8	10	12	16	19	23	23	16	99	240
100	5				0	7	5	10	12	15	19	20	23	23	16	99	254
120	5				0	7	8	11	16	17	19	20	23	23	16	99	264
140	5				0	8	9	14	16	17	19	20	23	23	16	99	269
160	5				0	8	13	15	16	17	19	20	23	23	16	99	274
180	5				7	3	13	15	16	17	19	20	23	23	16	99	276
200	5				7	5	14	15	16	17	19	20	23	23	16	99	279
220	5				7	6	14	15	16	17	19	20	23	23	16	99	280
240	5				7	9	14	15	16	17	19	20	23	23	16	99	283

(Table AI-10 continues on p. 88.)

Table AI-10 *(continued)*

Feet and minutes — columns 180 through 40.

Time of dive	Time to first stop	180	170	160	150	140	130	120	110	100	90	80	70	60	50	40	Total time
PARTIAL PRESSURE (310)																	
10	5	------	------	------	0	0	0	7	0	0	2	3	3	5	10	52	87
20	5	------	------	------	0	0	7	0	0	4	5	6	6	11	10	84	138
30	5	------	------	------	0	7	0	0	5	5	7	8	9	14	12	96	168
40	5	------	------	------	0	7	0	3	5	8	8	11	13	18	15	99	192
60	5	------	------	------	0	7	3	6	7	10	12	18	22	23	16	99	228
80	5	------	------	------	7	0	6	9	11	12	16	19	23	23	16	99	246
100	5	------	------	------	7	1	9	10	14	17	19	20	23	23	16	99	263
120	5	------	------	------	7	4	11	12	14	17	19	20	23	23	16	99	270
140	5	------	------	------	7	5	12	15	16	17	19	20	23	23	16	99	277
160	5	------	------	------	7	8	14	15	16	17	19	20	23	23	16	99	282
180	5	------	------	------	7	10	14	15	16	17	19	20	23	23	16	99	284
200	5	------	------	------	7	12	14	15	16	17	19	20	23	23	16	99	286
220	5	------	------	------	8	13	14	15	16	17	19	20	23	23	16	99	288
240	5	------	------	------	9	13	14	15	16	17	19	20	23	23	16	99	289
PARTIAL PRESSURE (320)																	
10	5	------	------	0	0	0	7	0	0	0	3	3	3	7	10	54	92
20	5	------	------	0	0	7	0	0	2	4	5	6	7	10	10	85	141
30	5	------	------	0	0	7	0	2	4	5	7	8	11	15	13	98	175
40	5	------	------	0	7	0	1	4	6	7	8	12	15	19	16	99	199
60	5	------	0	7	0	5	6	9	11	13	17	20	23	23	16	99	231
80	5	------	------	0	7	3	7	9	11	13	17	20	23	23	16	99	253
100	5	------	------	0	7	5	9	11	13	17	19	20	23	23	16	99	267
120	5	------	------	0	7	7	12	13	16	17	19	20	23	23	16	99	277
140	5	------	------	7	2	9	12	15	16	17	19	20	23	23	16	99	283
160	5	------	------	7	3	11	14	15	16	17	19	20	23	23	16	99	288
180	5	------	------	7	5	11	14	15	16	17	19	20	23	23	16	99	290
200	5	------	------	7	6	13	14	15	16	17	19	20	23	23	16	99	293
220	5	------	------	7	7	13	14	15	16	17	19	20	23	23	16	99	294
240	5	------	------	7	9	13	14	15	16	17	19	20	23	23	16	99	296
PARTIAL PRESSURE (330)																	
10	5	------	------	0	0	0	7	0	0	0	4	3	3	7	10	56	95
20	5	------	------	0	0	7	0	0	3	5	5	6	8	10	10	88	147
30	5	------	------	0	7	0	0	4	4	6	7	9	11	17	13	98	181
40	5	------	------	0	7	0	4	4	6	7	9	12	16	20	16	99	205
60	5	------	------	7	0	2	6	8	9	11	14	17	23	23	16	99	240
80	5	------	------	7	0	6	8	8	13	14	19	20	23	23	16	99	261
100	5	------	------	7	2	7	10	13	16	17	19	20	23	23	16	99	277
120	5	------	------	7	4	9	12	13	16	17	19	20	23	23	16	99	283
140	5	------	------	7	6	11	13	15	16	17	19	20	23	23	16	99	290
160	5	------	------	7	8	13	14	15	16	17	19	20	23	23	16	99	295
180	5	------	------	7	10	13	14	15	16	17	19	20	23	23	16	99	297
200	5	------	------	7	12	13	14	15	16	17	19	20	23	23	16	99	299
220	5	------	------	9	12	13	14	15	16	17	19	20	23	23	16	99	301
240	5	------	------	10	12	13	13	15	16	17	19	20	23	23	16	99	302

Table AI-10 *(continued)*

PARTIAL PRESSURE (340)

Time of dive	Time to first stop	180	170	160	150	140	130	120	110	100	90	80	70	60	50	40	Total time
10	5	------	0	0	0	7	0	0	0	2	3	3	4	7	10	59	100
20	5	------	0	0	7	0	0	2	3	4	6	5	10	10	10	90	152
30	5	------	0	0	7	0	1	4	5	6	8	8	13	17	14	98	186
40	5	------	0	7	0	1	4	5	7	7	10	12	17	22	16	99	212
60	5	------	0	7	0	5	6	8	9	11	15	20	23	23	16	99	247
80	5	------	0	7	2	7	8	10	13	15	19	20	23	23	16	99	267
100	5	------	0	7	5	9	9	13	16	17	19	20	23	23	16	99	281
120	5	------	7	1	7	10	13	15	16	17	19	20	23	23	16	99	291
140	5	------	7	2	9	12	14	15	16	17	19	20	23	23	16	99	297
160	5	------	7	4	10	13	14	15	16	17	19	20	23	23	16	99	301
180	5	------	7	5	12	13	14	15	16	17	19	20	23	23	16	99	304
200	5	------	7	6	12	13	14	15	16	17	19	20	23	23	16	99	305
220	5	------	7	8	12	13	14	15	16	17	19	20	23	23	16	99	307
240	5	------	7	10	12	13	14	15	16	17	19	20	23	23	16	99	309

PARTIAL PRESSURE (350)

Time of dive	Time to first stop	180	170	160	150	140	130	120	110	100	90	80	70	60	50	40	Total time
10	5	------	0	0	0	7	0	0	0	3	3	3	4	7	10	61	103
20	5	------	0	0	7	0	0	2	4	5	7	8	9	10	10	90	157
30	5	------	0	7	0	0	3	5	5	6	8	9	13	18	14	98	191
40	5	------	0	7	0	2	4	6	7	8	10	13	16	22	16	99	215
60	5	------	7	0	3	5	6	9	10	13	16	18	21	23	16	99	251
80	5	------	7	0	7	7	8	11	13	15	19	20	23	23	16	99	273
100	5	------	7	2	8	8	12	13	16	17	19	20	23	23	16	99	288
120	5	------	7	4	9	11	13	15	16	17	19	20	23	23	16	99	297
140	5	------	7	6	11	13	14	15	16	17	19	20	23	23	16	99	304
160	5	------	7	9	11	13	14	15	16	17	19	20	23	23	16	99	307
180	5	------	8	9	12	13	14	15	16	17	19	20	23	23	16	99	309
200	5	------	8	11	12	13	14	15	16	17	19	20	23	23	16	99	311
220	5	------	10	11	12	13	14	15	16	17	19	20	23	23	16	99	313
240	5	------	11	11	12	13	14	15	16	17	19	20	23	23	16	99	314

PARTIAL PRESSURE (360)

Time of dive	Time to first stop	180	170	160	150	140	130	120	110	100	90	80	70	60	50	40	Total time
10	5	0	0	0	7	0	0	0	2	2	3	3	5	7	10	64	108
20	5	0	0	7	0	0	0	4	4	5	5	7	9	13	10	94	163
30	5	0	0	7	0	1	4	4	5	7	8	11	13	18	14	99	196
40	5	0	7	0	1	3	5	6	7	8	11	14	17	23	16	99	222
60	5	0	7	0	5	5	8	8	11	12	16	19	23	23	16	99	257
80	5	0	7	2	7	7	10	11	13	17	19	20	23	23	16	99	279
100	5	7	0	6	8	9	11	15	16	17	19	20	23	23	16	99	294
120	5	7	1	7	9	12	14	15	16	17	19	20	23	23	16	99	303
140	5	7	3	9	11	13	14	15	16	17	19	20	23	23	16	99	310
160	5	7	4	10	12	13	14	15	16	17	19	20	23	23	16	99	313
180	5	7	5	11	12	13	14	15	16	17	19	20	23	23	16	99	315
200	5	7	7	11	12	13	14	15	16	17	19	20	23	23	16	99	317
220	5	7	9	11	12	13	14	15	16	17	19	20	23	23	16	99	319
240	5	7	11	11	12	13	14	15	16	17	19	20	23	23	16	99	321

The use of Tables AI-11, 12, 13, 14 and 15 is simple and identical to that described for the repetitive air tables AI-3, 5, 6 and 7. It is extremely important when using the helium-oxygen tables to realize that they are *computed on the basis of the partial pressures developed while using a gas mixture of 68/32 and are therefore valid only for this mixture.* Since the computations to determine decompression are not based upon linear functions, any attempt at direct extrapolation to other mixtures will produce erroneous schedules with concomitant probability of bends.

Commercial Diving Tables

The requirements of commercial diving and military operations for which the U.S. Navy tables were designed are quite different. The level of work is much higher on commercial diving operations and the standard tables will often not provide adequate protection.

Commercial diving systems are usually more versatile. Diving bell-deck decompression chamber combinations are normally used because decompression of divers in the water is inconvenient on a work site. The more complex system has the advantage of greater flexibility in gas supply percentages because the oxygen level can be changed during the dive. Gas mixtures with optimum rather than maximum oxygen percentages, commensurate with heavy work, may therefore be used without consideration of start of dive oxygen levels.

Table AI-11 (USN Table 1-17)
Decompression Table for Mixed Gas SCUBA
(68% Helium, 32% Oxygen supply)

Depth (ft)	Bottom time (min)	Time to first stop (min:sec)	Decompression stops (feet)					Total ascent time (min:sec)	Repetitive group
			50	40	30	20	10		
40	260	--------	--------	--------	--------	--------	0	0:40	L
50	180	--------	--------	--------	--------	--------	0	0:50	L
	200	0:40	--------	--------	--------	--------	20	20:50	L
60	130	--------	--------	--------	--------	--------	0	1:00	L
	150	0:50	--------	--------	--------	--------	20	21:00	L
	170	0:50	--------	--------	--------	--------	35	36:00	L
70	85	--------	--------	--------	--------	--------	0	1:10	J
	100	1:00	--------	--------	--------	--------	15	16:10	K
	115	1:00	--------	--------	--------	--------	25	26:10	L
	130	1:00	--------	--------	--------	--------	40	41:10	L
80	60	--------	--------	--------	--------	--------	0	1:20	I
	70	1:00	--------	--------	--------	5	10	16:20	J
	80	1:00	--------	--------	--------	10	15	26:20	K
	90	1:00	--------	--------	--------	10	25	36:20	K
	100	1:00	--------	--------	--------	10	35	46:20	K
90	45	--------	--------	--------	--------	--------	0	1:30	H
	60	1:10	--------	--------	--------	5	10	16:30	J
	70	1:10	--------	--------	--------	5	20	26:30	K
	85	1:10	--------	--------	--------	10	30	41:30	L

Table AI-11 *(continued)*

Depth (ft)	Bottom time (min)	Time to first stop (min:sec)	Decompression stops (feet)					Total ascent time (min:sec)	Repetitive group
			50	40	30	20	10		
100	35						0	1:40	G
	50	1:20				5	15	21:40	J
	60	1:20				10	20	31:40	K
	70	1:10			5	15	25	46:40	K
110	30						0	1:50	G
	40	1:30				5	10	16:50	H
	50	1:30				10	20	31:50	J
	65	1:20			5	15	25	46:50	L
120	25						0	2:00	G
	35	1:40				5	10	17:00	I
	45	1:30			5	10	15	32:00	K
	55	1:30			10	15	20	47:00	L
130	20						0	2:10	F
	30	1:50				5	10	17:10	I
	40	1:40			5	10	15	32:10	J
	50	1:30		5	5	15	20	47;10	L
140	15						0	2:20	E
	25	2:00				5	10	17:20	G
	35	1:50			5	10	20	37:20	J
	45	1:40		5	5	15	25	52:20	K
150	15						0	2:30	E
	20	2:10				5	10	17:30	G
	30	2:00			5	10	15	32:30	J
	40	1:50		5	10	15	20	52:30	K
160	10						0	2:40	E
	20	2:10			5	5	10	22:40	G
	35	2:00		5	10	10	20	47:40	K
170	10						0	2:50	E
	20	2:20			5	5	10	22:50	H
	35	2:10		5	10	15	20	52:50	K
180	5						0	3:00	C
	10	2:40				5	10	18:00	E
	20	2:20		5	5	10	10	33:00	H
	30	2:20		5	10	15	20	53:00	K
190	10	2:50				5	10	18:10	E
	20	2:30		5	5	10	20	43:10	H
	30	2:20	5	5	10	15	25	63:10	K
200	10	3:00				5	15	23:20	F
	20	2:40		5	5	10	20	43:20	I
	30	2:30	5	5	10	15	35	73:20	K

Table AI-12 (USN Table 1-18)
Decompression Table for Mixed Gas SCUBA
(68% Helium, 32% Oxygen supply)
Oxygen decompression at 30- and 20-feet stops

| Depth (ft) | Time (min) | Decompression stops | | | | | Repetitive group |
| | | He-O$_2$ | | | Oxygen | | |
		50'	40'		30'	20'	
60	170					20	L
70	115					15	L
	130					25	L
80	80					15	K
	90					20	K
	100					25	K
90	70					15	K
	85					25	L
100	50					15	J
	60					20	K
	70				5	20	K
110	50					15	J
	65				5	20	L
120	45				5	15	K
	55				10	20	L
130	40				5	15	J
	50		5		5	20	L
140	35				5	15	J
	45		5		5	20	K
150	30				5	15	J
	40		5		10	20	K
160	20				5	10	G
	35		5		10	20	K
170	20				5	10	H
	35		5		10	20	K
180	20		5		5	10	H
	30		5		10	20	K
190	20		5		5	15	H
	30	5	5		10	20	K
200	20		5		5	20	I
	30	5	5		10	25	K

(center column, vertical text: ALLOW 2 MINUTES TO COMPLETE BAG PURGE TO OXYGEN)

Table AI-13 (USN Table 1-19)
No-Decompression Limits and Repetitive Group Designators
for No-Decompression Helium-Oxygen Dives

Depth (ft)	No-decompression limits (min)	Repetitive groups (He-O₂ dives)											
		A	B	C	D	E	F	G	H	I	J	K	L
10	------------	70	190	720	------	------	------	------	------	------	------	------	-----
20	------------	25	60	95	145	215	335	720	------	------	------	------	------
30	------------	15	35	60	80	110	145	185	245	335	525	720	------
40	260	10	25	40	55	70	90	110	140	165	200	245	260
50	180	10	20	30	40	55	70	85	100	120	140	160	180
60	130	5	15	25	35	45	55	65	75	90	105	120	130
70	85	5	10	20	30	35	45	55	65	75	85	------	------
80	60	5	10	15	25	30	40	45	55	60	------	------	------
90	45	5	10	15	20	25	35	40	45	------	------	------	------
100	35	5	10	15	20	25	30	35	------	------	------	------	------
110	30	5	9	12	15	20	25	30	------	------	------	------	------
120	25	5	8	10	15	20	22	25	------	------	------	------	------
130	20	------	5	10	15	17	20	------	------	------	------	------	------
140	15	------	5	10	12	15	------	------	------	------	------	------	------
150	15	------	5	10	12	15	------	------	------	------	------	------	------
160	10	------	5	6	8	10	------	------	------	------	------	------	------
170	10	------	5	6	8	10	------	------	------	------	------	------	------
180	5	------	------	5	------	------	------	------	------	------	------	------	------
190	------------	------	------	------	------	------	------	------	------	------	------	------	------
200	------------	------	------	------	------	------	------	------	------	------	------	------	------

Table AI-14 (USN Table 1-20)
Surface Interval Credit Table for Helium-Oxygen Repetitive Dives

[Repetitive group at the end of the surface interval (He-O₂ dives)]

Repetitive group at the beginning of the surface interval from previous dive

	L	K	J	I	H	G	F	E	D	C	B	A
L	0:00 0:30	0:31 0:40	0:41 0:50	0:51 1:20	1:21 1:40	1:41 2:00	2:01 2:30	2:31 3:10	3:11 4:00	4:01 5:10	5:11 7:10	7:11 12:00*
K		0:00 0:30	0:31 0:40	0:41 1:00	1:01 1:20	1:21 1:50	1:51 2:20	2:21 3:00	3:01 3:50	3:51 5:00	5:01 7:00	7:01 12:00*
J			0:00 0:30	0:31 0:40	0:41 1:00	1:01 1:30	1:31 2:00	2:01 2:40	2:41 3:30	3:31 4:40	4:41 6:40	6:41 12:00*
I				0:00 0:30	0:31 0:50	0:51 1:20	1:21 1:50	1:51 2:20	2:21 3:10	3:11 4:20	4:21 6:20	6:21 12:00*
H					0:00 0:30	0:31 0:50	0:51 1:30	1:31 2:00	2:01 2:50	2:51 4:00	4:01 6:00	6:01 12:00*
G						0:00 0:30	0:31 1:00	1:01 1:40	1:41 2:30	2:31 3:40	3:41 5:40	5:41 12:00*
F							0:00 0:35	0:36 1:10	1:11 2:00	2:01 3:10	3:11 5:10	5:11 12:00*
E								0:00 0:40	0:41 1:30	1:31 2:40	2:41 4:40	4:41 12:00*
D									0:00 0:50	0:51 2:00	2:01 4:00	4:01 12:00*
C										0:00 1:20	1:21 3:10	3:11 12:00*
B											0:00 2:00	2:01 12:00*
A												0:00 12:00*

*Dives conducted after 12-hour surface interval are not considered repetitive.

Table AI-15 (USN Table 1-21)
Repetitive Dive Timetable for Helium-Oxygen Dives

Repetitive groups	Repetitive dive depth (ft) (He-O₂ dives)																
	40	50	60	70	80	90	100	110	120	130	140	150	160	170	180	190	200
A	13	10	8	7	6	6	5	5	4	4	4	4	3	3	3	3	3
B	26	21	17	14	13	11	10	9	8	8	7	7	6	6	6	5	5
C	40	32	26	22	19	17	15	14	13	12	11	10	9	9	8	8	8
D	56	44	35	30	26	23	20	19	17	15	14	13	13	12	11	11	10
E	74	57	45	38	33	29	26	23	21	19	18	17	16	15	14	13	13
F	93	71	56	47	40	35	31	28	26	24	22	20	19	18	17	16	15
G	115	86	67	56	47	42	37	33	30	28	26	24	22	21	20	19	18
H	139	102	79	66	55	48	43	39	35	32	29	28	26	24	22	21	20
I	168	120	92	76	64	56	49	44	40	37	33	31	29	27	26	24	23
J	203	141	105	87	72	63	55	49	45	41	37	35	32	31	28	27	25
K	248	165	120	98	80	71	61	55	49	45	42	39	36	34	32	30	28
L	305	191	137	111	91	79	68	61	55	50	46	43	40	37	35	33	31

Table AI-16: Helium/Oxygen Decompression Schedule
150-Foot Dive — Three-Hour Bottom Time

Minutes	Depth (Ft.)	Elapsed Time	Gas Mix
3	0-150	0:00-0:03	He 80%—O₂ 20%
177	150	0:03-3:00	
2	150-80	3:00-3:02	
9	80	3:02-3:11	Air
1	80-70	3:11-3:12	
14	70	3:12-3:26	
1	70-60	3:26-3:27	
14	60	3:27-3:41	
1	60-50	3:41-3:42	
30	50	3:42-4:12	
5	50-40	4:12-4:17	
15	40	4:17-4:32	
5	40-35	4:32-4:37	
10	35	4:37-4:47	100% O₂
5	35-30	4:47-4:52	
35	30	4:52-5:27	Air
5	30-25	5:27-5:32	
20	25	5:32-5:52	100% O₂
5	25-20	5:52-5:57	
55	20	5:57-6:52	Air
5	20-15	6:52-6:57	

(Table AI-16 continues on p. 96)

Table AI-16 (*Continued*)

Minutes	Depth (Ft.)	Elapsed Time	Gas Mix
30	15	6:57-7:27	100% O_2
5	15-10	7:27-7:32	
90	10	7:32-9:02	Air
5	10-5	9:02-9:07	
35	5	9:07-9:42	100% O_2
5	5-0	9:42-9:47	

Flow Rate: O_2 20% He 80% at 15 liters per minute.

Table AI-17: Helium/Oxygen Decompression Schedule
250-Foot Dive — Two-Hour Bottom Time

Minutes	Depth (Ft.)	Elapsed Time	Gas Mix
1	0-50	0:00-0:01	He 80%—O_2 20%
1	50-100	0:01-0:02	
1	100-150	0:02-0:03	
1	150-200	0:03-0:04	He 35%—O_2 15%
1	200-250	0:04-0:05	
115	250	0:05-2:00	
1	250-200	2:00-2:01	
1	200-150	2:01-2:02	
4	150	2:02-2:06	He 80%—O_2 20%
1	150-140	2:06-2:07	
4	140	2:07-2:11	
1	140-130	2:11-2:12	
9	130	2:12-2:21	
1	130-120	2:21-2:22	
9	120	2:22-2:31	
1	120-110	2:31-2:32	
9	110	2:32-2:41	
1	110-100	2:41-2:42	
14	100	2:42-2:56	
1	100-90	2:56-2:57	
14	90	2:57-3:11	
1	90-80	3:11-3:12	
20	80	3:12-3:32	Air
5	80-70	3:32-3:37	
25	70	3:37-4:02	
5	70-60	4:02-4:07	
40	60	4:07-4:47	
5	60-50	4:47-4:52	

Table AI-17 (*Continued*)

Minutes	Depth (Ft.)	Elapsed Time	Gas Mix
30	50	4:52-5:22	Air
5	50-45	5:22-5:27	
20	45	5:47-5:47	100% O_2
5	45-40	5:47-5:52	
40	40	5:52-6:32	Air
5	40-35	6:32-6:37	
20	35	6:37-6:57	100% O_2
5	35-30	6:57-7:02	
70	30	7:02-8:12	Air
5	30-25	8:12-8:17	
25	25	8:17-8:42	100% O_2
5	25-20	8:42-8:47	
80	20	8:47-10:07	Air
5	20-15	10:07-10:12	
30	15	10:12-10:42	100% O_2
5	15-10	10:42-10:47	
85	10	10:47-12:12	Air
5	10-5	12:12-12:17	
35	5	12:17-12:52	100% O_2
5	5-0	12:52-12:57	

Flow Rate: Helium/Oxygen at 15 liters per minute

Table AI-18: Helium/Oxygen Decompression Schedule
300-Foot Dive — Two-Hour Bottom Time

Minutes	Depth (Ft.)	Elapsed Time	Gas Mix
1	0-60	0:00-0:01	He 80%—O_2 20%
1	60-120	0:01-0:02	
1	120-180	0:02-0:03	He 85%—O_2 15%
1	180-240	0:03-0:04	
1	240-300	0:04-0:05	
115	300	0:05-2:00	
1	300-240	2:00-2:01	
1	240-180	2:01-2:02	
4	180	2:02-2:06	He 80%—O_2 20%
1	180-170	2:06-2:07	
4	170	2:07-2:11	
1	170-160	2:11-2:12	
9	160	2:12-2:21	
1	160-150	2:21-2:22	

(*Table AI-18 continues on p. 98.*)

Table AI-18 (*Continued*)

Minutes	Depth (Ft.)	Elapsed Time	Gas Mix
9	150	2:22-2:31	He 80% − O_2 20%
1	150-140	2:31-2:32	
9	140	2:32-2:41	
1	140-130	2:41-2:42	
9	130	2:42-2:51	
1	130-120	2:51-2:52	
9	120	2:52-3:01	
1	120-110	3:01-3:02	
5	110	3:02-3:17	
5	110-100	3:17-3:22	
20	100	3:22-3:42	Air
5	100-90	3:42-3:47	
10	90	3:47-4:07	
5	90-80	4:07-4:12	
20	80	4:12-4:32	
5	80-70	4:32-4:37	
15	70	4:37-4:52	
5	70-65	4:52-4:57	
30	65	4:57-5:27	
5	65-60	5:27-5:32	Air
30	60	5:32-6:02	
5	60-55	6:02-6:07	
30	55	6:07-6:37	
5	55-50	6:37-6:42	
40	50	6:42-7:22	
5	50-45	7:22-7:27	
15	45	7:27-7:42	100% O_2
5	45-40	7:42-7:47	
50	40	7:47-8:37	Air
5	40-35	8:37-8:42	
15	35	8:42-8:57	100% O_2
5	35-30	8:57-9:02	
55	30	9:02-9:57	Air
5	30-25	9:57-10:02	
20	25	10:02-10:22	100% O_2
5	25-20	10:22-10:27	
60	20	10:27-11:27	Air
5	20-15	11:27-11:32	
35	15	11:32-12:07	100% O_2
5	15-10	12:07-12:12	
70	10	12:12-13:22	Air
5	10-5	13:22-13:27	
40	5	13:27-14:07	100% O_2
5	5-0	14:07-14:12	

Flow Rate: Helium/Oxygen at 15 liters per minute.

Bibliography to Part I

Andersen, H.T. 1964. *Stresses imposed on diving vertebrates during prolonged underwater exposure.* Symposium, Social Experimental Biology 18:109.

———. 1965. Cardiovascular adaptations in diving mammals. *Am. Heart J.* 74:295.

Beckman, E. L. 1963. Thermal protection during immersion in cold water. *Proceedings, second symposium on underwater physiology* Nat'l. Res. Council pub. 1181. Washington: Nat'l. Academy of Sciences, Nat'l. Res. Council.

Beebe, C.W. 1934. *Half mile down.* New York: Zoological Society.

Behnke, Albert R. 1966. Cause and prevention of decompression sickness among compressed air tunnel workers. *Proceedings, National Safety Congress.*

Bennett, P.B. 1966*a. The aetiology of compressed air intoxication and inert gas narcosis.* London: Permagon Press.

        1966*b.* Performance impairment in deep diving due to nitrogen, helium, neon and oxygen. *Proceedings, third symposium on underwater physiology.* Lambertsen, C.J., ed. Baltimore: Williams & Wilkins.

Bert, P. 1878. *La pression barometrique.* Paris: Masson. Translated by Hitchcock, M.A. and Hitchcock, F.A. Columbus: College Book Co.

Bond, G.F. 1966. Medical problems of multiday saturation diving in open water. *Proceedings, third symposium on underwater physiology.* Lambertsen, C.J., ed. Baltimore: Williams & Wilkins.

Brauer, R.W. 1968. Seeking man's depth level. *Ocean Industry* vol. 3 no. 12.

Comroe, J.H. Jr. 1966. The lung. *Scientific American* vol. 214, no. 2.

Conference for National Cooperation in Aquatics. 1962. *New science of skin and scuba diving.* New York: C.N.C.A.

Cousteau, J.Y. 1953. *The silent world.* London: Hamish Hamilton.

———. 1965. *World without sun.* New York: Harper & Row.

CSL, U. of Fla. 1967. *Quarterly report of the Dept. of Speech, U. of Fla.* vol. 7, no. 1. Gainesville: U. of Florida.

Davis, R.H. 1962. *Deep diving and submarine operations.* 7th ed. London: Saint Catherine Press Ltd.

Diolé, P. 1953. *The undersea adventure.* Sedgwick & Jackson, L'adventure sous-marine. Paris: Albin Michel.

———. 1954. *4000 years under the sea.* New York: Julian Messner.

Donald, K.W. 1947. Oxygen poisoning in man. *British Medical Journal* 1:667-672.

Dugan, J. 1965. *Men under water.* Philadelphia: Chilton.

Fenn, W.O. 1969. The physiological effects of hydrostatic pressures. In *The physiology and medicine of diving and compressed air work.* Bennett, P.B. and Elliott, D.H., eds. London: Bailliere Tindall and Cassell.

Gilpatric, G. 1938. *The complete goggler.* New York: Dodd Mead & Co.

Grant, Alan H. 1969. Skindiver contact lenses. Personal communication.

Haas, H. 1947. *Drei Jäger auf dem meeresgrund.* Zurich, Orell Fussle, London: Jarrolds.

———. 1952. *Manta.* Berlin: Ullstein. (English: *Under the red sea.*) London: Jarrolds.

———. 1958. *We come from the sea.* London: Jarrolds.

Haldane, J.S. and Priestly, J.G. 1935. *Respiration.* Oxford: Claredon Press.

Hamilton, R.T., Bacon, R.H., Haldane, J.S., and Lees, E. 1907. *Report to the admiralty of the deep-water diving committee.* London: Stationery Office.

Haxton, A.F. and Whyte, H.E. 1969. The compressed air environment. *In the physiology and medicine of diving and compressed air work.* Bennett, P.B. and Elliott, D.H., eds. London: Bailliere Tindall and Cassell.

Hollien, H., Brandt, J. and Thompson, C. 1967 *a. Preliminary measurements of pressure response to low frequency signals in shallow water.* CSL/ONR Rpt. #9.

———, Brandt, J. and Malone, J. 1967 *b. Underwater speech reception thresholds and discrimination.* CSL/ONR Rpt. #7.

Hong, S.K. and Rahn, H. 1967. The diving women of Korea and Japan. *Scientific American* vol. 216, no. 5.

Johnson, C.S. 1968. *Countermeasures to dangerous sharks.* Undersea Technology Conference. ASME pub. 69-UNT-8.

Keatinge, W.R. 1969. *Survival in cold water.* Oxford: Blackwell Scientific Publications.

Keller, H. and Buhlmann, A.A. 1965. Deep diving and short decompression by breathing mixed gases. *J. of Applied Physiology.*

Krasberg, A.R. 1966. Saturation diving techniques. *4th International Congress of Biometerology.* New Brunswick, N.J.: Rutgers University.

Kylstra, J.A. 1967. Advantages and limitations of liquid breathing. *Proceedings, third symposium on underwater physiology.*

——. 1968. Experiments in water-breathing. *Scientific American* vol. 219, no. 2.

——. 1969. The feasibility of liquid-breathing and artificial gills. In *The physiology and medicine of diving and compressed air work.* Bennett, P.B. and Elliott, D.H., eds. London: Bailliere Tindall and Cassell.

Laffont, M. and Delauze, H.G. 1969. The Janus experiment—working at 500 feet. Special report: working in the sea. *Ocean Industry* vol. 4 no. 2.

Lanphier, E.H. 1955. Use of nitrogen-oxygen mixtures in diving. *Proceedings of the Underwater Physiology Symposium.* Goff, L.G., ed. Washington: Nat'l Academy of Sciences, National Res. Council, pp. 74-78.

——. 1963. Influence of increased ambient pressure upon alveolar ventilation. *Proceedings, second symposium on underwater physiology.* Nat'l Res. Council pub. 1181. Washington: Nat'l Academy of Sciences, Nat'l Res. Council.

——. 1969. Pulmonary function. In *The physiology and medicine of diving and compressed air work.* Bennett, P.B. and Elliott, D.H., eds. London: Bailliere Tindall and Cassell.

Linaweaver, P.G. 1963. Injuries to the chest caused by pressure changes, compression and decompression. *Am. J. Surg.* 105:514.

Los Angeles County. 1964. *Underwater Recreation.* Los Angeles: Dept. of Parks and Recreation.

MacInnis, J.B. 1966. Living under the sea. *Scientific American* vol. 214, no. 3.

Makiel, J. M. 1965. Thermal protection for the free swimming deep diver. *Westinghouse Research Laboratory Progress Report No. 2.* Research Memorandum 65-7D2-DEEP-M2.

Marx, R.F. 1967. *They dared the deep.* Cleveland: World Publishing.

Office of Naval Research. 1967. *Project Sealab Report, An experimental 45-day undersea saturation dive at 205 feet.* ONR rpt. ACR-124. Washington: U.S. Government Printing Office.

Paganelli, C.V., Bateman, N. and Rahn, H., 1966. Artificial gills for gas exchange in water. *Proceedings, third symposium on underwater physiology.* Lambertsen, C.J., ed. Baltimore: Williams & Wilkins.

Panel on Underwater Swimmers Committee on Undersea Warfare. 1955. *Proceedings, underwater physiology synposium.* pub. 377. Washington: Nat'l. Academy of Sciences.

Rahn, H. 1963. Respiratory effects of increased pressure. *Proceedings, second symposium on underwater physiology.* Nat'l. Res. Council pub. 1181. Washington: Nat'l Academy of Sciences - Nat'l. Res. Council.

Raymond, L.W. 1966. Temperature problems in multiday exposures to high pressures in the sea. Thermal balance in hyperbaric atmospheres. *Proceedings, third symposium on underwater physiology.* Lambertsen, C.J., ed. Baltimore: Williams & Wilkins.

Schaefer, K.E. 1969. Carbon dioxide effects under conditions of raised environmental pressure. In *The physiology and medicine of diving and compressed air work.* Bennett, P.B. and Elliott, D.H., eds. London: Baillier, Tindall and Cassell.

——, Allison, R.D., Dougherty, J.H. Jr. and Parker, D. 1968. Pulmonary and circulatory adjustments determining the limits of depths in breath-hold diving, *Science* 162:1020 (taken from USN Sub. Med. Ctr. Rpt. no. 584 Case Rpt.).

Stenuit, R. 1966. *The deepest days.* New York: Coward-McCann.

Strauss, M.B. 1969. *Mammalian adaptations to diving.* Naval Submarine Medical Center Report 562. Groton, Conn.

——, and Wright, P. 1969. *A diving casualty suggesting an episode of thoracic squeeze.* Naval Submarine Medical Center Report 584. Groton, Conn.

Terrell, M. 1967. *The principles of diving.* New York: A.S. Barnes & Co.

Thorne, J. 1969. *The underwater world.* New York: Thomas Y. Crowell.

U.S. Navy. 1968*a. Principles and applications of underwater sound.* NAVMAT P-9674. Washington: U.S. Government Printing Office.

——. 1968*b. Proceedings, first annual international naval diving conference.* Experimental Diving Unit. Washington: U.S. Government Printing Office.

——. 1970. *United States Navy diving manual.* part 1. NAVSHIPS 250-538. Washington: U.S. Government Printing Office.

Van Der Aue, O.E., Keller, R. J., Brinton, E.S., Barron, G., Gilliam, H.E. and Jones, R.J. 1951. *Calculation and testing of decompression tables for air dives employing the procedure of surface decompression and the use of oxygen.* Research report 1. United States Navy Experimental Diving Unit. Washington: U.S. Government Printing Office.

Weltman, G., Egstrom, G. H., Elliott, R.E. and Stevenson, H.S. 1968. *Underwater work measurement techniques: initial studies.* ONR Rpt. 68-11. Los Angeles: University of California, Biotechnology Laboratory.

Wilson, D.P. 1951. *Life of the shore and shallow*

Bibliography to Part I 101

sea. 2nd ed. London: Nicholson & Watson.

Workman, R.D. 1966. Recent Navy experiences in decompression. *4th International Congress of Biometerology.* New Brunswick, N.J.: Rutgers University.

_____. 1969. American decompression theory and practice. In *The physiology and medicine of diving and compressed air work.* Bennett, P.B. and Elliott, D.H., eds. London: Bailliere Tindall and Cassell.

Part 2
Diving Technology

3.
Air Diving

A diving system consists of all the equipment necessary to bring a diver from his natural atmospheric environment to a certain water depth, to provide a life-sustaining environment for him at that depth, to assure use of his normal sensory and motor capability at depth and to return him to his natural environment with no specific detrimental effects. Diving system complexity ranges from a simple set of swim fins and face plate for free diving to the most complex diving system ever developed, the $10 million Sealab III saturation diving experiment.

Classification of diving systems can be done both vertically and horizontally—vertically according to the depth of operation and horizontally according to system characteristics desired at the depth. The diver's task is the most important factor in diving system selection; for example, system requirements for a commercial diver working on a wellhead completion are different from those for a military diver engaged in covert reconnaissance or mine clearing operations. The commercial application, since it requires little horizontal movement, would be performed with a tethered life-support system; that is, the diver would be attached to an umbilical through which he would receive his breathing gas. The military operation would be served with Self-Contained Underwater Breathing Apparatus (SCUBA) equipment. The diver obtains his breathing gas from tanks carried on his back, so that his movement is not limited. The breathing medium would, of course, depend upon the depth of the dive; high pressure, normal-composition air for shallow diving to 200 feet, a mixture of oxygen/helium in the range of 200-1,200 feet and a mixture of oxygen/hydrogen for deeper than 1,200 feet.

Gas medium limitations are based upon the physiological responses of the human system to the mixtures as discussed in Chapter 1. Human tolerance to the mixtures presents a rather narrow "window" within which the system must perform. In deep diving systems, where a diver may go through a wide depth range, the life-support gas must be kept in a safe proportion. This makes the life-support unit a complex mechanism.

Breathing apparatus, although it is the core of the diving system, often represents a minor investment compared to the overall system, which could include the following subsystems: air or gas source with control and monitoring capability, protective clothing to minimize or replace heat loss from the

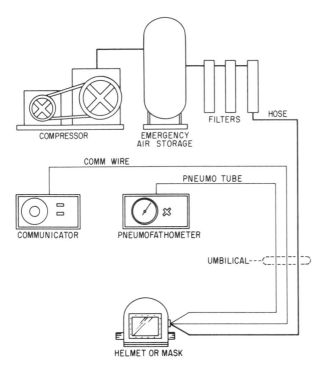

Figure 3-1. The basic tethered air diving system.

diver's body, facilities for decompression to atmospheric environment or recompression in the event of medical complications, diving bells for lowering and raising divers on deep work and pressurized surface or sea-floor living facilities for extended underwater missions.

Regardless of whether a breathing system provides air or mixed gas to the diver, the supply of breathing gas is delivered in one of two ways—by continuous flow or by demand. In a *continuous flow system,* a regulatable amount of gas continuously flows past the diver's face, completely independent of breathing rate. In a *demand supply system,* gas flows only when the diver inhales.

Life-support systems can also be categorized by the exhaust system. If the gas expired from the diver is vented into the water, the system is open circuit. If the gas expired from the diver is cycled through a purifying system to remove the expired carbon dioxide and oxygen is added for reuse, the system is a closed circuit system. If part of the gas is reused and part is vented, the system is classified as semi-closed circuit.

In summary, all breathing systems used in diving, both tethered and SCUBA, can be classified by their supply—free flow or demand, and by their exhaust—closed circuit, semi-closed circuit or open

circuit. The standard SCUBA diving rig, commonly called the "Aqualung," for example, would be classified as an open circuit demand system.

Air Breathing Systems

The air breathing system is the most common life-support system in use on diving operations today, and it will continue to be so for the next decade. Air breathing systems are the basic tool in commercial, military and scientific diving.

The limitation of the air system is the maximum depth to which it may be worked. The theoretical limit, based on an assumed maximum human tolerance for oxygen, is set at the point where the partial pressure of oxygen in the breathing gas reaches a concentration equal to 2 atmospheres. Since the oxygen content of atmospheric air is approximately 20%, air breathing would be limited to 10 atmospheres absolute or 297 feet.

The practical limit for air diving is, however, considerably shallower than the theoretical limit because of the physiological response to high partial pressures of nitrogen. Nitrogen narcosis limits the diver's effectiveness to such a point that most divers are incapable of performing useful work below 220 feet while breathing compressed air. Despite its limitations, air diving is the most economical and simplest technique for shallow water work. The aggregate of air diving time, in all major segments of diving, is at least ten times that of any other method.

Tethered Air Breathing Apparatus

The simplest of all standard diving rigs is the shallow water diving system. In a continuous flow, open circuit system, the basic components are normally a compressor, an umbilical hose and a mask with supply and exhaust valving systems. Figure 3-1 schematically shows a basic tethered air diving system. The diver may use any type of dress, or none at all, depending on the task. Shallow water diving equipment has several distinct advantages over all other systems in use today. It is inexpensive to acquire, equally inexpensive to operate and maintain and simple to the point that only minimum training is needed for operation. However, these attractive advantages are overshadowed by

Figure 3-2. The simplest of all commercial diving rigs is the "Jack Brown" shallow water gear built by DESCO. Although considered uncomfortable by most divers, the mask is still used by the U.S. Navy and for shallow water commercial work.

disadvantages in commercial usage. Changing safety requirements, increased demands for diver efficiency and a need for diver comfort during extended exposures are relegating shallow water diving equipment to a decreasing role in the diving field.

One of the more common shallow water diving systems in use today is based on the "Jack Brown" design. This is an open circuit system available in either free flow or demand configurations. The mask (Figure 3-2) is triangular in shape, covering the entire facial area. The control regulator on the demand configuration, or the air supply flow valve on the free flow system, is located on the right side of the mask at eye level. The exhaust valve, consisting of a simple rubber flapper with an external bubble diffuser, is located on the left side at about eye level. The high position of the exhaust valve increases the mask's tendency to flood. Noise level in the mask, caused principally by the intake valving system, is approximately +100 db and voice communication is restricted.

Since a large portion of diving performed in Naval repair and inspection activity is on ships in relatively shallow water, satisfactory shallow water

diving equipment has been considered mandatory by the U.S. Navy Supervisor of Diving Office. To fill the requirement, the Battelle Memorial Research Laboratory in Columbus, Ohio, was tasked with modernizing the existing "Jack Brown" design in 1969.

The resultant unit, which may be available to the Fleet by 1972, has removed many objectionable design problems. The exhaust system now employs a venturi at the bottom of the mask to keep it dry and to improve ventilation. Intake air has been quieted to a level where good communication is feasible, and the mask body has been faired with fiberglass to improve its hydrodynamic and aesthetic characteristics.

Several new shallow water masks of improved design have become available. The Kirby-Morgan band mask, with both demand and free flow capability, has gained wide acceptance for moderate depth diving. The mask, made on a fiberglass frame, has proven to be excellent for short duration dives of less than 150-feet. The mask is solidly built and has options for either free flow or demand operation. Ambient noise level within the mask is reasonably low and good communications are possible.

Hoses for shallow water diving gear are made of flexible reinforced rubber, usually with 3/8-inch inside diameter and terminating in standard oxygen fittings. The hose is made up with a minimum 1,200 pound tensile strength nylon lifeline. If communication equipment is to be used, the telephone wire is made up in the same bundle. All three are taped together with waterproof electrical tape at 36-inch intervals. Continuous taping or single construction of umbilical with integrated lifeline and communication wire is not feasible with standard hose construction techniques because the hose lengthens when pressurized. This extension varies with the manufacturing techniques used in the hose, but will average from 3 to 5 feet per hundred feet of hose.

The air source for shallow water diving systems is commonly a two-stage air compressor driven by a small diesel engine. Earlier compressors used gasoline engines, but restrictions imposed by the U.S. Coast Guard have limited their usefulness. The engine and compressor are mounted atop a receiver tank that serves the dual purpose of main-

Figure 3-3. Two U.S. Navy divers are lowered on a stage to an underwater work site. They are wearing standard dress and helium/oxygen hard hat equipment developed in the 1930s. A new generation of commercial diving equipment has replaced this cumbersome and limited capability gear. (Photo: U.S. Navy.)

taining a steady air flow and providing a volume of reserve breathing gas to permit the diver to surface should the compressor fail. The receiver tank volume is ideally a minimum of 60 cubic feet. Pressure in the system is maintained at least 50 psi above ambient water pressure at the diver's depth. Filtering to remove oil, particulate matter and carbon monoxide residue in the breathing air is provided at the outlet of the receiver tank. A check valve, which allows air flow in only one direction, is installed at the junction of the supply hose and the diver's mask to prevent mask flooding or squeeze if the air hose is severed or the air flow is stopped.

Diving Dress

For deeper work the standard diving dress is still in use, but is gradually being phased out for advanced equipment that more closely meets the operational needs of commercial diving companies. Several versions of standard diving dress are commercially available. The standard diving helmet, MARK V, is still used by the U.S. Navy Fleet; a modified version of this system is used in commercial construction and a ¾-scale version of the standard helmet, manufactured in Japan, is also available. Although the original design dates back to the nineteenth century, standard diving dress use continued up until the late 1960s because the equipment had definite advantages in the performance of heavy underwater tasks.

The standard deep sea dress, shown in Figure 3-3, consists of a spun brass helmet with glass ports, top, side and front, a brass breast plate for attaching the helmet to the suit and the diving suit made of heavy rubberized canvas. The suit was designed to keep the diver dry. The sealing is effected by rubber cuffs on the sleeves. Standard diving shoes with lead soles and hard wood upper soles and canvas tops, weighing 35 pounds per pair, are worn by the diver. Positive buoyancy of the system is overcome by use of a lead weight belt, varying according to the diver's size, but normally weighing about 80 pounds.

The design of standard deep sea dress varies little throughout the world and all versions show the influence of Augustus Siebe. The system was originally designed for the Royal Navy where the diver is not permitted to control the flow of air into the helmet, and the diving watch officer is responsible for setting the air flow rate. The helmet, therefore, does not have an integral flow control valve. On the American models, the helmet is equipped with a short length of hose, or "whip," that terminates at the diver's waist. A flow control valve is provided at this point. Buoyancy is controlled by the amount of air the diver maintains in his suit. With a minimum of air, the diver is negatively buoyant and can work on the sea-floor in a negative condition, making it possible to exert physical force. By increasing the volume of air in the suit, the diver can maintain neutral buoyancy and perform work on structures between the surface and the bottom.

Although buoyancy control provides distinct advantages for work, it presents a safety problem. If the diver allows himself to get in a feet-up posi-

Figure 3-4. Commercial diver Tom Devine adjusts air flow on his Advanced Diving helmet. One of the new generation of diving systems, it has made major contributions to diving safety.

tion, the standard dress has a tendency to "blow up"; the legs "balloon" to a point where the diver can no longer right himself. Once the diver loses attitude control, things happen very quickly. The lower portion of the suit acts as a lifting bag starting the diver on an uncontrolled ascent toward the surface. As the ascent continues, decreasing ambient pressure causes the trapped air to expand, further increasing lift and ascent velocity. The helmet exhaust system is not capable of passing this volume of air, so a positive pressure build-up occurs in the suit. If the suit is not in good condition, it may rupture. In addition to the problem of rapid decompression, only an attentive tender can prevent the diver from drowning. When the suit ruptures, the diver loses all his positive buoyancy and sinks rapidly, flooding the system and developing pressure squeeze. If the tender observes the trouble and quickly pulls in the diver's hose and lifeline, sinking can be avoided.

Standard dress had some other distinctive safety problems, particularly in underwater welding and burning operations. If a diver accidentally touched his helmet with a "hot" electrode while welding or burning, the probability was that he would burn a hole through the brass. In the near-zero visibility water common to most underwater construction projects, this was a commonplace accident.

New-Design Diving Systems

In the late 1960s several new-design diving systems became available which represented the first major steps in tethered diving equipment since 1910.

A complete system giving the diver the protection of a hard helmet, buoyancy control without the possibility of blow up, a more convenient suit arrangement and allowed an order of magnitude increase in mobility was introduced into the field. The Universal helmet (Figure 3-4), built by Advanced Diving Equipment Company of Gretna, Louisiana, consists of a fiberglass shell, with a large Lexan face plate and a top port. The Lexan face plate was tested and found capable of stopping a 38-caliber bullet at 25 feet without shattering. The helmet incorporates a noise-filtering air supply system that reduces noise level, permitting good communications and simultaneously giving effective wiping action for the removal of carbon dioxide from the helmet. The new ADEM helmet design incorporates a two-stage exhaust that prevents fouling and leakage common in diving equipment when underwater jetting operations are being performed in fine sand. The helmet is designed to be worn with any type of suit, including the standard Navy dress, lightweight dry suits, or the more convenient wet suit.

Buoyancy is provided by a heavy nylon breathing buoyancy bag that fits over the diver's shoulders and to the waist, front and back. By adjusting the helmet exhaust valve, the diver can maintain a volume of air in the buoyancy bag at any desired level. Because of its limited volume, the breathing/buoyancy bag attitude sensitivity is reduced and the blow up threat is eliminated. A nylon coverall is worn over the breathing/buoyancy bag and suit and acts as a restraint on the bag volume.

For use with standard diving dress, a fiberglass breast plate adapts the helmet to the diving suit. The helmet and breast plate combination, which weighs less than one-third of the Navy MARK V combination, lowers the diver's center of gravity and eliminates attitude sensitivity.

The helmet communication system consists of a noise-cancelling microphone, which eliminates most of the remaining helmet noise, and a pair of earphones for the diver which are either mounted in the helmet or worn in a nylon skull cap.

Lightweight Helmet Systems

Within the past few years, the lightweight diving helmet system has grown from only an innovation to the standard for the worldwide commercial diving industry. In addition to the Advanced system, lightweight helmets are now manufactured by Dragerwerk in Lubeck, Germany, General Aquadyne Corporation, DESCO, and Kirby-Morgan Corporation in the United States, and Societe Industrielle Des Etablissments Piel in France. Each system differs slightly in construction techniques, but all are designed to embody the same basic improvements: minimizing the weight and bulk of the diver's equipment, providing an effective wiping system to prevent carbon dioxide build-up, using gas economically, adapting to new developments in diving suits and protecting the diver's head from injury.

The Drager DM 200 helmet, although basically designed for mixed gas operation, can be used on air. The helmet and breast plate are of fiberglass construction and easily rigged. Gas flow control is provided on a whip that mounts on the diver's left side. Unlike American diving equipment, the exhaust valve is located on the right side of the helmet and, as characteristic of most European-manufactured diving helmets, the face plate may be opened while in a diving bell or on the surface.

The General Aquadyne Helmet is a fiberglass-base helmet. In addition to its functional capability, it has a well designed appearance. It features

Figure 3-5. The DESCO diving helmet is the only one of the new generation systems to retain the traditional spun brass construction. (Photo: Perry Submarine.)

a top port, a single-stage exhaust system located in the rear quadrant of the helmet, and uses an over-center cam locking neck ring arrangement for attachment to either wet or dry diving suits. The face plate and top port are impact-resistant Lexan.

The DESCO lightweight diving helmet, Figure 3-5, is the only unit of the newer styles that uses spun brass as the basic construction material. The helmet utilizes the same basic hardware as the shallow water diving mask. The standard jocking arrangement for the helmet is a nylon rope that passes between the diver's legs and attaches to the front and rear of the helmet.

The Kirby-Morgan Company of Santa Barbara, California, was one of the first to introduce fiberglass as the basic material in face masks and has in the past been responsible for many innovations in mask and helmet designs. Characteristic of all Kirby-Morgan helmet designs is an integral face mask. The cranial portion of the mask is usually free flooding. The most widely publicized unit has been the "Clamshell," a demand face mask with a hinged cranial protector that keeps the mask portion in place by tension. An oral-nasal mask minimizes the dead air space. The unit has been tested for operation to depths of 850 feet. In recent tests at the University of Florida, the Kirby-Morgan Clamshell, built for use with the U.S. Navy MARK I Diving System, scored highest of all helmets tested for communication intelligibility. During 1970, manufacture and distribution of Kirby-Morgan helmets and face masks were taken over by the Commercial Diving Division of U.S. Divers Company.

Depth Measurement

On the newer deep air diving systems, the umbilical consists of a ½-inch inside diameter air supply hose, a four-conductor shielded communication wire, a pneumofathometer tube, and a nylon lifeline. A "D" ring is provided on the chaffing gear to tie off the lifeline and umbilical.

The pneumofathometer system is the standard method of monitoring the diver's depth from the surface. If properly integrated into the system, this "kluge" line may also serve as an emergency air supply. The "pneumo" system operates very simply and with a surprising degree of accuracy. An air supply with a gage is connected to the plastic hose at the surface. The other end of the tube terminates at the diver's helmet. When air is allowed to flow through the kluge line, the pressure in the line is equal to the pressure at the diver's depth. By reading the gage pressure, the tender or dive master has an accurate depth indication for computing decompression procedures. The gage is normally calibrated to read out directly in feet, or the gage pressure indication can be calculated to foot equivalents by the simple formula, Depth = Gage Pressure ÷ 0.445 in sea water or 0.443 in fresh water. For example, if the gage pressure reads 100 psi, the diver would be at a depth of 225 feet.

Air Supply

Compressors used with deep diving equipment are considerably larger than those for shallow water gear. The ambient pressure at working depths may be as high as 100 psi, and most divers prefer to work with the air in their supply hose 100 psi over ambient. Compressor output pressure would therefore be about 200 psi on moderately deep

dives. The amount of air required at that pressure depends on many factors, but primarily on the limit placed on the partial pressure of residual carbon dioxide in the helmet. This limit is arbitrarily selected by the diving supervisor in commercial diving and by directive in military operations; the commonly accepted maximum is 0.03 AtA of carbon dioxide. Any higher percentage will begin to cause pulmonary distress.

The actual dead air space in a diving helmet and the effectiveness of the air path through the helmet control the air flow a specific piece of equipment requires to maintain the proper level. Great variances in these factors are found in system design. The most effective arrangement has proven to be a system that controls the flow so that approximately 80% of the input air is projected across the diver's face in a broad flat stream. The remaining air continually flushes the back and bottom of the helmet. Since expired carbon dioxide is heavier than air, it tends to sink to the bottom of the helmet. The exhaust system is most efficient when the outlet is slightly lower than the diver's mouth and in the main circulation path.

As with all underwater problems involving humans, the physiological variances from diver to diver assume importance. In determining the air flow required, the individual's tolerance to elevated partial pressures of carbon dioxide, his rate of carbon dioxide production, and the level of exertion and stress become important factors. On the average, the human system will produce about 0.01 cubic feet per minute of carbon dioxide at rest; during moderate exercise this will increase to between 0.04 and 0.05 cubic feet per minute.; under severe physical stress the output may reach as high as 0.10 cubic feet per minute. In flow calculations, the highest figure is always used.

In the standard dress, the air flow required for safe ventilation is 4.5 cubic feet per minute at ambient pressure. To supply a single diver with this flow rate at a depth of 99 feet, or 4 AtA, would require a compressor with a capacity of 4.5 x 4 or 18.0 cubic feet per minute of surface air. To supply a single diver with air at 200 feet, the machine should be able to compress 36 cubic feet of surface air to a pressure of 200 psi each minute. Since most compressor units are designed to supply both the submerged diver and an emergency stand-by diver, a suitable unit would be able to supply twice this amount.

Diving Suits

Protection of the diver against thermal stress has assumed great importance in the past decade. As technology has progressed, the diver's capability for longer underwater excursions has also increased. Increasing depth capability and the extension of ocean sciences to colder climates have focused attention on thermal stress. Protective clothing for immersion is a subject that is relevant to all areas of diving, regardless of purpose or the apparatus used. Certain factors such as high heat conductivity of the mediums used for mixed gas diving tend to make the problem more critical, but since it is a problem common to all diving, it will be treated in this chapter.

There are three basic techniques for keeping a diver warm in the water: insulate him and keep him dry, insulate him and keep him wet with minimum water circulation, or keep him wet or dry and apply heat from an external source.

Dry Diving Suits

The oldest method is to keep the diver dry and insulate his skin. Although this is the most effective method of thermal protection, it has major disadvantages when used with modern diving apparatus. Dry diving suit techniques require that suits be completely watertight and maintain a layer of air between the diver's skin and the suit material. If a layer of air is maintained between the diver and the suit, it will of course be subject to compression in accordance with Boyle's Law. Starting from the surface, the suit will gradually compress unless some means of adding air is provided. In the standard deep sea dress, where the technique was first used, this did not present a problem, since the suit and helmet are an integral unit and the amount of air within the suit may be varied by control of the supply and exhaust helmet valves. On the standard dress, the suit is sealed at the breast plate and the only other watertight seal required is at the cuffs. To insure watertight

integrity, these wrist cuffs must necessarily be tight. If fit is not good, the cuffs can reduce circulation to the hands. When this happens, even with protective gloves, a measurable loss in tactile sensitivity and a painful interlude result when the dress is removed. To avoid this, some suits are made with loose metal wrist rings which use rubber gloves to make the seal; although this increases comfort, watertight integrity of the suit can be destroyed by a small tear in the gloves. Most divers prefer cold hands to the risk of a flood-out.

The standard Navy diving dress is made from heavy rubberized canvas which even after repeated use is very stiff. A newer version of the suit, available from Japan, uses a nylon base material and is considerably more supple and comfortable. Standard suits are supplied in graded sizes designed to fit everyone, but generally provide a good fit for no one. The Japanese version of the suit, until recently, was adapted to the anthropometric characteristics of the oriental skeletal structure and did not provide a satisfactory fit for American divers. Later models have longer legs.

The function of the dry suit, as its name implies, is simply to keep the diver dry. Surrounded by a sheath of air, the diver has a warmth problem little different than he does on the surface; that is, sufficient insulation for the retention of body heat must be provided. The common solution is to use diving underwear. Construction of these garments varies from country to country, but generally they are of heavy-knit wool, not unlike the two-piece "long johns" used by our grandfathers. In moderately cold water (50 to 60°F), one set will provide comfort if moderate work is performed. For colder water or long periods of minimal underwater activity, two sets are worn.

Dry suits may also be worn with systems in which the helmet and suit are not an integral unit, but these require considerably different construction techniques. Usually they are made from a supple tough rubber so that when they collapse under water pressure they do not cause discomfort to the diver. Like the standard dress, the suit's function is to keep the diver dry. Insulation must be provided by undergarments.

Unlike the standard dress, sealing the suit is a major problem. In the standard dress, entry is made through the neck opening; on the compressible dry suit, since there is no seal between the helmet or life-support equipment and the suit, additional sealing must be provided. These suits are generally made in one piece with a front-entry opening tied to make a watertight seal. Watertight integrity is completed by seals at the wrist and face.

A modified version of this type dry suit, called the Constant Volume Dry Suit, is manufactured by U.S. Divers Corporation. It features a connection to the air supply that meters air into the suit to prevent collapse under pressure. Although the suit has gained a measure of popularity among European divers, it is seldom used in America by commercial divers.

Wet Suits

The wet suit (Figure 3-6) gained its original popularity during the middle 1950s. The suit allows a thin layer of water to enter between the suit material and the diver's skin. Theoretically, this layer of water does not circulate and the body heat warms it. The rubber suit acts as insulation between the warmed water and cold sea water. Wet suit material is thicker than dry suit material and is usually foamed neoprene rubber containing minute cells of trapped gas. Thickness of the material ranges from 1/8-inch for subtropical water to ½-inch for arctic water. The material surface is available in smooth rubber or nylon covered for rough service.

Fit of the suit is critical since only a thin layer of water can be effectively heated by the body. With proper fit, the suit provides a considerable increase in movement and motor capability over dry suits and is additionally attractive because of the absence of watertight integrity problems. Because the material is composed of unicellular gas bubbles it does have one major disadvantage: it is subject to the pressures of Boyle's Law.

As the diver descends, increasing pressure on these gas cells causes the material to become thinner, with two results. The thinner material provides considerably less insulation between the warm layer of body-heated water and the surrounding sea water and being more dense, the

Figure 3-6. The wet suit is the most common diving dress in use today. It is made in many configurations for all types of diving activity. (Photo: U.S. Divers Company.)

material provides less buoyancy. If the diver adjusts his buoyancy near the surface for neutral condition, loss of suit volume as he descends makes him negatively buoyant; for a free-swimming diver on SCUBA, this can present a difficult situation.

Externally Heated Suits

Under severe thermal stress it is often impossible to keep the diver warm even if he is kept dry. This is especially true when the diver is breathing a helium/oxygen mixture. If the suit is standard dress, the problem is further complicated because he is surrounded by an envelope of helium. Body heat conduction in the helium atmosphere is approximately seven times that of the air atmosphere. To compensate for this increased heat transmission, a synthetic heating arrangement must be provided. The earliest attempts to accomplish this used electrically heated underwear.

To provide a heater network that was electrically insulated from the diver's body and capable of withstanding the flexing caused by the diver's body presented a problem slightly beyond the reach of earlier technology. Resistance wires were brittle and electrical connections could not survive the flexing. A long list of failures resulted, with interesting side effects such as smoldering of garments within the suit. Divers came to believe that water and electricity were not a good combination, and the consensus within the diving community was that cold was better than fire, skin burns or electrocution. The problem has recently been solved by using capillary tubes of mercury as resistance wires. If one of the capillaries breaks, the liquid mercury runs out and there is no danger of electrical fire in the suit.

Because of its convenience and comfort, the wet suit is preferred in almost every phase of diving activity. To compensate for insulation loss under pressure and to provide additional heat for mixed gas diving, several new methods of heat addition are in use. In Europe, mercury resistance electrically heated underwear, similar to that used in the dry suit, is most widely used. Even with modern technology, the electrically heated garment is fragile and not generally considered suitable for the continuous heavy duty required in commercial underwater work.

The hot water wet suit is the most dependable system available today, primarily because it is the simplest. The diver wears a loose fitting wet suit with built-in channels similar to a garden sprinkler hose. Hot water is pumped through these channels which distribute the flow around the diver's body. The suit is open circuit; that is, the water is discharged after it performs its heating function. The hot water is carried to the diver through an insulated hose made up as part of his umbilical line. Several alternate methods of heating the water are available; a brute force surface oil burner is the standard.

SCUBA Air Breathing Apparatus

The idea of a diver surviving underwater from an air supply he carries with him is not new. Illustrations of divers using this type of apparatus appeared as early as the seventeenth century.

Compressed air SCUBA or, to use its common name, the "Aqualung," gained technological acceptance in the middle 1940s. In the intervening years it has caused a blossoming of underwater scientific investigation, created a new sport and changed the military diver from a salvor to an effective weapon system. Its growth as a tool in commercial diving has, however, been tenuous. Its commercial rejection has been based on both practical work situations and personal economic considerations within the commercial diving community.

The physical limitations of SCUBA are apparent from the apparatus and techniques.   Dive time is a function of the amount of breathing gas the diver can carry and the rate at which he uses it. The diver is neutrally buoyant and has minimal work capability. He is not connected to the surface, seldom has communication and is generally considered a higher safety risk than a tethered diver in the same work situation.

The economic considerations are less direct, but no less real. Less expensive equipment acquisition and maintenance has created a cadre of experienced sport divers eager for profitable application of their diving prowess. They are willing to work for less than scale diver wages and constitute a measurable economic threat, particularly for shallow water work.

Several years ago the mobility of air SCUBA seemed commercially attractive for inspection and survey operations requiring a high degree of lateral movement. However, development of lightweight commercial gear that a diver could actually swim with snipped this blossom in the bud. Today air SCUBA is the basic tool for scientific, sport and military diving, but is not accepted nor shows any probability of acceptance in the commercial diving business within the next few years.

All air breathing SCUBA in use today is based on the demand breathing system and is of the open circuit type; that is, air flows only when the diver inspires and is vented to the water after it is expired. Configurations vary widely, but all systems consist of the same basic units: a storage cylinder to carry the high pressure air supply, a valving system to reduce the high pressure air to the ambient water pressure and a demand mechanism to deliver air to the diver with minimal respiratory effort.

Based on an operational history of twenty-five years, the design of compressed air SCUBA has progressed to a point where differences between commercially available equipment are not measured by major innovation but rather by quality control and miniscule improvements in performance. The rate of air consumption in these systems and depth capability with compressed air are not functions of properly built equipment. They are directly related to diving physiology and diver experience.

The Air Storage System

Compressed air for SCUBA systems is normally carried in high pressure cylindrical vessels designed for air pressures between 1,800 psi and 3,000 psi. This "working pressure" is the pressure to which the tank can be repeatedly filled without causing abnormal metal fatigue. In the United States, regulations controlling specifications for material, design and manufacture of these cylinders are under control of the Federal Transportation Commission. Tanks are rated in size, not by their physical dimensions, but by the volume of standard atmospheric air they contain when pressurized to their rated working pressure. For example, a 70-cubic foot tank (Figure 3-7) filled to its rated 2,250 psi would contain enough air to fill an enclosure with a volume of 70 cubic feet to a pressure of 1 atmosphere, or 14.7 psi; the actual internal physical volume of such a tank would be only about 2.0 cubic feet.

A high pressure valve completes the air storage system. Valve configuration varies with the manufacturer, but the mechanism is one of two types—a straight-through "K" valve or a reserve "J" valve. The "K" valve is a simple "on/off" mechanism; the "J" valve contains a spring-loaded shutoff that automatically closes when pressure within the tank storage system reaches a preset minimum. Air to the diver is cut off until he operates a lever which restores the air flow. Of course, the diver knows at

Figure 3-7. The 70 cubic foot tank with a working pressure of 2,250 psi is the standard SCUBA air storage system. A wide variety of valve configurations are available. (Photo: U.S. Divers Company.)

this time exactly how much air remains in the system.

The Regulator

The pressure reducing system and the demand valving system that deliver air to the diver are integrated into one unit called the regulator. Although they can have entirely different configurations, all regulators embody these two functions. The two general configurations in use today are double-hose (Figure 3-8A) and single-hose (Figure 3-8B) regulators. The double hose, although it has been refined from an engineering standpoint, is still the basic unit designed by Cousteau and Gagnon. The single-hose regulator, the most popular unit today, was developed in the late 1950s.

The regulator unit is attached to the tank system by means of a "yoke," and the seal between the regulator and the tank valve is effected by an O-ring. When the tank storage valve is turned on, high pressure air is supplied to the regulator,

Figure 3-8. The regulator is the heart of the SCUBA system. The two-hose regulator (top) was the first dependable demand SCUBA regulator to be commercially available; improved versions are still popular in military diving activity. The single-hose regulator (bottom) was perfected in the 1950s and is in general use today. (Photo: U.S. Divers Company.)

the demand system is always equal to the ambient or surrounding pressure; for example, if the system is being used at sea level, or 1 atmosphere, the pressure of the air supply to the diver would be 14.7 psi. If the diver were at a depth of 33 feet, the pressure of the air supplied to the diver through the demand system would be equal to the surrounding water pressure, or 29.4 psi.

The criteria for determining the regulator's effectiveness is its capability to provide the large quantities of air required for heavy underwater exertion with a minimum of suction effort. If for some reason, either poor design or improper maintenance and adjustment, the regulator does not flow freely, the diver could become distressed from carbon dioxide build-up in the respiratory system or have uncontrollable spasms of the respiratory muscles from over-exertion.

In summary, the regulator portion of the compressed air SCUBA is available in two general configurations—single hose or two hose. The function of the regulator system is to reduce the high pressure air in the air storage system and to deliver it to the diver upon demand.

Air is made available to the diver in any of several different ways, most commonly through a bite-type mouthpiece designed to be clenched in his teeth. On the two-hose regulator, the mouthpiece forms part of the hose system, and the entire regulator mechanism is built into a single unitized housing that attaches directly to the air storage system valve. When the diver inspires, the pressure reduction within the hose activates the mechanism in the regulator. In the single-hose configuration, the mouthpiece forms part of the demand system itself. Demand valving and diaphragm are contained in a small housing with the mouthpiece attached. A single interconnecting hose carrying air at intermediate pressure connects the demand mechanism, or second stage, to the pressure reducing first stage, which is attached to the tank valve.

but there is no flow until the diver inspires. The vacuum created in the demand system by this inspiration causes movement of a diaphragm which is levered to the air flow valve as flow begins. The high pressure reducing system lowers the pressure of the air being supplied to the demand portion of the regulator to approximately 100 psi. Pressure in

4. Mixed Gas Diving

Since the early 1960s, there has been a growing need for extended diving operations beyond the safe depth limitations of compressed air. To fill this need, a completely new technology, mixed gas diving, has developed within the diving community.

The use of synthetically mixed gas is not a new idea in diving. Helium/oxygen mixtures, hydrogen/oxygen and specially compounded mixtures of oxygen/nitrogen have been proposed, tested and used on special operations since 1920. The development of deep diving mixed gas equipment has undergone an exponential expansion in recent years. About 10% of commercial diving operations using mixed gas now account for approximately 28% of the revenue of the commercial diving business. Indications are that both these percentages will double in the next 5 years.

The use of mixed gas has solved many deep diving problems — and at the same time created a completely new set. Helium/oxygen diving is the most well known and widely applied technique in use today. In deep diving, the system must supply the diver with a level of oxygen high enough to support his life functions and ergometric requirements, but not so high that it will cause oxygen toxicity.

To carry oxygen to the diver, an "inert" gas that causes a minimum of abnormal physiological responses must be selected. Nitrogen, considered an inert gas, causes dramatic narcotic effects beyond a depth of 200 feet. Helium, on the other hand, behaves well at that depth; in fact, it continues to do so down to 1,200 feet.

By mixing oxygen with helium, the first gas barrier—nitrogen narcosis—is passed, but not without paying a price. Helium is much lighter than nitrogen and has entirely different heat conduction characteristics. The lightness has its effect by changing the speech output to a high pitched, Donald Duck sound, which worsens as depth increases. In recent test at 850 feet, with the divers breathing a helium/oxygen mixture, the University of Florida Communications Sciences Laboratory found that the intelligibility level of the divers' speech was less then 5%. Helium speech converters designed to reconstruct, or normalize, the divers' speech output have been unable to significantly improve intelligibility at these depths. The thermal conductivity of helium has proven to be another major problem area. Conducting heat away from the diver's body at a rate seven times greater than air, helium breathing places the diver under thermal stress that requires an external heating source to sustain life.

The detrimental physiological effects of helium under pressure are not apparent until a depth in excess of 1,200 feet is reached. With rapid compression to depth, the effects can occur sooner, but with good procedures the symptoms called High Pressure Nervous Syndrome (HPNS) are not seen in most divers until 34 (AtA) pressure. Like oxygen toxicity and nitrogen narcosis, the effects vary from diver to diver and seem to be independent of diving experience. HPNS occurs dramatically and very little data has been gathered about the physiological process.

The 1,200-foot depth helium limit is, however, not an impenetrable barrier. In fact, experimental dives to 2,000 feet are now underway. During 1969 animals were brought to 3,500 feet and returned, with no apparent physiological modifications, breathing a mixture of oxygen/hydrogen. Exhaustive experiments have been conducted by the British Navy, at shallower depths, using oxygen/hydrogen mixtures. There has been reluctance to use these gases in operational activity because of their explosive reaction with each other, but concentrations that would be used below 1,200 feet (oxygen less than 2%), danger of explosion is negligible. With proper handling and equipment, the safety factor is as high as with other synthetically mixed gases used in diving procedures.

It is technically possible to use oxygen/helium or other synthetic mix gases in standard air breathing apparatus—possible but economically unfeasible. The cost of the constituent gases, which must be provided in large pressure vessels, makes it mandatory to recycle the breathing mixture. Most mixed gas life-support systems are therefore considerably more complex than air breathing equivalents. When a diver inspires the breathing mixture, a gas exchange takes place in his lungs. The inspired gas is oxygen/helium/carbon dioxide. The oxygen concentration of the expired gas is less than the inspired gas and the amount of oxygen

depletion is dependent on the diver's metabolic rate. The helium, used only as a carrier, is little changed in quantity or quality. To reuse the gas, the carbon dioxide must be removed. Oxygen equal to the amount absorbed by the diver must be added, and the purified, reoxygenated mixture must be delivered to the diver's breathing intake.

Mixed Gas Apparatus

The Semi-Closed System

Two methods of performing the necessary functions are in general use today. The first and most widely used, both in SCUBA and tethered configurations, is the semi-closed system, in which gas mixture is actually a function of flow. In this system, a premixed gas, usually oxygen/helium, is provided to the diver by either continuous flow or demand. His expired gas is partially vented to the water and the remainder passes through a canister containing a chemical compound which absorbs carbon dioxide. The purified gas from the canister, or scrubber, is then added to the gas supply system. The recirculated gas, which has a lower oxygen content than the fresh gas supply, is mixed with the fresh gas and provided to the diver.

The sum of the oxygen content of the recirculated gas and the fresh gas must give a level of oxygen sufficient to prevent anoxia but low enough to avoid oxygen toxicity or hypoxia. The controlling factors are, therefore, the content of the fresh gas supply, the amount of flow rate of the purified gas and the overall flow rate through the system.

The Closed Circuit System

The second type of mixed gas apparatus is the closed circuit system, which vents none of the gas to the water and controls the oxygen level in the gas supplied to the diver by measuring the oxygen partial pressure. Oxygen is automatically metered into the recircualting gas to maintain the proper level. By far the most operationally economical, this system has only recently become feasible because of the complex measuring and control electronics required. With dependable controls, the method is preferable to the mass flow method because the oxygen partial pressure is the function that must be accurately controlled; the closed circuit partial pressure control system is a direct method.

Tethered Mixed Gas Diving

The tethered mixed gas, continuous flow, semi-closed circuit apparatus is the most widely used mixed gas diving equipment in commercial diving. The basic system for moderate depths consists of a surface gas supply of oxygen and helium cylinders containing premixed gas, a regulator, a control console, interconnecting hoses and diver equipment, usually a fiberglass helmet and carbon dixoide absorbing backpack. Although several configurations of equipment are available, the basic operating principles are similar.

In the Advanced Diving mixed gas system, shown in Figure 4-1, the deck tank storage supply is fed through a pressure reguaitor to reduce it from the 2,000-3,000 pound supply pressure to an intermediate pressure of approximately 500 psi. The regulator is usually located at the gas supply so that piping on the deck of the diving barge or ship contains intermediate pressure gas. Whenever possible, high pressure piping in the vicinity of work sites is avoided. The intermediate pressure gas is piped to a control console. The control console performs several basic functions. Primarily, it takes the mixed gas and puts it through a regualtor to adjust it for adequate flow so that it is approximately 100 psi above ambient pressure at the diver's helmet.

The console also provides the capability of delivering standard compressed air or oxygen to the diver during the decompression cycle or in emergencies. A pneumofathometer is normally located at the console. Therefore, at one critical location, the console operator or dive master can monitor the type of gas, gas pressure, the diver's depth, and has the capability of changing the diver's gas supply. Most consoles are equipped with a stand-by air supply in the event of failure of primary mixed gas supply, the certified air or the oxygen.

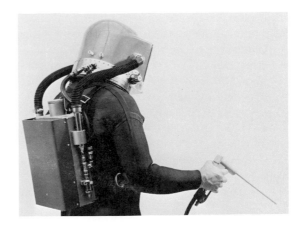

Figure 4-1. The Advanced Diving mixed gas rig is widely used for deep commercial diving. Constant gas recirculation is accomplished by a venturi/mass flow system. Backpack mounted cannister purifies recirculated gas for up to 7 hours. Diver is holding an oxyarc burning torch, one of the basic underwater tools. (Photo: Advanced Divers Supply.)

From the console the gas is carried in a ½-inch inside diameter hose to the diver. The hoses are normally made up in the same manner as the deep diving air hose system; that is, a single bundle containing the supply hose, armored communication cable, the pneumo, or kluge line and a 1,200-pound test life line.

The mixed gas diving helmet is identical in basic construction to an air diving helmet, with the addition of two hoses that connect to the backpack for recirculating a portion of the gas. The gas supply is brought from the surface to the backpack where it is introduced through a jet nozzle that forms part of a venturi system. The fresh gas supply is brought into the helmet through the right-hand side interconnecting hose. The gas is passed in front of the diver's face at up to 300 liters per minute when the pressure in the supply hose is set at 100 psi over ambient pressure. The exhaust valve on the left-hand side of the helmet is set with sufficient back pressure so that when the diver inhales it is in the closed position. Under these circumstances, all the gas injected into the helmet is brought out through the return recirculating hose into the backpack.

As the diver exhales, pressure within the helmet builds up. By the time the diver has finished approximately 50% of his exhalation, the exhaust valve opens and gas containing a high proportion of carbon dioxide is vented into the water. As the exhalation is completed, the exhaust valve closes and the gas is again recirculated to the backpack. The exhaust valve is normally adjusted so that it will open when the pressure inside the helmet is approximately 3/4-pound above ambient pressure. Set at this point, approximately 10-20% of the total gas flow is lost through venting into the water.

Gas recirculated to the backpack is brought in on the left side of the canister and travels laterally through the carbon dioxide absorbent material. The material can be soda lime, baralyme, or any of the common carbon dioxide absorbent materials. As the gas passes through the granules, an exothermic chemical reaction takes place and the material absorbs all the exhausted carbon dioxide from the diver's helmet. On the right side of the backpack, the gas supply is being injected through a nozzle over the venturi. This provides a vacuum on the right-hand side of the backpack and causes the expired recirculated gas to flow evenly through the absorbent material. When operating properly, the canister will effectively scrub the recirculated gas for up to 7 hours.

The used, repurified gas is added to the gas supply flow at this point and recirculated past the diver. The loss of oxygen in the used gas depends on several factors, such as the ergometric situation of the diver, and the flow rate through the helmet. If the exhaust valve on the helmet is set so that most of the diver's exhalation is exhausted and the diver is working at a normal level, oxygen depletion in the recirculated gas is normally minimal and is compensated by a more richly oxygenated primary gas supply. The partial pressure of the oxygen supplied to the diver is therefore dependent on the oxygen partial pressure in the fresh gas supply and the mass flow rate of the gas through the recirculating system. The residual level of carbon dioxide in the system depends on the effectiveness of the carbon dioxide absorbing system. This in turn is a function of the manner in which gas is brought through the chemical, the temperature of the canister and the length of time the material has been used.

The effectiveness of the canister and scrubbing of the carbon dioxide are directly related to the mechanical configuration. If the cansiter is not properly designed, a "channeling" effect can take place in which the gas to be scrubbed follows two or three major paths and does not pass through all the scrubbing material. Even in a well designed canister, the channeling can sometimes take place because of packing of the material; in this case a sharp rap on the canister will usually resettle the granules into a more acceptable structure.

The temperature of the canister and the gas are important factors in scrubbing efficiency. The removal of carbon dioxide from the gas takes place in an exothermic or heat-producing reaction. If the canister or the gas is too cold, the effectiveness drops appreciably and can, in fact, reach a point where carbon dioxide build-up in the system becomes a serious problem. For operations in extremely cold water, external heat from either recirculating hot water or an electrically operated helium blanket is necessary as part of the canister system to keep the exothermic reaction in process.

Scrubbing chemical effectiveness can also be adversely affected by excess moisture. If the material becomes wet, scrubbing action is reduced. With some chemicals, wetting produces a caustic gas which could be recirculated into the diver's helmet. In the event of foreign gas generation in the canister, or loss of effectiveness due to channeling or depletion of the chemical, a by-pass is provided which allows the diver to flow the fresh supply gas directly through the helmet.

Except for very brief work periods, the process of lowering a diver to a depth over 300 feet, having him perform work, and then bringing him up through his decompression cycle has proved to be operationally impractical and economically unfeasible. The most common technique used today for other than a "bounce" dive is to lower the diver or divers in a diving bell maintained at atmospheric pressure until the divers are near the work site. At that time the bell is pressurized to the ambient pressure at a controlled rate. The diver exits, using tethered mixed gas continuous flow semi-closed circuit apparatus for life-support equipment and moves a short distance to the work site.

The diver's gas supply can either come from the surface to the bell and diver, or it may be self-contained on the diving bell itself. If the supply is self-contained, the bell is of course larger, and the storage tanks are mounted on the outside and piped to a flow control system inside. This is the most commonly used method and, in fact, nearly all bells used today are completely self-contained, with the exception of electrical power. Use of surface-supplied gas to the bell through an umbilical presents an expensive configuration and increases handling problems during lowering and ascent.

Dragerwerk of Germany manufactures an autonomous system. The plant is designed for simultaneous supply to three divers using semi-closed circuit equipment. The gas bank consists of a series of vertically mounted cylinders on the bell that are manifolded together in up to three premixed combinations for work at depths to 600 feet. With premixed gas, it is necessary to have several mixtures available so that the limit of oxygen partial pressure is not exceeded as the diver descends to the bottom. If the working depth is less than 600 feet, fewer gas mixtures may be required. In this system, the Drager MVU III, mixed gas of the proper proportion is manifolded to the inside of the chamber where it is piped to a separate control for each diver.

The Drager MVU III incorporates a special warning feature that alerts the diver if the gas supply is interrupted. A pneumatic/electric flow control sensor is mounted in the diver's equipment directly in the gas flow path. An electrical signal is fed back from this device to the control unit in the diving chamber. A series of lights is mounted in the diver's helmet or mask, and if the system is operating normally a green light burns continuously. Should the gas be interrupted, the green light in the face mask or helmet goes out, a red light comes on and an audible signal is actuated in the diving bell. In normal operations three divers are seldom used simultaneously; one diver usually acts as tender within the bell to monitor gas flow and communications and to assist the other divers in and out of the bottom hatch.

Tethered mixed gas equipment used by the U.S. Navy and some commercial diving companies outside the United States operated on a slightly different principle than the Advanced Diving equipment. Breathing bags form part of the system and gas is provided to the diver through a bite-type mouthpiece or a full face mask. In the U.S. Navy MARK IX mixed gas underwater breathing apparatus, two nylon fabric breathing bags are fitted like a vest as part of the equipment harness. The bags extend from the front of the diver over the shoulders. The gas supply goes through a control block that has three selectable gas flow rates and then into the inhalation or right-hand bag. The gas passes from this bag through a hose to the mouth-piece. Exhalation from the diver goes through a check valve system into the left-hand or exhaust bag, which is fitted with an adjustable pressure-relief valve for dumping a portion of the exhaust gas.

Gas flows continually from the exhalation bag to the carbon dioxide absorbent canister and back through the control block, where it is mixed with the incoming gas mixture from the umbilical. The canister and a pair of high pressure emergency storage tanks are mounted in a fiberglass-faired housing carried on the diver's back. The emergency tanks serve as an alternate supply in the event of failure in the umbilical. They also provide extra gas during descent or periods of exceptional exertion. The length of time the emergency supply may be used is of course dependent on the flow rate and the depth. The MARK IX system is equipped with an emergency demand regulator so that the diver can provide breathing gas to his diving partner if an equipment failure occurs.

The Dragerwerk SMS I system operates on similar principles. However, the unit is considerably less complicated and bulky than the U.S. Navy MARK IX unit because it was principally designed for commercial divers who would always be working in close proximity to their diving bell. The MARK IX was designed for use on heavy salvage projects and laboratory experiments such as Sealab, where the diver might be operating with a rather long (up to 600 feet) umbilical hose from a diving bell or an underwater habitat.

In SMS I system, gas flows from the control unit in the diving bell through a valve on the backpack directly into the inhalation bag. Input is monitored by a flow control alarm system. From the inhalation bag the gas goes through the inhalation hose and the diver inhales it through the bite-type mouthpiece. On exhalation the gas goes to an exhalation bag fitted with a pressure-relief valve and from there into a soda lime canister where carbon dixoide is removed. A water trap is provided in the canister to remove any excess moisture or water that might inadvertently enter the system. A single emergency tank containing approximately an 8-minute supply of premixed gas for use at 600 feet forms part of the diver's system. If the primary gas supply is stopped, the diver opens the emergency valve on the lower left-hand side of his backpack and gas is provided to him at a flow rate of approximately 20 liters per minute. In the Drager system, all life-support components, including the breathing bags, are housed in a fiberglass-reinforced polyester backpack.

The high operational cost of helium diving has placed pressure upon technology to produce a system which permits complete or nearly complete recovery of all helium. There are, at this time, closed circuit SCUBA systems in which helium loss is minimal for operations down to 1,000 feet. Most commercial companies, however, prefer using umbilicals in their diving operations, since tethered equipment generally has superior communication capability. In addition, the use of a tender improves the safety factor in deep diving operations.

Several closed circuit umbilical systems now used in commercial diving have been produced by in-house research and development. Details of these systems are highly proprietary and not available for publication. However Drager, of Lubeck, Germany, has developed several commercially available bell diving systems of the closed circuit type. The first, capable of supporting three men in the diving bell (one tender and two divers), is completely self-contained. A gas supply consisting of oxygen and helium is mounted externally on the bell. The gas is not premixed and the oxygen and helium are contained in separate cylinders and piped into the bell individually. The inside of the diving bell is pressurized with a helium/oxygen mixture selected according to the operating depth. Helium is admitted to pressurize the bell, and suffi-

cient oxygen is metered to maintain the oxygen partial pressure at a safe level.

An oxygen sensor system in the bell continuously monitors the oxygen level and automatically meters oxygen into the chamber as it is required. The bell acts as a large gas reservoir for the diver's breathing apparatus. Ambient gas is sucked in through a low pressure regulator to a compressor. From there the gas goes through a two-element carbon dioxide absorption unit to a high pressure regulator.

The regulator is set to maintain a positive pressure above the chamber and the diver's gas supply is manifolded at this point. The hose used by the diver is a two-element unit including both a supply and a return. The diver's exhaust, which is under a vacuum, enters the system just beyond the low pressure regulator and continual flow with minimum breathing resistance is assured. Within the bell itself, the carbon dioxide, being heavier than either the oxygen or the helium, sinks to the bottom of the chamber.

The intake for the closed circuit system is located in the chamber's lower portion so that both the chamber and the system are continuously purged. During operations it is unnecessary for the tender inside the bell or the stand-by diver to use auxiliary breathing apparatus, which adds considerably to their mobility and comfort.

Because the entire chamber is pressurized with the breathing mixture, the system is not suitable for very short periods of diving activity. As the bell is brought back to the surface, the gas within the chamber is vented, so that there would be a considerable loss of gas during ascent.

In the second system designed for closed circuit bell operations, the entire bell is not pressurized with the helium/oxygen mixture, and the tender and divers wear life-support equipment. The diver is equipped with a backpack containing inhalation and exhalation breathing bags, a carbon dioxide absorbent canister and a supply of helium. The exhalation bag is equipped with an over-pressure relief valve, which normally does not dump gas at a preset rate, as in semi-closed apparatus, but only on ascents made during the diving operations. Oxygen is supplied from within the bell. The umbilical consists of an oxygen supply hose and oxygen-sensor electrical conductors. The sensing element, which controls the oxygen flow, is located in the inhalation bag of the diver's life-support equipment. Sensor output is transmitted back along the cable to control and oxygen metering devices located within the bell.

As the diver inhales and exhales, operation of the unit is very similar to the semi-closed circuit unit except that the only gas vented is by leakage. Carbon dioxide is removed from the recirculating gas in the soda lime absorbent and the gas is re-enriched with oxygen to compensate for depletion by the diver's respiratory system.

Since the helium is acting only as a carrier for the oxygen, loss of helium in this closed circuit is minimal. If helium is inadvertently dumped, the diver's breathing bags are reinflated by operation of the supplementary helium valve which is fed from the self-contained helium tank and pressure reducer system. In addition to the advantage of economical helium usage, the system has the added advantage of small size and, therefore, minimal restriction of the diver's movements during excursions from the bell.

Mixed Gas SCUBA

Mixed gas self-contained underwater breathing apparatus is not new to diving technology. Only within the last 5 years, however, has it gained any popularity in general diving usage. Its commercial diving applications have been limited for several reasons. First, gas supply is limited to what the diver can carry. At mixed gas depth, the rate of usage gives short operational time. Second, in most commercial applications the diver is not required to make long distance excursions from the work site; generally, operations require vertical ascent and descent with an excursion of not more than 50 feet at the work site.

The principal advantage of self-contained underwater breathing apparatus over tethered diving is mobility. When mobility is not required, there is very little to recommend it. With the latest advances in technology, however, it is now possible to build a completely closed circuit SCUBA life-support system that is lightweight, easily maintainable and with sufficient diving time to make it a practical tool for commercial applications.

In military diving, mixed gas SCUBA has been widely adopted. The military has, in fact, provided much of the impetus and research funding for the development of this apparatus. Its effectiveness in military operations is obvious. At moderate depths, semi-closed circuit mixed gas SCUBA provides considerably longer diving time per pound of equipment than ordinary air SCUBA. Also, considerably less gas is vented into the water so on covert operations fewer bubbles reach the surface, and the possiblity of divers' positions being given away is more remote. For mine clearing operations such as those performed by Explosive Ordinance Disposal Units (EDU), semi-closed circuit mixed gas SCUBA operates more quietly than air SCUBA. It therefore provides an additional safety margin in disarming underwater acoustic ordnance.

Little mixed gas SCUBA has been used in scientific diving applications and this situation will probably not change in the very near future. Mixed gas diving in any form is generally more expensive than air diving and most scientific projects are unable to support the costs involved. Excluding some geological diving, few scientific diving investigations require diving deeper than 150 feet. Even if underwater time could be extended by the use of mixed gas SCUBA, the practicality of equipping for and performing the decompression schedules involved in longer exposures is not in keeping with current scientific diving practices.

In the next decade, as man's deep underwater activity increases, mixed gas SCUBA equipment will assume a more important role in the general diving community. Rapidly improving technology promises to make the units dependable, economical and simpler from the standpoint of operation and maintenance.

Mixed gas SCUBA is classified as either closed circuit or semi-closed circuit. The use of mixed gas in standard open circuit SCUBA equipment, although possible, is impractical. The high cost of the gas and the consumption rate at depth are prohibitive. On some sea-floor laboratory projects such as Sealab II, open circuit SCUBA charged with a helium/oxygen mixture has been used for excursions from the underwater laboratory. During the Sealab II project, aquanauts using twin 90-foot cylinders filled with a gas mixture of 85%

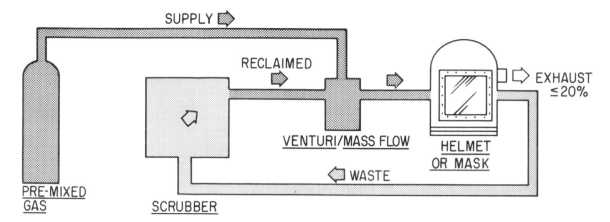

Figure 4-2. All semi-closed mixed gas systems operate on the same principle. In tethered diving the gas supply is located at the surface and the diver is fed through an umbilical; with SCUBA, the mixed gas supply is carried in back-mounted tanks.

helium and 15% oxygen were able to get a maximum of 25 minutes diving time at 205 feet. This represents about 50% efficiency compared to a semi-closed circuit SCUBA rig.

Semi-closed Mixed Gas SCUBA

The operating principles of most semi-closed mixed gas SCUBA available today are similar although they differ in configuration, method of construction and performance. The U.S. Navy MARK VI semi-closed circuit underwater breathing apparatus, manufactured by Scott Aviation Corporation, and the Model FGG III mixed gas diving apparatus manufactured by Dragerwerk are the most widely used today.

Figure 4-2 shows a typical operating schematic for semi-closed circuit mixed gas diving apparatus. The system gas, generally helium/oxygen, is premixed for operation within a specific depth range. Gas is conducted from the high pressure storage cylinders through a flow control or "constant dosage" valve. These arrangements usually have three selectable flow rates. In some equipment the flow rate can be changed by the diver during operations; in other equipment, it must be preset prior to start of the dive. After passing through the metering or flow control valve, the gas is conducted to a breathing bag. These systems are provided with two breathing bags—one inhalation, one exhalation. The fresh gas is injected into the inhalation bag and flows from there through a hose to the diver's mouthpiece or face mask. From there it is conducted through another hose to the exhalation bag. The exhalation bag is equipped with a preset exhaust valve to permit venting of excess gas. This valve will normally dump from 8 to 20% of the circulating gas, depending on the pressure setting.

From the exhalation bag, the used gas is conducted to a canister where it flows through the carbon dioxide absorbent material. This material removes all the expired carbon dioxide and returns the purged gas to the inhalation bag. The recirculating gas is, of course, depleted in oxygen. The degree of depletion is a function of how hard the diver is working and the rate of flow through the system. In the inhalation bag, the recirculated gas is mixed with the fresh gas flowing from the supply and the cycle is repeated.

The U.S. Navy MARK VI was the first mixed gas SCUBA accepted for general use by the Navy. It consists of a twin tank block with a carbon dioxide absorbent canister mounted in between. The breathing bags are contoured to fit over the shoulders close around the neck. They are usually contained in a zippered vest that also has weight pockets for adjusting buoyancy.

The Drager FGG III (Figure 4-3) is a newer design breathing apparatus. All major components

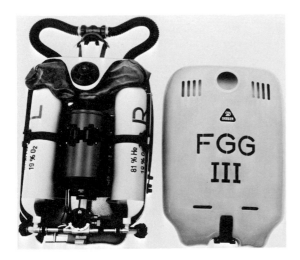

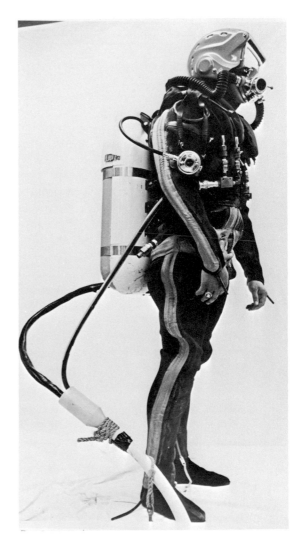

Figure 4-3. The Drager FGG III semi-closed circuit SCUBA is one of the most advanced designs commercially available. Its operating principles are similar to the new U.S. Navy MK 11 SCUBA currently under evaluation for fleet use. (Photo: Dragerwerk.)

including the twin high pressure mixed gas tanks, carbon dioxide absorbent canister, flow control valves, regulator and by-pass valve are housed in a protective case that mounts on the diver's back. Gas is supplied to the diver through two hoses that may be connected to either a standard bite-type mouthpiece or a full face mask. The FGG III is equipped with connections so that the unit may be operated on an umbilical from a supplementary supply. When operated in this manner, the fresh gas supplied from the bell or from the surface is injected into the inhalation bag. Flow control must therefore be provided at the source.

A unit with similar capability to the Drager FGG III, the U.S. Navy MARK VIII underwater breathing apparatus (Figure 4-4) was designed and scheduled for tests on the Sealab III project. The unit used two large 90-cubic foot cylinders for the helium/oxygen supply. Operation of the system is similar to the Drager FGG III, with the addition of a standard demand regulator for emergency breathing procedures. Provision was made for connection to an umbilical at the flow control valve. An oxygen sensor was provided on the output of

Figure 4-4. This U.S. Navy aquanaut is dressed out in the Sealab III life-support system consisting of MK VIII semi-closed circuit mixed gas system, hot water suit and Kirby-Morgan helmet. More practical gear has replaced the complicated umbilical and cumbersome MK VIII. (Photo: U.S. Navy.)

the carbon dioxide absorbent canister so that the effectiveness of the scrubbing action could be measured while the aquanaut was on an excursion from the habitat. The unit weighed approximately 30 pounds more than the standard MARK VI or the FGG III and was considered unsatisfactory by divers because of its bulk and handling characteristics.

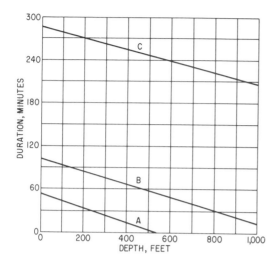

Figure 4-5. Comparison of diving duration capability of: A, open circuit SCUBA, B, semi-closed circuit SCUBA, and C, closed circuit SCUBA.

The duration of dives using any semi-closed circuit SCUBA without a supplementary gas supply is dependent on work load, flow rate and the amount of gas stored in the supply tanks. The average diving time under light work conditions for the Drager FGG III in the depth range of 165 to 360 feet is approximately 40 minutes. In the range of 360 to 525 feet, the duration is about 30 minutes, and at its deepest design range, 460 to 655 feet, the maximum diving time is approximately 20 minutes.

To extend diving time on mixed gas SCUBA equipment without using supplementary gas supplies, it is necessary to go to a completely closed circuit system where oxygen partial pressure is controlled directly rather than through preset mass flow. With this capability, none of the gas is dumped (except by accidental leakage or during ascent) and oxygen is used only as needed. Figure 4-5 shows a comparison of duration versus depth between an open circuit SCUBA, an efficient semi-closed circuit SCUBA rig and a completely closed circuit SCUBA.

Closed Circuit Mixed Gas SCUBA

A new closed circuit underwater breathing system was introduced by Bio Marine Industries in

1968. This system, shown in Figure 4-6, incorporates an entirely new control principle for mixed gas diving apparatus. In all the systems previously described, the oxygen level in the system was controlled by regulation of a continuous flow of premixed gas through the system. The Bio Marine system operates on a different principle. Duration of dive time is nearly independent of the operating depth. For example, a completely charged unit will give approximately 5 hours of dive time in very shallow water and 4¼ hours near 1,000 feet. The difference in dive time as a function of depth reflects the unavailability of residual gas pressure left in the bottles as ambient pressure increases with depth.

The Bio Marine operates on the principle of controlling the oxygen partial pressure supply to the diver. Therefore, the flow of oxygen—rather than the flow of dilutent gas—is the regulated function. Based on a requirement for a working diver of 1.5 liters per minute at the surface, the requirement in actual oxygen does not increase as the depth increases, so if the diver were swimming at the surface and descended to 300 feet, the rate of oxygen supply compared to the surface would be the same. The system electronically controls the oxygen partial pressure within ±10% of the level selected. This point, based on the anticipated work load, can be preset anywhere between 0.2 and 1.2 atmospheres of oxygen.

As an additional feature, a high oxygen and low oxygen setting can be incorporated within the unit. This permits operation at some normal level and switching to a higher oxygen level during decompression so that the process is shortened. Provision is made in the oxygen control system for both manual and automatic operation. Total system redundancy is provided within the equipment, and no single electronic component failure can cause complete failure of the apparatus.

The carbon dioxide scrubber unit uses baralyme and is designed to operate efficiently in water temperatures as low as 28°F. Operation of this portion of the system is similar to that of the semi-closed circuit. Exhaled gas from the diver passes through a hose into the carbon dioxide scrubber unit. Absorbent material removes excess water vapor from the gas at this point. Oxygen sensors in the scrubber chamber sense the level of oxygen in

Figure 4-6. Technological developments in oxygen partial pressure sensing devices have made completely closed circuit mixed gas SCUBA practical. Light, efficient and economical, this type of equipment will be standard in deep diving operations of the future. (Photo: Bio Marine Industries.)

the gas mixture. The three galvanic oxygen sensors are arranged around a triangular bracket and put out an electrical singal to a solenoid valve on the oxygen supply circuit. Oxygen is metered into the output of the carbon dioxide scrubber where it is mixed with the oxygen-depleted recirculating gas. The gas, after being enriched with oxygen, passes into a diaphragm assembly which accepts the gas in the same manner as the inhalation breathing bag of the semi-closed circuit SCUBA equipment.

As the diver descends, water pressure exerts a force on one side of the diaphragm and mechanically causes the diaphragm to operate a valve (similar to a two-hose air SCUBA regualtor). When the valve is tripped, the dilutent gas is metered in to the dry side of the diaphragm until the pressure is equalized. During ascent this process works in reverse and gas is exhausted or vented into the water.

The "sample data" control technique, which samples and periodically introduces fixed pulses of oxygen when necessary, reduces fluctuation in oxygen partial pressure and makes the ±10% control specification feasible. The entire system, excluding the connecting hoses to either face mask or mouthpiece, is contained in a fiberglass backpack measuring 8x14x23 inches. The assembly is neutral in water and weighs 54 pounds in air.

A special monitoring gage is worn by the diver that permits him to constantly monitor the condition of the system. Failure of any of the major components results in both audio and visual alarms.

Several other systems with operating principles similar to the Bio Marine closed circuit unit are currently in development. The economy and simplicity of these units and their ability to be connected to a small-diameter umbilical hose for supplemental gas supply and communication makes their acceptance as general purpose diving rig for commercial, military and scientific purposes a distinct possibility.

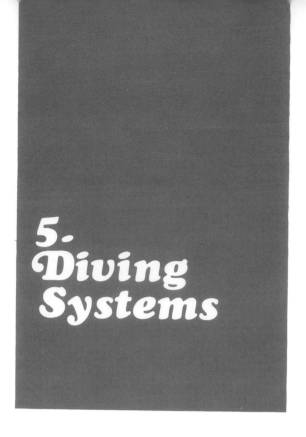

5. Diving Systems

Chambers

The pressure chamber is a basic component of the diving system. Major chamber functions are simulating pressure environments for experimental diving, serving as pressurized underwater transportation vessels and as pressurized vessels for the decompression process and recompression for medical treatment. Chamber size and complexity vary from small portable units, just large enough to hold one man, to 30-foot long multiple complexes used in saturation diving techniques. To appreciate their function as operational components, it is necessary to follow them through their evolution from medical tools to their present use as diving apparatus.

Treatment Chambers

The most common "accident" suffered in professional diving is a "hit," or a case of the bends. Incidents of the bends were quite common in early diving procedures and a great effort went into developing treatment procedures. The treatment found to be most suitable for relieving symptoms was to place the victim in a pressurized environment until symptoms disappeared. In most severe cases, dramatic symptoms ceased by the time the patient was pressurized to the equivalent of 165 feet.

Early chambers were quite crude, but as the techniques for treatment of pressure-incurred ailments improved, the chambers improved with them. Today, all major diving facilities are equipped with recompression chambers for treatment of the bends; all commercial diving activities performed in 60 feet or more of water have a recompression chamber available at the site.

The simplest of all chambers is the portable rigid recompression chamber (Figure 5-1). This is a steel pressure vessel slightly less than 3 feet in diameter, 8 feet long, with a hatch at one end through which the patient is passed. The "portable" chamber, completely plumbed and ready to operate, weighs about 1,000 pounds. It is equipped with a high pressure storage tank for pressurization life support of the patient.

Normally, standard oxygen fittings are provided so that oxygen may be used to speed treatment in the final decompression stages. Accurate pressure gages are provided to monitor the internal

Figure 5-1. The portable recompression chamber provides emergency treatment and serves as a pressurized transportation vessel for bends victims.

The one shown, by Galeazzi of Italy, is equipped with medical lock, communications and exceptionally good creature comfort facilities.

pressure of the vessel. Automatic constant flow of high pressure air is usually provided for breathing and venting. A small "lock," or double hatch, is provided so that food or medical supplies may be passed in to the diver. Portholes for patient observation and a communication system are included. Portable decompression chambers are not known for their creature comforts, but the comfort of being relieved of the bends symptoms is usually sufficient compensation for the diver.

Portable recompression chambers are used principally for emergency treatment. In most cases it is necessary to transport the patient in the chamber to a more complex facility to receive medical attention. These large chambers, although they operate on principles identical to the portable chamber, are multiple-compartment arrangements with more control equipment. They are fitted out for a prolonged stay by the patient and attending physician.

Figure 5-2 shows a stationary decompression chamber. Stationary chamber installations are 54 to 72 inches in diameter and, depending on the

number of compartments, up to 12 feet long. The normal stationary treatment chamber consists of a cylinder with a pressure hatch at one end for access, an equalizing compartment and a treatment compartment long enough to accept a prone patient and his attendant. At the farther end of the treatment chamber, facilities are provided to accept a portable one-man recompression chamber while under pressure. This is accomplished in the following manner. The portable recompression chamber and stationary chamber are fitted with mating O-ring sealed flanges. The portable chamber is attached to the end of the treatment chamber while under pressure. The attendant enters the treatment compartment prior to its pressurization. The treatment chamber is then sealed and brought up to a pressure equivalent to that in the portable chamber. Two hatches, the one on the treatment chamber and the pressure hatch on the portable chamber, are opened and the patient is transferred to the treatment chamber. The hatch on the treatment chamber is then closed and the portable chamber may be removed.

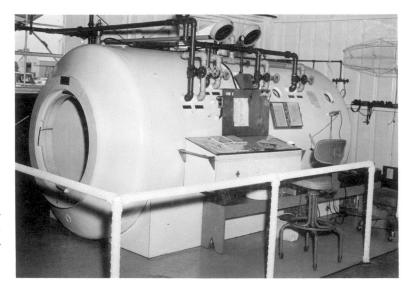

Figure 5-2. The station decompression chamber, with lock in/out capability, is designed for several men. The one shown doubles as a hyperbaric chamber for testing and research. (Photo: U.S. Navy.)

For subsequent entry and exit of personnel into the treatment chamber without loss of pressure, the outer entrance lock to the system is sealed and pressurized equal to pressure within the treatment compartment. The interconnecting hatches then open and personnel may move between the treatment chamber and the equalization chamber. The hatch is then sealed and the treatment chamber can remain at pressure while the equalizing chamber is brought back to atmospheric pressure.

Treatment chambers are equipped with a medical lock so that food, supplies and small instruments can be passed into the chamber without going through the process of pressurizing the larger equalization lock. The treatment chamber is pressurized with air, but in some treatments it is necessary for the diver to breathe 100% oxygen. Oxygen under pressure, especially in the presence of hydrocarbons, is explosive. Therefore, a recompression chamber filled with relatively high pressure oxygen is a volatile environment.

Although not common occurrences, fires and explosions in pressure chambers have caused many deaths. In 1965, two Navy divers lost their lives in a flash fire at the U.S. Navy Experimental Diving Unit at Washington, D.C. The two divers had just completed a test dive in the facility's wet chamber at a simulated depth of 250 feet. They entered the main recompression chamber, which was pressurized with a mixture of 28% oxygen, 36% nitrogen and 36% helium at a pressure of 41 psi. A portable carbon dioxide scrubber with an integral circulating fan was operating in the main chamber. Several minutes after the divers entered, the oxygen level within the chamber dropped to 27%. Several cubic feet of pure oxygen were added and the scrubber energized. One minute after the oxygen was injected, a diver reported a fire in the chamber and immediately after this, observers saw a flash that engulfed the inner lock. Pressure inside the chamber instantly rose to 130 psi. Investigation showed that the scrubber motor had caused the fire.

Today, when oxygen breathing is required, it is not uncommon to use a special life-support system within the treatment compartment. A special pressure-vacuum life-support system manufactured by Scott Aviation Corporation for use in chambers uses two regulators with an oral-nasal mask. Breathing gas is brought to the diver on an intermediate pressure hose similar to that used on a SCUBA regulator. Expired gas from the diver, which still contains a relatively high concentration of oxygen, passes through the second regulator and is brought out from the chamber through a vacuum system. In addition to avoiding contamination of the chamber environment, the system also minimizes the use of relatively expensive breathing gas.

Submersible Decompression Chambers

As diving techniques progressed, operations requiring deeper diver work and longer exposures became commonplace. The decompression procedures required for deeper diving are lengthy. For instance, if a diver makes a single dive to a depth of 190 feet and remains on the bottom for over 50 minutes, the time needed to decompress him would be 3 hours. In commercial diving operations, a man going through underwater decompression for 3 hours can be costly, uncomfortable and can interfere with work on the site.

To minimize the complexity involved in decompressing a diver, the submersible decompression chamber (SDC) was developed. With this system, the diver ascends to a depth that would normally be his first decompression stop. The submersible decompression chamber is waiting for him at that depth.

The *submersible decompression chamber* is a single-lock cylinder approximately 7 to 8 feet high and 4 feet in diameter, with a single hatch at the bottom for entry and exit of the diver. The diver enters and removes his diving equipment. Any water in the chamber is blown out and the lower hatch is sealed. The SDC is then hauled to the surface and placed on the deck of the barge or ship. Still under pressure equivalent to the first decompression stop, the chamber is connected to surface gas supply and the decompression cycle is performed on deck. In addition to being considerably more comfortable for the diver, decompression is performed at the surface and the diver's reactions can be observed directly.

Diving Chambers

As working depths become deeper, it is increasingly impractical to send a diver from the surface to the work site and return him through decompression, even with a submersible decompression chamber. Over the past several years, a new type of chamber has evolved—the diving chamber. Actually, the diving chamber is a versatile multi-purpose unit. It serves as an elevator for bringing the diver to the bottom and as a submersible decompression chamber on the return trip. In

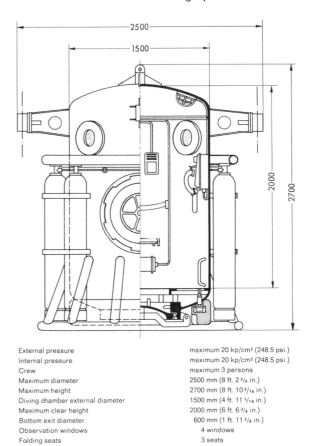

External pressure	maximum 20 kp/cm² (248.5 psi.)
Internal pressure	maximum 20 kp/cm² (248.5 psi.)
Crew	maximum 3 persons
Maximum diameter	2500 mm (8 ft. 2 ³/₄ in.)
Maximum height	2700 mm (8 ft. 10 ⁵/₁₆ in.)
Diving chamber external diameter	1500 mm (4 ft. 11 ¹/₁₆ in.)
Maximum clear height	2000 mm (6 ft. 6 ³/₄ in.)
Bottom exit diameter	600 mm (1 ft. 11 ⁵/₈ in.)
Observation windows	4 windows
Folding seats	3 seats

Figure 5-3. The Dräger TK 200 diving bell is designed for work to 650 feet. Drawing shows some of the construction details. (Photo: Drägerwerk.)

addition to providing a higher degree of safety, it improves diver efficiency and lowers decompression requirements.

Figure 5-3 shows a cutaway drawing of a typical diving chamber, the Drager Model TK 200. The unit is designed to transport and support three persons—normally, two divers and a tender. The system can be set up to operate from a surface gas supply or independently from an externally mounted manifold consisting of eight 50-liter tanks. The chamber is built to work with either an external or internal over-pressure of 285 psi. It can therefore be lowered to 650 feet with an internal environment of 1 atmosphere, or it can be raised from a depth of 650 feet fully pressurized to bottom depth. The chambers are provided in several configurations compatible with either side or bottom mating capability for deck decompression chambers.

Operationally these "bells" are used with many techniques. One of the most common is to seal the divers in at the surface and lower the chamber to the work site on one or more guy wires. On descent, the internal atmosphere can be maintained at normal surface atmosphere, reducing bottom time. At the site, the internal atmosphere is pressurized with the breathing mixture to the external ambient pressure. The hatch is then opened and one or both of the divers exit to perform their work tasks. The tender remains in the chamber to monitor the divers' activity and serve as communicator. When the work is completed, the divers return to the chamber, which is once again sealed. The chamber is then winched to the surface.

During ascent, decompression is started and proceeds independently of the external pressure. At the surface, the decompression process, if not extensive, may be completed in the diving chamber. When long decompression schedules must be followed, the diving chamber is usually mated to a deck decompression chamber and the divers transferred under pressure for decompression in more comfortable quarters.

Deep Diving Systems

Among the standard equipment inventory of every major diving concern is at least one deep diving system. The larger contractors may operate and maintain a dozen systems. The deep dive system is usually a "portable" complex consisting of a deck decompression chamber, gas supply and diving chamber. The deck decompression chamber and diving chamber, or bell, are normally mounted on a skid arrangement for easy handling and transfer to the work boat. The differences between system concepts used in commercial diving are very few; the only major one is whether the diving chamber gas supply is self-contained at the chamber, or is maintained at the surface and transmitted to the chamber through an umbilical cable.

Since many of the deep diving systems are designed by the people who use them, they show a considerable variety of mechanical innovations that result from the personal experiences of these divers. The most obvious of these innovations is the method of mating the diving chamber to the deck decompression chamber and the locking mechanism between the two chambers. Details of plumbing and rigging used in the system usually reflect the work specialty of the company, and no industry standard has been set. The most successful systems do have, however, several common configurational and operational characteristics:

1. They are self-contained, with all machinery required for operation (except gas supply) mounted on a single skid.
2. They are compact and can be placed upon the work vessel in such a manner as to make them operational without interfering with other phases of work being performed at the site.
3. They contain a minimum of complex subsystems that could, if they failed, disable the system.
4. All life-support and surfacing subsystems on the diving bell are redundant.

The depth capability of these systems varies with each company's requirements, but most systems placed into operation over the past several years have been built to work at depths between 400 and 1,000 feet. Although several companies have performed one-time jobs deeper than 600 feet with these systems, operations for the most part have been to demonstrate capabilities, and such depths are not the industry's common working range.

During the 1960s several diving equipment companies undertook the problem of designing standard deep diving systems. Reading and Bates Offshore Drilling Company of Tulsa, Oklahoma, designed and produced a complete series of diving systems for the commerical diving industry. This line was later taken over by Wilson Marine Systems of Houston, Texas. Figure 5-4 is a drawing of the standard diving system built by Wilson. It uses a round diving chamber 66 inches in diameter that has an externally mounted emergency diving gas supply. The bell is positively buoyant with 2,000

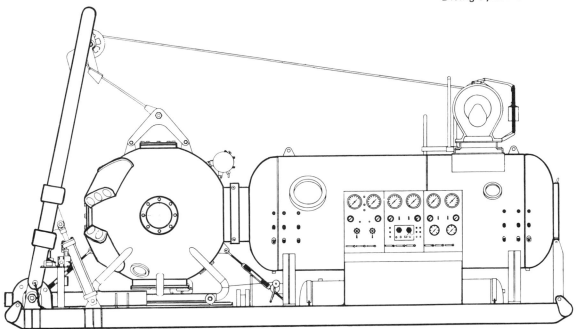

Figure 5-4. The Standard Commerical Diving System, manufactured by Wilson Marine Industries of Houston, Texas, is used for deep diving.

pounds of ballast for descent. In an emergency, the ballast can be released by operation of a lever inside the bell. A carbon dioxide scrubber unit is provided in the bell for decontamination of the gas.

The diving chamber end connects to the 54-inch inside diameter deck decompression chamber. The davit system is hydraulically operated and the main winch for raising and lowering the diving bell is mounted on top of the decompression chamber. The primary breathing gas supply is carried through an umbilical cable that also includes a four-wire telephone line and pneumofathometer lines.

The system designed and used by COMEX of France (Figure 5-5) uses a different concept. The bell is self-contained and carries the breathing gas supply below the spherical bell. The umbilical cable used to hoist the bell to the surface contains only the strength members and electrical cables for the underwater floodlights. Two guidelines, kept under constant tension by winches at the surface, guide the bell to the work site. The bell is positively buoyant and two releasable weights are used

to lower it. The entire system is designed for 600-foot operation.

Underwater Vehicles

During the late 1960s a need developed for highly mobile commercial diving. Pipeline inspection was one of the main areas requiring new developments. For years, the normal inspection method was for the diver to walk the line using an umbilical attached to a compressor on a live moving boat. "Live boating" at its best is a risky business. The boat operator cannot see the diver, so he must rely on the position of the diver's bubbles at the surface to determine his position on the bottom. A severed umbilical was not an uncommon occurrence; if the water was shallow, the possibility of the diver reaching the surface was fair, but the probability of survival decreased with increasing depth.

To safely perform pipeline inspection and similar tasks, a number of small submarines with lockout capability were designed and built. One of the first to successfully demonstrate its capability in this area was the *Shelf Diver,* built by Perry Oceanographics of Palm Beach, Florida. Although a submarine may seem quite different from a diving bell, it possesses a lockout capability and is, in

Figure 5-5. Diving bell design varies throughout the world, but all work on the same basic principles. The bell shown is typical of the design used by the French diving firm, COMEX. (Photo: COMEX.)

reality, a self-contained, steerable underwater diving bell.

The *Shelf Diver* has been used for scientific and commercial applications and has performed lockout tasks as deep as 700 feet. The boat consists of two main pressure-proof enclosures. The first, which normally remains at atmospheric pressure, is located forward and houses the operator, observer and all operational controls. The second, located midship, is similar to a cylindrical diving bell and has a bottom egress hatch. Operation of this compartment is identical to that of a self-contained diving bell, since it can be pressurized to ambient pressure to allow the diver to exit; after the diver completes his task, decompression proce-

dures can be commenced. After surfacing, the submarine can be mated to a standard deck decompression chamber to continue decompression. Figure 5-6 shows a diver live boating in front of the *Shelf Diver*. Using this technique, the diver is always in view of the submarine operator so the possibility of accidents is greatly reduced. On relatively deep jobs, the efficiency of the lockout submersible makes its use economically feasible.

Lateral movement of divers has not been a chronic concern of diving except for military missions. In offshore oil production, however, the situation will begin to change during the 1970s. Well fields will be completely handled underwater and divers will require transportation for the

Figure 5-6. A Perry submarine with diver lockout capability serves a mobile diving chamber for this diver. Equipped for mating to a deck decompression chamber, the submarine is finding many applications where diving is deep and a wide area must be covered.

operation, maintenance, inspection and repair of subsea installations. The vehicles providing this transportation will be wet; that is, they will be free-flooding to ambient pressure, and divers will wear life-support equipment while in them. Although they are not actually diving bells, they will be self-contained vessels carrying all equipment necessary to sustain divers on extended missions. The first of these vehicles designed specially for commercial application is the *Totalsub 01* used by COMEX. The vehicle carries four men, 880 pounds of equipment and tools, and complete life support for 12-hour operation at a depth of 200 feet. Cruising speed of the *Totalsub* is 4 knots, with a maximum speed of 6 knots. Military ver-

sions of wet submersibles have been in use, with varying degrees of success, for the past 10 years.

Saturation Diving Systems (SDS)

The saturation diving system provides access to the work area, maintains life-support functions at depth or depth equivalent pressure, provides creature comforts during pressurization and controlled compression and decompression capability for access to and from the diving operation over extended periods of time. The elimination of decompression procedures between each work period, especially when the working depth is over 200 feet, makes deep diving economically feasible.

Two SDS types, the surface habitat system and the underwater laboratory system, are now in operation. In the surface habitat system, divers live in a habitat on the surface. They are maintained at near equivalent working depth and use a pressurized capsule for descent and return from the work site. In the underwater laboratory system (UWL), or "Sealab," the divers live in a habitat on the sea floor. Each system has merit and the controlling factor in their uses to date has been application. The sea-floor habitat has proven satisfactory for sustained deep ocean scientific work. The surface habitat has proven to be most effective for commerical applications.

Surface Habitat Systems (SHS)

Surface habitat systems cover a wide range of equipment designed to bring divers to and from deep diving work sites and to maintain them in a saturated environment for extended periods. The saturation time limit is seldom dependent upon the system capability; it is normally related to the complexity of the work task and physio-psychological diving conditions.

SHS systems are usually portable equipment complexes that can be set up on the deck of a vessel or barge. Several basic configurations are in general use, but all have the same basic components: the habitat with support equipment on deck, the deck decompression complex, and mating diving chamber or personnel transfer capsule for transporting divers to and from the bottom. The SHS system differs from short-term deep diving systems in that it is designed to support divers in saturated conditions. The deck complex, therefore, is larger and capable of providing the basic creature comforts necessary for extended missions. The systems are being used in both commercial offshore operations and military salvage work. However, commercial systems tend to be "leaner," principally because the space available on construction and drilling rigs is severely limited.

The latest version of an SHS system used by the military is the U.S. Navy's Deep Dive System (DDS), MARK I. This DDS was designed and built for a working depth of 850 feet and emergency operations to 1,000 feet. Capable of supporting four men operating as two-man diving teams, the complex requires a seventeen-man support crew. Using the maximum work schedule, diving time for each team may be as long as 4 hours at 850 feet. The sequence, with alternate teams, can be repeated for up to 14 days. Mission time, which includes compression and decompression time, can be as long as 29 days.

The deck decompression complex (Figure 5-7) and its associated main control console (MCC) are the heart of the system. The complex consists of two 1,000-foot rated decompression chambers, each with living accommodations and separate life-support functions. The chambers can be operated independently, permitting the alternate teams to be separately compressed to, or decompressed from, saturated condition. Food, medical supplies and small equipment can be passed into the system through medical locks located in each chamber. Comfort has been maximized in the chambers to include even piped in hi-fi music and TV.

The MCC contains controls and monitoring instrumentation readouts for operating the entire complex, including the deck decompression complex and personnel transfer capsule. All electrical power, communications and video monitoring, system pressurization and system atmosphere composition are operated from this point.

The PTC (Figure 5-8) is a diving chamber that serves as a pressurized elevator for transferring saturated divers to the work site and as the diving station on the bottom. With the exception of the lifting cable, which doubles as power and communication cable, the PTC is completely self-contained. Breathing gas, premixed for the saturation depth, is carried in storage tanks mounted outside the pressurized sphere. The capsule can comfortably carry two divers completely dressed in their diving gear with working tools and equipment. Two umbilicals are provided connecting the divers to the PTC breathing gas and communication system for working on site. The capsule mates with the deck decompression complex at the entrance lock. A double door O-ring seal, located in the bottom of the capsule, connects to the top of the lock, allowing divers to move to and from the DDC.

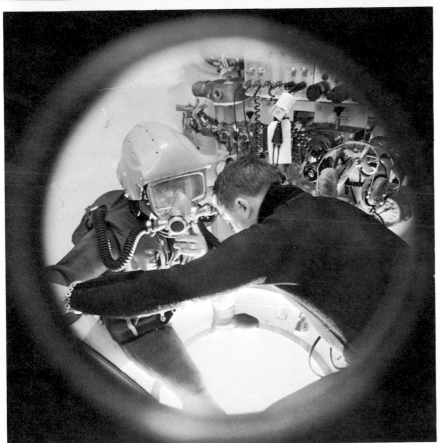

Figure 5-7. (Above). Artist's concept of U.S. Navy Deep Dive System deck decompression chamber. This type of system is capable of keeping four men in saturated condition for 29 days. (Photo: U.S. Navy.) Figure 5-8. (left). Divers test operations in the personnel transfer capsule (PTC) of the U.S. Navy DDS MARK I. Capsule is designed to support four divers at 850 feet. (Photo: FMC.)

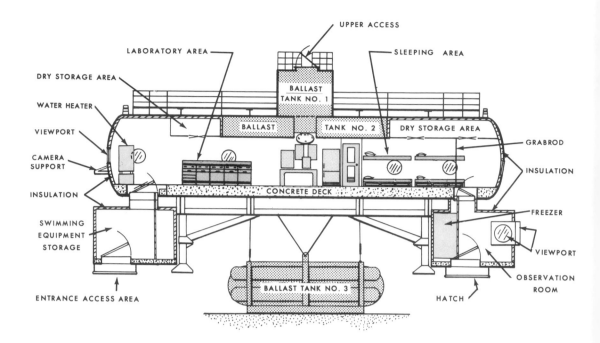

LABORATORY AREA
DRY STORAGE AREA
WATER HEATER
VIEWPORT
CAMERA SUPPORT
INSULATION
SWIMMING EQUIPMENT STORAGE
ENTRANCE ACCESS AREA
UPPER ACCESS
BALLAST TANK NO. 1
SLEEPING AREA
BALLAST TANK NO. 2
DRY STORAGE AREA
GRABROD
INSULATION
CONCRETE DECK
FREEZER
VIEWPORT
OBSERVATION ROOM
BALLAST TANK NO. 3
HATCH

Figure 5-9. (Above). Simplified cross section of the Sealab III habitat. In figure 5-10 (Opposite page), the Sealab III habitat undergoes tests for *watertight integrity. The 57½-foot sea-floor habitat was the U.S. Navy's last effort at deep sea-floor living in the 1960s. (Photo: U.S. Navy.)*

Sea-floor Habitat Systems

The basic sea-floor habitat saturation diving system consists of a surface support ship, a bottom habitat, an elevator to transport devices between the support ship and the habitat. This system is much more complex than the surface habitat system and, in fact, contains a complete surface habitat system as a subsystem. On deep operations, support functions for such a complex system are difficult. The economics are not in keeping with commercial diving use and the expense can barely be justified on the basis of scientific experimentation. Many major experimental series using sea-floor habitat systems were successfully completed in the 1960s. Among them were Cousteau's Conshelf series, the U.S. Navy's Sealab series, the Tektite Project and West Germany's Helgoland.

Sealab Habitat

The Sealab III system, designed for a 60-day stay at 600 feet, was the most costly diving system ever built, as well as the most complex. Figure 5-9

shows a simplified cross section. The habitat was a 57½-foot long cylindrical structure with capped ends and a small conning tower and had the appearance of a railroad tank car without wheels. The main cylindrical structure contained the controls, electronic equipment, galley, bio-medical/marine research laboratory and berthing facilities for eight aquanauts. Figure 5-10 is a photograph of the habitat.

The habitat was essentially the same used in the previous Sealab II experiment. The cylindrical structure was 12 feet in diameter with two rooms, each 12 feet square by 7½ feet high, installed under the main structure.

Aquanauts entered the habitat through a hatch in the diving station, a small room attached to one end of the main habitat structure. The diving station provided space for the aquanauts to don and doff their diving gear, to check and adjust their equipment, to charge SCUBA tanks (from gas banks attached to the habitat), to control breathing gas supplied by the umbilical to divers in the water and to clean and stow swimming gear. There was a 12-inch view port in each side of the diving station. Directly above the diving station a utility

1.6% oxygen. Relative humidity was maintained at about 70% with a temperature of 90°F. The ventilation system was modeled after the standard submarine system with lithium hydroxide (LiOH) carbon dioxide scrubbers and charcoal filtration.

Because of the higher thermal conductivity of helium and the resulting heat loss to the cold surrounding sea water, a "comfortable" temperature of about 90°F was considered a necessity in the habitat. Two-inch cork insulation covered all metal in the habitat exposed to the sea. The deck of the main structure, diving station and observation room were made of concrete with embedded heating cables. There were also overhead radiant heaters. Eight bottles were mounted on the habitat proper and eight additional gas bottles mounted on the "clump" under the habitat.

Two external oxygen cylinders were provided to replenish the habitat's atmosphere with oxygen. A sensor monitored the partial pressure of oxygen in the atmosphere, and if a loss was detected, the oxygen supply would bleed in automatically and an acoustic alarm would be actuated. The oxygen was cut off when the habitat's partial pressure normalized. In the event of failure of the automatic sensor device, a visual readout sensor was included and, should both oxygen sensing systems fail, a manual oxygen makeup system was provided. An emergency built-in breathing system (BIBS) was provided in the event the habitat's atmosphere became contaminated. BIBS manifolds were installed throughout the habitat, including the diving station and observation room. SCUBA mouthpieces, with hoses and regulators to fit the manifolds, were provided at each bunk and in the diving station.

The habitat complex, positively buoyant to facilitate handling and towing, used a variable ballast system for lowering and anchoring. The ballast system consisted of three water ballast tanks plus about 2,000 pounds of portable lead pigs located inside the habitat, in trays outside the habitat and on the clump suspended under the habitat. Ballast tanks were located in the small conning tower of the habitat, in the overhead of the habitat and in the clump under the habitat.

The habitat was made some 9,000 pounds heavy for lowering by flooding the upper ballast

space was located at the stern which had a shower stall, a combination clothes washer-dryer, hot water heater and space for additional gear storage.

The laboratory area, located forward of the utility space, served as operations center for the habitat. It contained instrumentation for monitoring and controlling the atmosphere in the habitat, power and lighting system, communications and special instrumentation. Going forward, the next area contained the habitat's galley with a refrigerator, electric range, infrared oven, sink and storage space. The berthing area, occupying the forward end of the main chamber, contained eight bunks, lockers and additional stowage space. Directly below the berthing area was the observation room. This was equipped with a large bottom hatch which could be opened to provide direct access to the sea in the event of emergency.

The habitat was equipped with eleven 24-inch plexiglass view ports in the main cylinder, providing observation of almost all the surrounding water. (The number and size of the view ports was decided after a near-fatal accident in Sealab I. An unconscious aquanaut outside the habitat was rescued only when his bottles bumped the side of the structure. He was not visible from inside the habitat.)

The atmosphere of the habitat at 600 feet consisted of 92.4% helium, 6.0% nitrogen, and tanks. After lowering to the ocean floor by crane and with the clump resting on the bottom, the lower ballast tank was flooded, providing an anchor. In the event the ocean floor was uneven, the habitat, which was attached to the clump by a series of cables, could be maintained in a level attitude by means of automatic tensioning wires and by shifting the portable lead ballast.

Located beneath the diving station and observation room were two air-tight skirts, 18 inches in depth. The skirts provided a variable volume below each hatch to accommodate fluctuations in the water level due to tidal variations and changes in the internal pressure of the habitat.

The subsystem within the system used for compression, decompression and transfer of divers in pressurized condition from the surface was the U.S. Navy Deep Diving System, MARK II. It consisted of two deck decompression chambers and two personnel transfer capsules.

Deck Decompression Chambers (DDC).

Mounted on the support ship, the DDCs were designed to support four divers for prolonged periods at an internal pressure up to 378 psi (the equivalent of an 850-foot depth). Normally the DDC would accommodate four divers for up to fourteen days plus the necessary decompression time. In emergency situations more men could be decompressed with the duration dependent only upon available gas supplies.

The chamber was divided into three sections: entrance lock, living area and sleeping area. The entrance lock was 5 feet long and separated from the rest of the chamber by a pressure hatch. This arrangement enabled personnel to enter or leave the main chamber while the main chamber was under pressure or provided with a special gas mixture. This outer lock also contained the DDC sanitary facilities and was fitted with toilet, wash basin, shower and small cabinet for the storage of personal articles and first aid equipment.

A non-structural bulkhead separated the living and sleeping areas. The former was fitted with a table, which could be folded out of the way when not in use, and seats. The sleeping area contained four submarine-type bunks and lockers for stowage of personal gear. At the sleeping end of the

DDC was a small lock for passing in medical supplies, food and other small items.

The life-support system for each decompression chamber consisted of two loops, either of which could be operated independently or simultaneously when there were increased demands on the system. Each loop consisted of a carbon dioxide absorber, condenser, filter, heater, control valves, and associated gages and piping, all of which were external to the DDC. The system was designed to meet the normal requirements of four men at a depth range of from 0 to 850 feet sea water; inside temperature ranges were maintained between 75°F and 95°F.

Each DDC had an emergency built-in breathing system which could supply oxygen or helium-oxygen gas mixtures to breathing masks in the event the chamber's atmosphere became contaminated.

Personnel Transfer Capsules (PTC).

The PTCs transported divers between the DDC and the ocean floor. The PTC could also support divers working on the ocean floor, providing them with breathing gas through umbilicals and a rest station. After the divers completed their UWL mission, the PTC would return them to the surface under pressure (to 378 psig) and mate to the DDC, enabling the divers to pass freely into the chamber for decompression.

The PTC contained seating for four divers and could support four divers for 8 hours during normal operations and for 24 hours in an emergency. This support included breathing gases, rations, drinking water and provision for body warmth. Safety equipment in the PTC included seat belts and crash helmets to prevent body injury when the PTC was being handled in rough sea conditions.

Breathing gas for the PTC was provided by ten gas bottles containing oxygen (1), helium (5), and helium-oxygen (4). Each bottle held 1,200 cubic feet of gas. Four stowage racks were provided outside the PTC for 200-foot hoses to enable divers to work in the proximity of the capsule using the bottled breathing gases.

Access to the PTC's interior was through the large bottom entrance hatch. The hatch was large enough to enable divers with SCUBA rigs to have easy entrance. Four view ports in the PTC enabled divers inside the chamber to view divers working nearby. The PTC could also be used as a dry, 1-atmosphere observation chamber to carry observers to the ocean floor for inspections of the experiment area.

During operations the PTC had positive buoyancy. A winch was provided in the base of the capsule for hauling the PTC to the bottom. The winch cable was attached to an anchor, or clump. Explosive release studs were fitted to both the haul-down cable connection and the strength-power-communication cable, which raised and lowered the PTC, so that either or both cables could be blown clear in the event they were severed or fouled. (The haul-down cable was 3/8-inch galvanized steel, 1,200 feet long.)

The strength-power-communication cable was designed to provide the strength, electric power and communications connection between the support ship and personnel transfer capsule during a diving operation. The 1,400-foot long cable had an inner core of electrical wires which carried communication and instrumentation signals and electrical power. The outer sheath was made up of galvanized plow steel wires. The cable had sufficient strength to lift a flooded PTC from the ocean floor to the surface and an internally dry PTC from the ocean floor to the deck of the surface support ship.

Integrated Command and Medical (IMC) Vans.
Located on the support ship's deck, the IMC served as command, communication and medical monitoring centers. The command van had dual communication stations which served as monitors and switchboards for all communications with the sea-floor habitat. These facilities enabled the Sealab III on-scene commander and his staff to communicate with the habitat, aquanauts in the water, shore and ship support facilities and the diving system on the support ship. The engineering evaluation console was also installed in the command van.

The adjacent medical van had complete facilities for monitoring the atmosphere in the Sealab III complex and the psychological condition of divers. The gas chromatograph in the medical van, which could sample gas in the habitat and anywhere in the diving system, was manned at all times because of the potential fatal consequences of variation in the prescribed breathing gas mix for aquanauts.

Other equipment in the medical van was designed to allow Navy doctors to perform complete laboratory tests on aquanauts while they were on the ocean floor (using blood samples brought up in a dumb-waiter system), and while the aquanauts were on the surface before and after a dive. The biological parameters which could be measured by equipment in the medical van included electrocardiography, blood chemistry, urinalysis, gas uptake and elimination, culture studies and respiration.

Constructed of aluminum, both vans were 26 feet long, 8 feet wide, and 8 feet, 7 inches in height. As mounted on the support ship, they were connected by a flexible passageway. The vans had independent heating and air conditioning systems, which used electrical power from the ship. During Sealab III the vans were welded to the deck of the support ship and were designed for rapid removal for use ashore or afloat in later experiments and research.

For a complete list of communications and medical equipment used in the system, see the Appendix.

6. Diver Tools

The diving system is the total equipment necessary to bring a diver to a specific depth, to provide life support, and to return him to his natural environment with no specific, or cumulative, detrimental effects. The diving system then is a means of transportation that provides the diver with a safe environment at the work site and returns him to his normal atmospheric surroundings. It has little to do with his function underwater. In performing his underwater work, the diver relies on a number of diving tools. A diving tool is any apparatus—electronic, electrical or mechanical— that assists the diver in performing his primary diving function.

Diver tools cover a wide range of apparatus and instrumentation, including communications equipment that enables the diver to speak with those on the surface and to other divers, sonar equipment that enables him to extend his vision range and special instrumentation used for measurement. Finally, it includes tools that enable him to perform specialized journeyman functions underwater. A word that cannot be spoken, an object that cannot be seen, the length mark that cannot be read and the bolt that cannot be secured can contribute equally to the abortion of an expensive and long-planned underwater operation. Diver tools help avoid these problems.

Our hand tools are regarded as natural evolutions that are the best for the job. A key, the simple device that tightens the chuck of a drill, is considered the optimum for performing its function in air; underwater it is next to useless. To use a drill in air, the key is inserted into the chuck and turned clockwise to grip the drill bit. This is a simple operation. However, attempting this in diving gear, with heavy gloves, leaves something to be desired. The solution is to weld a 6-inch stem onto the key so that it can be felt. This may seem miniscule, but if a $5,000-an-hour operation is waiting at the surface for the diver to drill a specific hole, it becomes very important.

Underwater tool designers face a group of very stringent parameters. They must design tools that are usable by people who are under narcosis stress, encumbered by diving gear, and who have lost the greater part of their tactile sensitivity through cold exposure. Worst of all, they cannot see the tool. Although this may seem an impossible combination of requirements to meet, underwater work is done and the degree of effectiveness of the underwater technician is directly propor-

Figure 6-1. Navy aquanaut Paul Wells uses a special tool for injecting plastic foam into a sunken aircraft during a practice salvage operation off Sealab II. Underwater tool usage is complicated by poor visibility, cold, lack of a foothold and restrictions of the diver's bulky dress. (Photo: U.S. Navy.)

tionate to the usability of the tools. Figure 6-1 shows a diver using a special tool during a salvage operation.

Underwater Communications Equipment

As important to safety as it is to efficiency, underwater communications equipment ranks next to life-support equipment in the underwater environment. Only within the last 5 years has dependable equipment with a reasonable level of intelligibility been commercially available for both the tethered and the free-swimming diver. Regardless of the development rate of underwater life-support systems, man can never be an effective underwater operator without the capability of voice communication—communication between both himself and other divers and the operational control center.

Underwater communication equipment for divers, however, has been suspiciously like the weather. There are many complaints about prob-

lem areas, and everyone using the apparatus agrees that something more can be done, but the basic problem remains. At a recent University of Florida Communication Sciences Laboratory symposium, some of the reasons for this impass came to light.

Underwater Communication Link

Although not commonly recognized, the critical problem areas in dependable communication are now the same for either free SCUBA divers or tethered divers. To see the commonality, we need only look at a typical underwater communication link.

A link consists of three main subsystems: the input transducer that converts the diver's speech into an equivalent electrical signal, the transmission system that delivers the signal to the receiver transducer system and, finally, the receiver that converts the signal to a form suitable for human perception. Severity of the problems in each of these subsystems is directly related to how closely subsystem performance is associated with human physiology. In increasing order of dependence and problems, we have the transmission link, the receptor system and the input transducer system.

The transmission link, whether it is a simple pair of telephone wires or a complex single sideband acoustic transmission system, is completely independent of the underwater environment's effects on the human body. The transmission link can be considered a straightforward engineering problem—a problem that has been resolved in several satisfactory ways.

Wired Communications

In tethered diving the solution is a waterproof telephone line driven by an audio amplifier. System complexity varies from a single amplifier with a two-wire telephone line that allows talking in one direction at a time to more complex audio systems suitable for use in multi-element diving systems.

One of the greatest problems with hardwire communication systems is that very often the system is not designed as a *system*. The electrical and electronic problems involved are straightforward, and the difference between a good system and a poor system is related directly to the mechanical/marine environmental engineering. The system's components must be extremely rugged in design, capable of withstanding physical abuse, water splashing, mud and corrosion.

Communication systems designed for deep diving systems embody the same principles, but require a greater number of subsystems stations to tie the complex together. The communications system used on the U.S. Navy MARK I Deep Diving System was a custom engineering system built by Aquasonics of San Diego. The system consists of the following major components: master control console communication station containing the central controls for the entire communications system and equipped with monitor speaker, volume controls for pressure chamber speakers, volume controls for microphone outputs, volume controls for deck speaker, four 6-watt power amplifiers, power supply, two head sets with noise-canceling microphones, input and out-

put terminals for various speakers, microphone and other components, deck speaker and interconnecting cable.

The PTC communications station includes a microphone, intercom loud speaker, connection to a stand-by acoustic system, on/off switches, volume controls, and plugs for headsets and microphones. The DDC communications panel and entry lock (EL) communications panel each consist of a monitor microphone, loud speaker, two noise-canceling microphones, two headsets and the plugs.

All units are designed to operate at a temperature range of 0 to 50°C and a relative humidity of 100%. Components in the PTC, DDC and EL operate in ambient pressures as high as 450 psi.

Wireless Communications

Several types of wireless transmission links are commercially available, ranging from simple voice amplifiers to rather complicated single sideband suppressed carrier transmission systems. The modes of transmission most common today are acoustic transmission, electromagnetic transmission, electric field transmission, AM and FM transmission and single sideband transmission.

Acoustic Transmission

Direct acoustic transmission of voice signals underwater is the simplest wireless method. It involves amplifying the voice and injecting the voice energy into the water with a suitable transducer. Maximum range for use with minimum usable intelligibility is restricted to several hundred feet in the water, under ideal conditions. Transmission is complicated by underwater noise and the diver's ability to receive the audio signal. As discussed in Chapter 1, the normal sound reception system is desensitized when occluded by water. In addition, the neoprene wet suit normally worn by a diver provides sufficient discontinuity between the diver's head and the water to cause great loss in the sound level reaching the bone structure. The net result of the hearing loss due to water occlusion and insulation provided by the neoprene wet suit is that the acoustic system becomes a very questionable transmission link. The system is, however, simple and cheap to manufacture and has found some application in diving.

Electromagnetic Radiation Communication

Electromagnetic radiation, or low frequency radio, is now used in and around the water. Its drawback is that energy absorption by the water in that portion of the frequency spectrum is great, and range is therefore limited. Although it is a convenient tool for communication between surfaced divers, it does not provide a satisfactory solution for underwater communication.

Electric Field Communication

Electric field transmission has been tried as a transmission medium for underwater communication systems. The principle of electric field transmission involves generation of an electric field, usually between two plates, by amplifying the diver's voice and generating a potential between the plates. The electric field generated in this manner can then be picked up by an equivalent receiver. It has the advantage of being a particularly quiet transmission channel, when not in the vicinity of other electronic or electrical equipment generating high level fields. Transmission distance is, of course, a function of the power of the transmitter and sensitivity of the receiver. As power is increased, however, the diver perceives the electric field as a mild shock. Since very little is known about the effects of electric fields on submerged divers, the system has not gained much popularity within the diving community. New electric field transmission techniques are developing, however, and it is possible that within the next two years this method may offer a safe, convenient mode of underwater transmission.

Modulated Carrier Communications

The most successful wireless underwater transmission systems developed to date are the modulated sonar frequency carrier systems, both AM and FM. The modulated sonar frequency link is capable of transmitting a voice signal with the same fidelity as a telephone line. The type of carrier system used is selected as a function of the application. The controlling parameters are ambient noise level in the area where the equipment is to be used and the transmission range required.

The greatest source of underwater noise is marine organisms, particularly bivalves. If the communicator is to be operated near heavily encrusted piling, the carrier frequency must be high enough to eliminate most of this background noise. In practice, a frequency of 40 kHz provides a good tradeoff between the range to power ratio and ambient noise. A typical communicator of this type is the Aquasonics 420, an amplitude modulated carrier system. Range under normal adverse conditions is about 150 yards diver to diver, and up to 500 yards in the open sea where noise level is less.

Single Sideband Transmission

Where relatively long range is required, a lower frequency with better transmission characteristics must be used. The most common choice is the 8-11 kHz range, the frequency used for U.S. Navy ship-to-ship underwater telephone systems. Operating at this frequency range and using single sideband transmission techniques, diver to diver communication ranges up to 1,500 yards can be obtained while still maintaining an equipment package size compatible with free-swimming diver operations. The Hydro Products 811 suppressed carrier single sideband communicator is typical of this type of equipment. It provides a minimum of 1,000-yard transmission link on the standard AN/UQC frequency of 8.0875 kHz. In the open sea, ranges of over 1,500 yards are normal. The system is also provided for use in small submersibles and is commonly applied today as an emergency stand-by system on diving bells.

Sound Reception

Solutions to reception of voice communication underwater have been available for the past several years. They do not represent new breakthroughs, but rather are applications of known technology and quantified data on how man actually hears underwater.

When the diver's head is in the gas environment of the diving helmet, the normal sound sensing system is operative; hearing is simply a function of a favorable signal to noise ratio. The best performance is obtained by mechanically reducing gas injection noises and placing a moderate signal level close to the ear rather than a higher power signal several inches from the ear.

When the diver's head and ears are exposed to the water, occlusion of the ear by water mechanically damps the system to a point where acuity loss and distortion occur. Bone conduction transducers in contact with the head restore near-normal hearing sensitivity.

Input

It is impossible to discuss the input system without considering the source generator, man. The actual process of converting a sound signal to an electrical equivalent is simple. Helmet noise-canceling microphones, moisture-proof and built to military specifications, are available off the shelf. For free-swimming divers, several microphone configurations with suitable frequency response are available, including a boot covered dynamic microphone from Aquasonics.

Specifically, from an engineering standpoint, it is possible with existing equipment to pick up an available voice communications signal and transmit it up to 1,500 yards to a receptor with an overall fidelity of ±3 decibels. The key to understanding the relatively low intelligibility scores in underwater communication is the term *available voice communication signal.* The signal that a diver can generate under high ambient pressures is quite different from his normal output at the surface.

The factors that distort underwater speech before it reaches the communication system's input are ambient pressure in the human voice system, unnatural resonant chambers created by life-support equipment and the physical characteristics of the breathing gas. Recently Dr. Harry Hollien of the University of Florida Communication Sciences Laboratory tested the effect of high ambient pressures on speech output. These tests showed that a loss of approximately 4% intelligibility occurs for each atmosphere of pressure. The mean score for a depth of 190 feet, or nearly

7AtA, was only 68.8%. These tests were conducted in an acoustically adjusted compression chamber and did not take into account damping of mechanical action of the human noise-producing mechanism by the pressure of external water.

The external distortive effects of life-support equipment, especially minimal back pressure in the breathing system, and unnatural resonant cavities is dramatic. The most common talking chamber used by free-swimming SCUBA divers is the Nautilus mouthpiece, available in either single or double hose configuration. In the single hose version, which includes the second stage regulator, the exhaust valve is located in a position approximately 1½ inches lower than in the two-hose unit. This small distance gives a difference of 0.9 ounce in exhaust or back pressure. Translated into distortion, however, this small increment in pressure, combined with the other distortive effects of the regulator mechanism in front of the diver's mouth, result in a 6% drop in intelligibility.

The effects of different breathing gases on a diver's ability to generate understandable signals is less clearly understood. Chapter 1 discussed tests run by scientists at the University of Florida Communication Sciences Laboratory under elevated ambient pressures of oxygen/helium. The results showed a loss of 50% intelligibility for each 200 feet equivalent pressure. Starting with 90% at the surface, intelligibility decreases to 50% at 200 feet, 20% at 450 feet and 10% at 600 feet. How much of this distortion is a result of the direct effects of pressure is still not known. If the elevated air pressure tests are extrapolated to these depths, they yield distortion errors of the same order of magnitude.

Locating Objects Underwater

Operating on land, man can use his perception and instincts to follow a direct path to nearly any location he wishes. Modern man expects highways and transportation between locations and generally finds it difficult to appreciate the problems involved in locating an object underwater. Consideration of the restricted vision, the absence of long distance landmarks and the presence of underwater currents reveals the problem of underwater object location in its proper complexity.

The problem is real, both physically and psychologically. In discussions with Sealab aquanauts after their underwater tours of duty, psychologists found that one of the most stressful psychological considerations for the aquanauts using SCUBA gear on excursions is a fear of not being able to find their way back. Being lost on land can be discomfiting and in rare instances dangerous—but being lost underwater on a saturation dive results in death.

Locating underwater objects also involves economic considerations. The diver's inability to find a wellhead can be a costly operation for a commercial crew. The obvious solution is to mark the object with a surface buoy; this undoubtedly would be a straightforward solution to the problem, with one exception. For some reason, people who traverse the ocean surface have a passion for recovering objects and buoys. Fortunately, this instinct for recovery does not include navigational buoys. Nevertheless the attrition rate on buoys used to mark underwater objects is awesome. What these collectors do with the buoys after removing them from the location is beyond comprehension.

Pingers and Pinger Locators

Today the most effective way to mark a location is to tag it underwater. This tag is usually an encapsulated electronic package that emits an acoustic signal or ping at regular intervals. The power from these sources varies and so does the length of time they will operate. Acoustic pingers that mark wellheads are capable of operating for years. Normally, these long term pingers are set up to operate once each 2 seconds with a power output detectable underwater at a distance of several miles.

For shorter period operations, less expensive devices are used. These low power pingers, generally designed for one-time use, are relatively inexpensive and considered expendable. Picking up the signal from an acoustic pinger requires a sensitive underwater receiver with directional capability. Equipped with a pinger receiver, a diver can submerge and swim directly to the marked object.

Typical of the equipment used is the pinger receiver built by Burnett Electronics Laboratory. The unit is a hand-held acoustic receiving device

Figure 6-2. Divers use a pinger receiver to return to a specific location. When they are close, they will submerge with the instrument and swim directly to the previously marked spot. (Photo: Burnett Electronics.)

operating in the frequency range of 30 to 45 kHz. The battery powered unit is contained in a watertight housing designed to operate to 200 feet. In this frequency band signals transmitted by acoustic pingers, or beacons, are converted to audio signals in the receiver; this signal is presented to the diver on a set of earphones. Designed for locating underwater objects such as missile cones, instrument packages, research rockets, torpedoes and wellheads that are intentionally marked with pingers, the unit is compact, lightweight and slightly positively buoyant underwater. An adapter is provided so that the unit can be used from the surface to bring the diving boat into the general vicinity of the beacon. Figure 6-2 shows divers using a pinger receiver.

Pingers, of course, are the simplest underwater location markers, but they have one drawback—their use requires preplanning. The object must be marked with a beacon and the search must take place within the life of the beacon. As with most underwater problems, especially in emergencies, preplanning is not characteristic, so much more sophisticated instrumentation must be used.

Diver-Held Sonar

In military diving such as explosive ordinance disposal (EOD) activities, the name of the game is finding the object that wasn't supposed to be found. The object is usually a mine, and is certainly not acoustically marked. To fill both needs, that is, to find objects that are supposed to be found but are not marked and to find objects that are not supposed to be found, Burnett Electronics Laboratory builds the Detecting-Ranging set, SONAR, AN/PQS-1C. The equipment is a self-contained unit designed for use by free swimmers in locating underwater objects. It has several modes of operation; in addition to its sonar capability it has facilities for locating acoustic pingers.

In the *search mode* of operation, the diver-held sonar set transmits an ultrasonic wave which sweeps through a specific bandwidth. A transducer and conical reflector direct the transmitted waves into a narrow beam to provide a precise angular sensing of the target location. The returning ultrasonic signal, which has bounced off the object, combined with a sample of the transmitted signal, produces a different frequency proportional to the distance from the target. This difference is an audio frequency that can be detected and amplified. The minimum and maximum frequencies in this range correspond to the minimum and maximum range of the scale selected. The lower the tone, the closer the operator is to the reflecting objects. Scale ranges of 20, 60 and 120 yards are normally provided. In the hands of an expert rater, the unit can provide a great deal of information. In addition to telling the experienced operator how far the object is, he can normally tell what it is as he approaches. Though this may seem rather unimportant, it becomes extremely important in an area where there are two or three objects—especially if one of them is an armed mine.

In the *marker-receiver mode,* the equipment can locate pinger transmitters operating in the range of 30 to 40 kHz. When operating in this manner, the equipment transmitter power amplifier is disabled and the unit operates in a manner very similar to the standard pinger receiver. Since the unit is receiving a signal from a fixed transmitter in this mode, the operator can only tell the direction to the source, not the distance.

Underwater Welding

One of the diver's major underwater tasks particularly in offshore well production or pipeline activity, is the joining and cutting of metal. These underwater activities can be performed in either *wet* conditions or *dry* conditions. Most welding performed in salvage or emergency repair is done in *wet* conditions with the shielded metal arc process, or as it is more commonly known, stick electrode welding.

Stick Electrode Welding

Welding of the stick electrode type is accomplished by maintaining an electric arc between the work and a coated electrode 12 to 15 inches in length. The arc vaporizes chemicals in the electrode coating and the gases formed shield the arc from the water and help to stabilize it. Figure 6-3 shows a diver welding by the stick electrode method.

The normal joint is made by laying weld beads in a V-shaped groove by lapping or butting the two pieces of warp to be joined. In the near-zero visibility encountered on most salvage or construction projects, the groove acts as a guide and the welder can actually complete the process without ever seeing the groove.

Wet welding is a slow and costly process. Even with the most skillful operator, underwater welding requires nearly four times as long as an equivalent job would take on the surface. If time consumption and expense were the only considerations, wet welding would still be a satisfactory process. But in addition to its cost, wet stick electrode welding underwater produces a generally unsatisfactory weld. The surrounding water acts as a heat sink, reducing heat penetration and causing unsatisfactory fusing. The weld and the areas near the weld can suffer under bead cracking caused by too rapid cooling, and from excess hydrogen caused by the arc's disassociation of hydrogen from the surrounding water. With today's technology, wet welding can be used only on mild and low carbon steels.

Herbert W. Mishler and Milton D. Randall of the Battelle Memorial Institute in Columbus, Ohio, suggest that it may be possible to produce high quality underwater wet welds if a technique can be devised for excluding water from the joint area. They suggest that high velocity jets of an inert gas might be used to create a 3- to 5-inch diameter water-free weld region. Where high quality welds on pipelines and drilling structures are to be performed underwater now, a welding technique that brings a dry environment to the underwater location is necessary.

Dry Welding

Underwater welding habitats are now part of the standard diving technique in the offshore oil industry. With these facilities the diver works in nearly the same conditions as he would at the surface.

After the welding habitat is secured at the work site and the water forced from inside, the divers are free to begin work. The atmospheric volume within the habitat is limited and the normal procedure is for the divers to wear life support equipment within the habitat to avoid breathing the contaminant gas produced by the welding process.

The dry welding process does produce a high quality weld underwater, but it is extremely expensive. A single welded tie-in pipe joint now costs between $150,000 and $250,000. The high costs are caused mainly by the amount of personnel support equipment and the time required to accomplish the task. But even with these high costs, the process is economically feasible compared to other methods of joining.

Two methods of dry welding in underwater habitats are generally used today: the gas metal arc (GMA) and the gas tungsten arc (GTA). The equipment for these processes is considerably more complex than those required with the stick electrode process. But the stick electrode method, because of the output of smoke and fumes from electrode coatings, is unsatisfactory in the closed environment of the welding habitat.

Figure 6-3. "Wet" welding by the stick electrode method has many quality control problems. Commercial diving firms now use "dry" welding methods which, though considerably more expensive, give quality welds. (Photo: T.A. Bennett.)

Gas Metal Arc Welding. In the GMA process a small consumable electrode is fed continuously through the welding torch as fast as the wire is melted by the arc. The end of the electrode arc and molten weld metal are shielded from the atmosphere by a flow of inert gas which is conducted through the torch under pressure, escaping from a nozzle that surrounds the electrode. The GMA system includes an electrical power source, a welding gun, a means of feeding the electrode wire from a spool and suitable controls. In underwater dry welding all the equipment except the electrical power source is located within the diving habitat. The electrical power supply is located at the surface and connected to the habitat by an umbilical cable.

Rate of deposition of metal by either the stick electrode or the GMA process is about the same in underwater habitats. However, some problems are associated with the GMA process at depth. As ambient pressure increases, the arc in the GMA system becomes more intense and melts the filler wire faster. The molten pool gets larger and the welder-diver is hard pressed to control it. This excess metal can lead to weld defects such as overlap and improper fusion. In addition, the shielding gas becomes dense under increasing pressure, and flow rates up to ten times those required at the surface must be used.

Mishler and Randall at Battelle suggest possible solutions to these problems. One is a modified GMA welding technique that would require smaller inputs of heat and thus avoid creating the excess molten metal. According to them, the dip-transfer and pulsed-arc methods require 20 to 30% less heat. In the *dip-transfer technique,* the arc "shorts out" when the filler wire dips into the weld pool but the resulting surge of current melts the wire and quickly reestablishes the arc. These shorts occur 50 to 70 times a second and metal is transferred each time. They also suggest that GMA underwater welding might be improved by reducing the filler wire melting rate. The melting rate can be lowered significantly by adding chemicals to the filler wire or the arc atmosphere.

Gas Tungsten Arc Welding

The Gas Tungsten Arc welding technique uses a tungsten electrode which does not melt, rather than the low melting point wire used in the GMA process. The arc melts the edges of the metal pieces being joined, and filler metal, in the form of bare wire, is fed into the weld pool as needed. Gas flow in the GTA apparatus is similar to that in the GMA. Since the filler wire used in the GTA system does not melt as fast as in the GMA, the process is considerably slower. The GTA process does, however, yield a high quality weld, and few difficulties have been experienced using the GTA technique at depth.

Underwater Cutting

Two cutting processes, flame and arc cutting, are in general use today. Flame cutting has progressed little since its introduction by the U.S. Navy in the 1920s. Surface flame cutting uses an oxygen/acetylene flame, but acetylene is extremely explosive above pressures of 15 psi. Its use

underwater, therefore, is limited to 25 feet, and even at this depth it is relatively volatile.

The common oxygen fuel flame used in underwater gas burning is oxygen/hydrogen. It has limited effectiveness and cannot cut stainless steel or non-ferrous metals. Cuts can be performed in slow oxidizing metals by melting instead of burning, but the process is extremely crude and slow.

Arc Cutting

Compared to arc cutting, flame cutting has little to recommend it and has gradually been discontinued on commercial operations. Arc cutting, or oxygen burning, employs an electric arc and oxygen to accomplish the process. The basic equipment is very similar to that in stick electrode welding, except that the electrode is hollow. After an arc is struck between the electrode and the work, oxygen flows through the electrode to oxidize the hot metal and blow it away. It is two to three times faster than flame cutting. Because only one gas is used, arc cutting equipment is simpler, easier to handle and easier to operate. Its main advantage is not its simplicity, but that it cuts all metals, although stainless steel and non-ferrous metals cannot be cut as quickly as ordinary steel.

Arc cutting is the most widely used cutting method on construction today and will continue to be for the next few years. Low voltage electricity is used, and although shocks are possible they are relatively uncommon with experienced workers. The electric field generated by the electrode is not normally of sufficient intensity to cause serious electric shock. The chief disadvantage is that the electrical wiring must be kept absolutely watertight and oxygen must be kept from feeding back into the torch. If an electrode is inserted and is not properly seated, it is possible for oxygen to get back into the torch handle and cause an explosion. With properly designed and maintained equipment, however, this is a relatively uncommon occurrence.

Plasma Arc Welding and Cutting

A new method of welding and cutting, plasma arc, is now under development and promises improvements over both the wet and dry methods. Plasma arc welding is similar to GTA except that heat is provided by a constricted plasma created by an electric arc within a stream of ionized gases. The standard welding arc can be considered a plasma, but in plasma arc welding the arc gas mixture is restricted to a much narrower diameter. The plasma arc method has several major advantages over GTA; it has a higher heat concentration as a result of the constricted plasma, it is less sensitive to changes in arc length and shows better arc stability. Deeper penetration into the weld is possible because of the plasma's high velocity.

Plasma cutting is essentially a modification of the GTA welding process, but more electrical power and higher gas flows must be maintained to assure cutting rather than welding. The high energy and velocity of plasma makes it an excellent cutting medium. The chief drawback to plasma cutting underwater is electrical shock. Successful cutting requires currents of approximately 600 amperes at above 80 volts, where as simple arc cutting requires 300 amperes at 6 volts. Researchers at Battelle Memorial Institute feel that these disadvantages can be overcome. If so, plasma arc welding and cutting will be one of the underwater tools of the near future.

Power Tools for Divers

Several power sources, basically the same as those used on the surface, are available for underwater use in diver tool systems. They include electrical, pneumatic, hydraulic and high velocity impact types. Power sources for diver tools have been a continuing problem over the years, and it has only been recently that these problems seem to be approaching a solution.

Electrical Tools

Electrically powered hand tools for divers have been under development and testing for many years. So far few, if any, successful electrically operated diver tools have been marketed. First, insulation problems and the relative power advantage obtainable with electric tools have slowed their acceptance. Generally, electric power tools must be operated from high voltage power sources, and the possibility of electrocution of the diver has been a primary consideration. Inadequate knowledge of high voltage electric field effects upon the human body in a salt water environment, integrity requirements within the electrical system, and the competitive pressure of both pneumatic and hydraulic tool systems preclude the possibility of electrical tools assuming great importance in underwater work.

Pneumatic Systems

The earliest successful underwater tools employed pneumatic power operation. Pneumatic tools have been commercially available for surface use for many years. Since diving operations normally require compressed air, pneumatic tools would seem to be good choices for diver support. In most cases high quality pneumatic units are usable underwater with little or no modification. As diving work moves deeper, however, long runs of hose for tool air supply become a problem. The hose diameter must be increased to deliver the proper air flow at the tool power unit, but as the diameter increases the hose becomes more unmanageable.

In the corrosive underwater environment, pneumatic tools require a comprehensive program of maintenance. Without it, tool life is extremely short and excessive costs result. For diving in water shallower than 100 feet, the pneumatic power unit is perhaps the most economical choice. Beyond that depth, particularly beyond 150 feet, it begins to lose its advantage.

Hydraulic Systems

In the last several years commercial diving companies have been leaning heavily towards hydraulically powered units in order to systemize their tools for all diving jobs. Because of the versatility of hydraulic systems, this tendency will probably continue. Except for specialized functions, the power tool kit of all diving companies will eventually be based on hydraulics.

At the Naval Civil Engineering Laboratory, Port Hueneme, California, a continuing program of commercial tool evaluation and special hydraulic tool development has been underway for over a year. At the 1970 Symposium, Equipment for the Working Diver, Stanley Black and John T. Quirk presented a paper describing the work at NCEL which, although concerned with their specific projects, is indicative of the state-of-the-art in underwater hydraulic tools.

According to Black and Quirk, two basic problem areas are involved with the development of tool power systems for diver use. The first is supplying hydraulic power to the unit in the diver's hand, and the second is providing the diver with the tool and associated hardware that is safe and easy to operate. Although these may seem to be rather obvious parameters, few tool systems have managed to perform the two functions and simultaneously provide the versatile service necessary for underwater work.

There are two methods of getting hydraulic power to the diver's tools: surface-supported systems that require an umbilical from the surface to the work site, and bottom-supported systems where the total system is submerged.

The hydraulic umbilical system is most adaptable to work accomplished in depths less than 150 feet. As with the pneumatic system, hydraulic lines become larger in diameter and create both topside and underwater handling problems as working depth increases. In addition, the input power required at the surface pump is a function of line loss in the hoses and therefore increases with hose length. To provide hydraulic power at deeper depths, the most effective system uses a hydraulic power unit located at the diver's depth with electrical power provided to the hydraulic pump from the surface.

The majority of hydraulic tools available today have been designed and developed for surface use. These designs give little consideration to the human factors and ergometric capability of divers under near-zero gravity. Standard tools are often difficult to use underwater because of visibility, cold, diver buoyancy, and the lack of adequate footing and hand holds. Divers' muscles have an increased tendency to become cramped and painful in periods of thermal stress if the triggers and handles are not properly designed, or if the tools produce too much torque or vibration. Excessively stiff hoses or cables can become practically unmanageable in normal underwater currents.

The basic hydraulic system consists of a drive unit, hydraulic pump and reservoir, and a dual hydraulic hose. Black and Quirk categorize all hydraulic systems under two basic types: the open centered hydraulic system and the closed center hydraulic system.

In the open center system, hydraulic oil is continuously cycled from the hydraulic pump through the high pressure hose to the tool. When the tool is not in operation, the oil flow by-passes the motor sections and returns to the low pressure reservoir. When the tool is activated, the oil is directed through the motor section of the tool. As high loads are applied with the tool, hydraulic pressure increases in the high pressure line.

In the closed center system, flow in the hydraulic hoses occurs only when the tool is in operation. Closed center hydraulic systems are normally used in applications where energy conservation is necessary, such as in a battery powered tool system. When the tool is not in use, the oil flow by-passes hydraulic lines and goes directly from the pump to the low pressure reservoir. This eliminates power losses caused by flow resistance in the hydraulic lines. Pressure losses in hydraulic lines are a function of the flow rate and hydraulic circuit and may vary from 10 to 200 psi. Translated into available power, this may be as high as 2 hp.

Each system has definite advantages and disadvantages. When operating in low temperature water (40° F or below), the open center system is preferable. The heat generated by the frictional loss of the hose and fittings tends to reduce the oil viscosity and therefore prevents sluggish action or stalling of hydraulic tools. With the closed center system, the viscosity of the hydraulic oil can increase to a point where tool operation is extremely sluggish or prevented altogether.

One of the main advantages of a closed center system is that energy storing accumulators can be used in conjunction with a pump that delivers less power than the tool requires. This system is useful where a tool operates on an intermittent basis, such as in drilling and nut running operations. In tests performed at NCEL, the average tool run time for drilling and tapping operations was less than 50% of diver work time. In the closed center system, hydraulic oil is stored in accumulators maintained at line pressure while a tool is not in use. When a tool is operating, the flow required beyond pump capacity is drawn from the line accumulators.

Hydraulic Power Drive Units

Several basic power drive units are available for the hydraulic system; diesel hydraulic power units, electro-hydraulic power module, and a completely self-contained drive unit.

Diesel Hydraulic Power Units. The diesel hydraulic power unit is normally an air cooled diesel engine coupled directly to a hydraulic pump. The high pressure side of the pump is coupled to an unloading valve set to relieve system pressures exceeding pump output characteristics.

Electro-Hydraulic Power Modules. The electro-hydraulic power module differs from the diesel hydraulic unit in that the drive unit is an electrical motor. The units may be either shipboard mounted or completely submersible. Where the unit is submersible, the electric motor must be of the submersible type or housed in a protective container, and either pressure compensated or designed to withstand the hydraulic pressures. Normally, the hydraulic pump is enclosed in a pressure compensated container filled with hydraulic oil; the container also serves as the reservoir for the hydraulic system.

Self-Contained Units. The Naval Civil Engineering Laboratory recently designed, fabricated and evaluated the battery powered closed center electrical electro-hydraulic power module. The unit consists of battery package, switching circuit and power unit. The unit is completely submersible and can be used directly at the work site.

Hydraulic Tools

A wide variety of standard hydraulic tools adaptable for underwater use is available. In most cases, the basic operating mechanism is directly adaptable, but requires additional sealing for continuous operation in the corrosive underwater environment.

The most complete line of hydraulic tools, sealed for underwater use, is produced by Ackley Manufacturing Company of Clackamas, Oregon. Typical sealing used on these systems is seen in their Underwater Hydraulic Impact Wrench. Three O-ring seals are used at the anvil or socket end, and all other interfaces are O-ring sealed, with at least one capable of preventing seepage of fine "sugar sand" into the moving parts. Ackley was one of the first major tool manufacturers to realize the potential for good working tools in the diving industry and has moved into a commanding proprietary position in this expanding market.

Ackley and other manufacturers produce impact wrenches, chain saws, rams for pushing, pulling or jacking, pipe and cable cutters and abrasive cutters. Most of these tools are available for use with either open or closed center hydraulic systems. Typical of these tools is the Ackley Lightweight Impact Wrench. Operating at a flow rate of 6 gallons per minute, the tool develops 1,000 rpm. A ball and cam impact mechanism gives two impacts per chuck revolution. Weighing under 6 pounds in air, the tool is well balanced and causes a minimum of fatigue in long term operations.

Velocity Power Tools

As on land, velocity impact tools have limited underwater application, but properly used they are extremely effective and productive. These tools are primarily used to fire special purpose projectiles into steel plate, concrete and wood. The projectiles are most often fasteners that leave an exposed threaded portion, or stud, when impacted into the work. As underwater construction becomes more economically important, more tools of this nature will probably be marketed. But to date, most development effort has been expended under the auspices of the U.S. Navy Supervisor of Salvage.

In 1969 two velocity impact tools developed by the Naval Ordinance Laboratory at White Oak, Maryland, in conjunction with the Mine Safety

Appliance Company of Pittsburg, Pennsylvania were put into use. Functionally the tools are identical, differing only in projectile penetration capability. They operate in a manner similar to the normal "stud gun." The barrel breaks open to receive a projectile, pre-loaded with a firing cartridge. To imbed the projectile into the work, the muzzle of the tool is pressed against the work to release the first safety mechanism. With the muzzle pressed against the work, the barrel of the tool is then rotated to release the second safety. With both safety mechanisms released, the trigger can then be pulled, firing the projectile into the work. The double safety arrangement undoubtably makes the tool a minimum risk device, but concomitantly makes it a minimum use tool. Few work situations exist where a diver can use both hands to operate a tool and still find a point of leverage to exert the force necessary for operation.

Getting a Foothold

If it were not for the viscosity of water, the problems experienced by divers attempting to perform underwater work would be identical to those of our astronauts in space—there is simply no place to dig the heels in. NASA spent millions of dollars designing, implementing and conducting experiments concerned with the ergometric capability of men in a zero gravity environment. It was not until the first "walk" in space that the full implications of not being able to get a foothold became evident. This is a daily problem for the working diver. Even if tools are properly designed to transmit minimum torque, even if power is transmitted to the tool by intermittent impacts, how can sufficient power be transmitted to the cutting or ripping edge of a tool? This is a "better mousetrap" problem that to date has not solved.

The problem is not merely a mechanical one. Divers generally fit the full range of anthropometric types, from those who must clench their fist to avoid stepping on their fingers to those capable of exerting a reasonable amount of physical force when required. Even under ideal test conditions with physically toned divers, the reasonable exertable flow is low. In recent tests conducted at the Naval Civil Engineering Laboratory at Port Hueneme, results showed that the force

exerted by a diver could be expected to vary from 30 pounds to over 100 pounds for a 1-minute, two-arm operation, and from 9 to 33 pounds for a 6-minute, one-arm operation. These tests were made under conditions where the diver could get a foothold or a bight to exert the force. On most work operations there is no way to do this without mechanical assistance.

The obvious solution to the foothold problem is to provide the diver with a harness that will attach him directly to the work. This gives a restraining force that allows him to transmit the necessary force to the tool edge. To provide the real and pyschological margin of safety, the harness must be equipped with quick-release mechanisms. This too requires only relatively simple solution. The core of the problem is *how does the diver attach this harness?* In most cases, the single diver cannot put it around the leg of a drilling rig or find any place to secure it on a salvage vessel. Both magnets and suction cups have been tried. These devices require a relatively flat surface, free of any marine growth, to be effective. Even if these ideal conditions are found or made by scraping the attachment surface, the ratio of the magnet weight to holding power and the attitude sensitivity of suction cups, coupled with the difficulty of disengagement, have proven to be major deterrents.

Visual Recording

One of the diver's most important functions is that of an inspector. In this capacity, his job is to relay information to the surface about underwater situations that can be the basis for action decisions. Often these decisions require the attention of a specialist, or they may be directly involved with regulatory operations and legal decisions. Where the diver is not capable or legally qualified to make an *in situ* judgment, he must transmit visual information to the specialist. There are several methods available to accomplish this function.

Underwater Photography

The most obvious visual recording method is the still photographic cámera. In the middle '60s

Figure 6-4. The Nikonos underwater camera has made underwater photography an economic possibility for all divers—sport, scientific or commercial. (Photo: Dave Woodward.)

underwater photography for industrial purposes was a highly specialized art. Today every commercial diving company has in-house photographic capability, which may range from a $195 Nikonos camera to the PC775 Hydro Products camera system. Tool selection depends on the desired end object. These are the only two cameras now commercially available that are specifically designed for underwater use. They are separated by considerable capacity and price.

Underwater photography is not restricted to cameras designed for underwater use. Between the Nikonos and the Hydro Products PC775 there is a wide range of cameras originally designed for land use but which can be enclosed in commercially available pressure proof housings. Regardless of how well these surface cameras are adapted to the housings, their construction is still a tradeoff.

This book contains samples of photographs taken with both cameras (and many others).

However, selection of a camera for underwater work is a controversial and personal matter and it is not within the scope of this book to evaluate camera systems. The minimum standard is definitely a Nikonos, the maximum, the Hydro Products PC775. Below minimum standards fall such equipment as snapshot cameras in plexiglass housing. It is an operation in maximum futility for a diver to go the trouble of suiting up, diving to the work site, and then using a camera capable of taking only medium quality snapshot photos.

The Nikonos underwater camera (Figure 6-4) was one of the first cameras specially designed by divers. The camera body is anodized aluminum and completely waterproof. O-ring sealing is used throughout, and the camera has proven to be a versatile underwater tool capable of giving excellent photographic results in the proper hands.

Two lenses are available for the unit—a general purpose F 2.5, 35mm lens and an F 3.5, 28mm

Figure 6-5. Dr. Andrew Rechnitzer uses a Hydro Products camera system to photograph reef formations during an underwater study. The system can take up to 400 frames without reloading. (Photo: Chuck Nicklin.)

underwater lens. The general purpose lens is usable both above and in the water, although the field of view is restricted underwater. The 28mm underwater lens compensates for underwater distortion and provides a considerably wider field of view. The lenses are interchangeable (change must be performed at the surface) and use O-ring sealed, bayonet-type mountings. On both lenses, an optically flat window forms the lens/water interface. Focus and aperture are controlled by knobs on the side of the lens. These settings, plus depth of field, can be read through the front lens window.

The waterproof camera body houses a focal plane shutter system with speeds ranging from 1/30 second to 1/500 second. Synchronized flash is provided at a connector on the bottom of the camera body. The camera uses a single lever system for film advance, shutter cocking and shutter release. A thumb-operated safety is provided to avoid accidental tripping of the shutter. Frame number is read on an automatically resetting indicator on the bottom of the camera.

As with all 35mm camera systems, the small negative format, 35 x 24mm, restricts the usefulness for high resolution photography. Standard film cassettes of 36 exposures can also be a problem on underwater operations since the diver must surface to reload the camera. This can be inconvenient and, in the case of commercial operations, expensive.

The Hydro Products PC775 diver-held camera system (Figure 6-5) is one of the most modern and efficient underwater cameras available today. It is based on a design by the late Carl Shipek, one of the pioneers of modern deep sea photography. Although the system appears rather cumbersome compared to surface cameras, it is easy to manipulate underwater and produces photographs of the highest quality. The negative format, 57 x 57mm, is nearly four times larger than the 35mm camera system and high resolution photographs can be obtained.

The camera uses a 43.7mm, F 2.8 water contact lens system, fully corrected for air/glass interface, with a 60° field angle in water. Focus range is 0.5 meters to infinity and is adjustable by a control on the end of the housing. The focus settings of 0.5, 0.6, 0.9, 1.2, 1.8, 3.1m and infinity have detent stops that can be felt when using the camera for underwater night photography. 70mm film is cartridge-loaded and the camera has a capability of up to 30.5m (100 feet), or 400 exposures. Film may be loaded in daylight or in the darkroom.

The shutter system used in the camera is a Prontor with full "x" synchronization. Speed range is 1 second to 1/125 second, with the shutter motor controlled and operated when the trigger is pulled. Recocking and film advance is automatic. A frame counter mounted on the end of the camera permits easy viewing of the frame number being shot. Recycle time between operations is 5 seconds.

The camera housing is aluminum with a hard anodized finish. It weighs 11.3 killograms (25 pounds) in air, but is neutrally buoyant underwater. Housing and controls are designed for operation to 305m (1,000 feet). The system is powered by an internal 28-volt rechargeable cadmium battery pack that will operate the camera through three full film loads between charges.

The light source for the Hydro Products diver-held camera is the PF710 underwater strobe light which has construction similar to that of the camera. The strobe may be operated from the camera's internal 28-volt power pack or from a separate power source. Output light intensity is continuously variable over the range of 50 to 200 watt seconds by means of a rear mounted control. Recycle time of the unit is 3 seconds. The strobe may be operated in two modes—"camera" or "slave." In the camera mode, the synch output from the camera triggers the unit. As a slave, the unit can be used as an auxiliary light source in conjunction with another strobe being operated in the camera mode. When the camera mode strobe flashes, it photoelectrically triggers the slave unit. On complicated underwater lighting problems, multiple flash units can produce dramatic results.

Although underwater photographic recording has been used for many years, real time visual recording is gaining importance in commercial diving. The perfection of underwater television and videotape recording has had a great effect on underwater technology. Typical of new applications for real time visual observation and recording is the method of certification and cleaning of drilling rigs instituted by the American Bureau of Shipping.

Periodic inspections to renew loadline certificates normally require that a drilling rig be removed from service and towed into relatively shallow water. Here, ballast is adjusted to raise one side at a time clear of the water for visual inspection. With the new method, a team of divers cleans the area to be inspected with rotary underwater cleaning tools at the work site. Divers using underwater television cameras at closeup range then transmit TV pictures topside. At the surface, American Bureau of Shipping inspectors can see details of the structure's condition and the inspection is permanently recorded on video tape.

Underwater television inspection systems, like camera systems, may be either specially designed, integrated systems for underwater use, or they may be adapted from standard closed circuit surface television systems. Characteristically, closed circuit TV systems for surface use, when packaged in underwater housings, require considerably more space. In addition, control functions are arranged so that interconnecting cables for adapted systems are cumbersome and expensive. Underwater systems generally use compact housings and have control functions built into the underwater camera to minimize conductors in the underwater cable.

One of the most widely used diver inspection underwater television systems is the Hydro Products model 303. It consists of an underwater cam-

Figure 6-6. Thallium iodide light penetrates murky water as a diver makes an underwater TV image to the surface where it is monitored and recorded. (Photo: Hydro Products.)

era with a 250-watt thallium iodide light, interconnecting cable, surface monitor/control unit and video tape recorder.

The underwater TV camera is 3 inches in diameter and weighs approximately 3½ pounds in water. The thallium iodide light (Figure 6-6) is fixed to the camera housing to permit easy underwater handling. Energy output from the light falls in the part of the color spectrum that has maximum transmission capability in the water. The vidicon tube, the image sensing portion of the camera, has its maximum sensitivity in the same portion of the spectrum. This combination of sensing tube and light output gives the TV camera 25% more vision sensitivity than the diver's eyes. Control of the camera focus, which has a range of 3 inches to infinity, is from the surface monitor. Aperture is automatically controlled at the camera. Recording of the diver's voice and that of the surface observer is included on the video tape. The voice record, combined with the high resolution image recording, gives a permanent record that could be obtained in no other way.

Appendix 2

Equipment Operation

The Scuba Regulator

Most regulators manufactured in the United States use two stage mechanisms. In recent years, preference has been for two stage, single hose units, but two hose regulators still enjoy wide usage and will continue to for some time. Each configuration has specific applications where it can do the better job.

Two Hose, Two Stage Regulators (Fig. AII-1)

Two hose, two stage regulators are manufactured by several equipment companies. The following is an operational description of the U.S. Divers Company standard two hose unit:

High pressure air from the storage cylinder enters the first stage and is reduced to an intermediate pressure of approximately 100 psi. The first stage employs an "upstream valve" arrangement which uses the high pressure cylinder air as the closing force to seat the valve. Counteract-ing the closing force of the cylinder air is a large spring linked to the valve through a small diaphragm and stem, which forms part of the input valve. This heavy spring is preadjusted to hold the input valve open against the pressure of a fully charged cylinder. When the intermediate chamber reaches design pressure, the small diaphragm moves, compressing the large spring and allowing the input valve to be closed by the force of the cylinder pressure.

Spring tension on the first stage diaphragm is designed to give the proper intermediate pressure when the high pressure air cylinder is fully charged. When the pressure within the cylinder drops as air is used, the force available to seat the input valve also drops. This allows the valve to remain open longer, giving more air flow into the intermediate stage and a rise in intermediate air pressure. The rise in intermediate pressure is a design function and is directly controlled by the size of the orfice between the first and intermediate stages. The variation is proportional to the size of

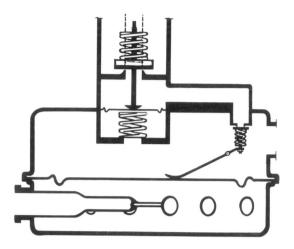

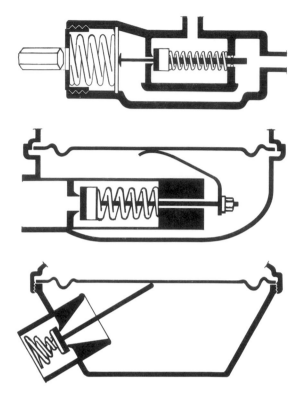

Figure AII-1 (above). Schematic diagram, two hose regulator. Figure AII-2 (right). Schematic diagram single hose regulator components: Top—balanced first stage; Middle—downstream second stage; Bottom—tilt valve second stage.

the orfice, that is, the larger the orifice the larger the variation. In the two stage, two hose regulator, a wide range of intermediate pressure can be tolerated without an increase in breathing resistance because of the large second stage diaphragm.

The first stage is depth compensated because air pressure in the second stage housing is always equal to the ambient water pressure. As depth increases, differential pressure is added to the spring tension on the small diaphragm. Intermediate pressure is always equal to the preset intermediate pressure of approximately 100 psi plus the ambient water pressure.

The second stage valve is also designed on the downstream principle. Increasing intermediate pressure caused by dropping cylinder pressure has no detrimental effect, but rather allows the valve to open more easily with concommittant lower breathing resistance.

Single Hose Regulators (Fig. AII-2)

Single hose regulators are made by all SCUBA manufacturers and are the most common choice of equipment in sport, scientific and commercial diving. As in the two hose configuration, there are many variations in design that can provide satisfactory performance. This description deals with the U.S. Divers Calypso, a single hose, two stage regulator that features a balanced first stage and a downstream second stage valve.

In the balanced first stage, operation is independent of cylinder air pressure. Pressure in the input chamber will be equal on all parts of the spring-loaded valve and no differential pressure can be established. As in the two hose system, input valve operation is controlled by the spring-loaded diaphragm which is connected to the input valve through a mechanical linkage. Spring tension can be set to give the exact intermediate pressure desired. This pressure will be constant as cylinder pressure drops. By using a large orifice balanced by an equal size penetrating valve stem at the opposite end of the input chamber, a large volume of air can be passed with no increase in breathing resistance. The first stage is depth compensated so that absolute pressure within the intermediate stage is always the design plus ambient pressure.

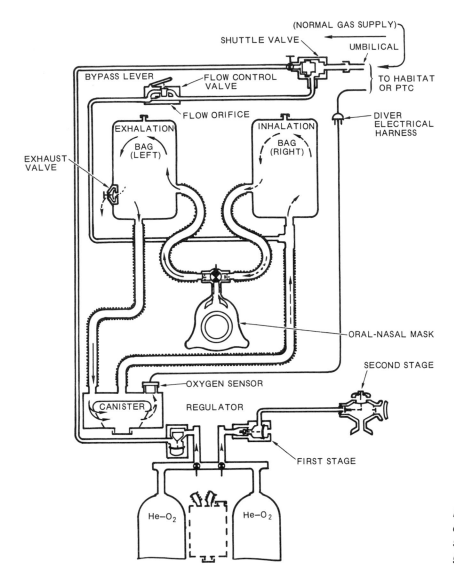

Figure AII-3. Schematic diagram, U.S. Navy Mk 8, semi-closed circuit, mixed gas breathing apparatus.

The Calypso-type second stage opens down-stream with the flow of air. An over pressure in the hose tends to make the valve open more easily. In the event of a first stage leak, the air is harm-lessly vented, whereas if the second stage valve was an upstream device such as a tilt valve, the system would require a safety pop-off valve to prevent hose rupture. In a straight opening valve such as used on the Calypso regulator, the seat must move a distance equal to only ¼ the diameter of the orifice to obtain full flow.

Mixed Gas Apparatus

Semi-Closed Circuit Mixed Gas Apparatus (Fig. AII-3)

Semi-closed circuit mixed gas equipment is available commercially from several manufacturers and will probably be the choice for relatively deep diving operations in the near future. Each unit available from commercial companies incorporates proprietary design information that cannot be disseminated. Although the U.S. Navy MK 8 SCUBA equipment does not represent the optimum in design characteristics, it incorporates design concepts similar to commercially available equipment. The MK 8 SCUBA is well documented and is considered in detail here to demonstrate general operational principles.

U.S. Navy MK 8 Mixed Gas SCUBA

The MK 8 SCUBA provides life support during dive missions, primary and emergency modes of operation through a semi-closed circuit system and an additional open circuit emergency system.

The primary helium/oxygen gas supply is provided to the semi-closed system at the dive pressure through an umbilical cord from the PTC. The umbilical is connected to the diver's breathing apparatus through a quick-disconnect at the shuttle valve located on the right-hand side of the vest. The ball valve allows the diver to select gas from the primary umbilical supply or from the emergency backpack cylinder supply.

From the shuttle valve, the gas goes to the flow control valve where a constant gas flow is maintained by a sonic orifice. A lever actuated by-pass provides a high gas flow for initial pressurization and any additional gas required by the diver. This gas is injected into the semi-closed breathing circuit at the inhalation bag canister hose fitting. Gas is inhaled by the diver from the bag through the inhalation hose and check valve. Exhaled gas passes through the tee assembly valve, exhalation check valve and hose to the exhalation bag.

During exhalation, the exhalation bag is inflated and pressurized. About 20% of the exhaled gas is vented to the sea through an exhaust valve mounted on the top of the exhalation bag. This valve is adjustable to relieve pressure from ¼ to 1 psig as desired by the diver.

The pressure differential build-up between the exhalation and inhalation bags causes gas to flow through the canister where baralyme (barium hydroxide) absorbs the carbon dioxide. The canister outlet contains an oxygen partial pressure sensor which transmits an indication of oxygen level to an indicator at the diving monitor station.

The emergency gas supply is provided by a cylinder in the backpack charged to 2,250 psig with the helium/oxygen mixture. The cylinder is connected to an outlet manifold containing two shutoff valves. The outlet farthest from the regulator is delivered through a gas supply hose to the shuttle valve. A ball valve (normally closed) is provided to prevent loss of emergency supply gas. This valve is opened by the diver to supply emergency gas to the semi-closed breathing circuit upon loss of the primary umbilical supply.

The open circuit system is also known as the demand regulator assembly or buddy system. The manifold outlet closest to the cylinder supplies gas to a first-stage regulator. The first stage regulator delivers gas at 155 psig nominal pressure through a hose and quick-connect coupling to a combination second stage demand regulator, mouthpiece and purge valve. The purge valve is depressed to supply extra gas to clear water from the mouthpiece. The demand regulator assembly is normally retained by a strap on the right-hand side of the vest at the waist. With the Kirby Morgan helmet or face mask, gas is supplied to a second stage demand regulator located in the helmet or mask

Backpack Unit. The backpack unit consists of a molded fiberglass support board with cover and attaching hardware, making a package approximately 26 inches high by 15 inches wide and 7 inches thick (front to back). A gas cylinder and manifold assembly are held in place by two stainless steel cylinder straps. The canister unit is mounted alongside the cylinder and held in place by two elastic straps which engage slots molded into the case. Handling and access holes are cut into the unit at upper and lower ends.

The backpack unit is attached to the diver's vest by a rod which engages loops in the vest and support board at the shoulder. In addition, two straps on each side of the support board engage buckles on the front of the vest.

Manifold and Cylinder Unit. The manifold and cylinder unit supplies mixed breathing gas to the regulator and demand regulator assembly. The galvanized steel 6.4 liter cylinder supplies gas at an initial pressure of approximately 2,250 psig. The cylinder is coupled by a manifold assembly with a common connecting line and two outlets controlled by manual shutoff valves. Each outlet may be operated independently of the other and does not interfere with gas flow from the cylinder.

The manifold consists of two sections connected by hex nuts and O-ring seals. The input section provides and O-ring seal with the cylinder. The output section contains the outlet ports and shutoff valves.

The input section includes a safety plug and blowout disc, rated for burst pressure of 3,400 psig. The end sections are provided with a vent hole to warn against removing the manifold while under pressure. Sufficient thread engagement remains to hold the manifold in the cylinder when venting occurs.

Gas enters the input section of the manifold through a filter hole in the dip tube and is routed to the shutoff valve housings. Gas passes through a circular groove around the outside of the housing into a diagonal hole and up to the seat orifice. With the valve in the open position gas passes around the valve plug into a port to the outlet located in the valve head. From the outlet, gas is discharged into the breathing apparatus pressure regulators. The demand regulator assembly in attached to the outlet closest to the cylinder only during an actual dive excursion. The outlet is also used for charging the cylinder by first removing the demand regulator assembly.

Canister Unit (Fig. AII-4). The canister unit is a container for the carbon dioxide absorbent chemical, baralyme, and is mounted in the backpack unit along side the mixed gas cylinder. The canister is a hexagonal stainless steel box divided into three interior chambers by two perforated stainless steel screens and a solid end plate. Four threaded ports are provided: baralyme filler, gas inlet, gas outlet and oxygen sensor fitting. Two low pressure hoses connect the gas outlet and inlet ports to the inhalation and exhalation breathing bags, respectively. The canister has a water jacket so that warm water may be passed over the surface of the canister to extend the service life of the baralyme.

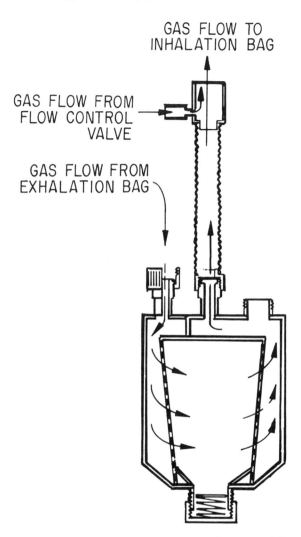

GAS FLOW TO
INHALATION BAG

GAS FLOW FROM
FLOW CONTROL
VALVE

GAS FLOW FROM
EXHALATION BAG

Figure AII-4. Schematic diagram, canister unit, Mk 8 SCUBA.

Approximately 4 quarts (8 pounds) of baralyme are packed in the interior chamber and held under compression by the filler end plate and spring. Since baralyme breaks into smaller particles and powders under handling, a vacuum is pulled on the gas hoses after the fill operation. This removes small particles and dust, precluding their later entry into the breathing system. During filling, the canister is gently pounded on the sides to assure complete filling. The filler port is then closed hand-tight with the cap, end plate and spring.

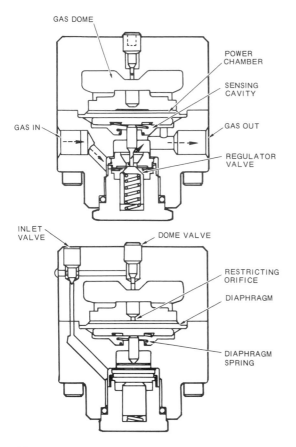

GAS DOME

POWER CHAMBER

SENSING CAVITY

GAS IN

GAS OUT

REGULATOR VALVE

INLET VALVE

DOME VALVE

RESTRICTING ORIFICE

DIAPHRAGM

DIAPHRAGM SPRING

Figure AII-5. Cross section, regulator unit, Mk 8 SCUBA.

The canister hose farthest from the oxygen sensor fitting is attached to the outlet of the vest exhalation bag. The canister hose closest to the oxygen sensor fitting is attached to the inlet port of the vest inhalation bag. Exhaled gas containing carbon dioxide, under ¼ to 1 psig pressure, depending on the exhaust valve setting, is forced through the screen and across the baralyme. Carbon dioxide is absorbed in the interior chamber and the remaining gas passes into the outlet chamber. A sensor samples the oxygen content at this point. The circulating gas is mixed with fresh gas from the flow control valve within the outlet hose and passes into the inhalation bag.

The baralyme is adequate for several hours depending on diver activity levels during diving operations. When the carbon dioxide saturates the baralyme, the color of the granules changes form

pink to blue. The oxygen sensor provides electrical signals to a remote monitor, and the diver may renew the gas mix in the breathing bag by using the flow control valve by-pass lever.

Regulator Unit (Fig. AII-5). The regulator unit, located on the manifold outlet farthest from the cylinder, reduces the initial 2,250 psig cylinder gas pressure to a predetermihned pressure for the dive. The gas then flows through the shuttle and flow control valves to the inhalation bag. The regulator uses an internally loaded dome chamber to balance inlet and outlet pressures. The design allows outlet pressure adjustment to 1,500 psig with an inlet pressure of 3,500 psig. Outlet pressure varies about 1% with a 5°F change in temperature and increases about 2.6 psi per 100 psi drop in inlet pressure due to mechanical rebalance of spring forces.

There are five cavities in the regulator unit which control its operation: The inlet, outlet, and sensing cavities, the power chamber, and the gas dome.

The inlet cavity contains gas at cylinder pressure. Gas is metered through the single-seated regulating valve into the outlet cavity and out to the shuttle valve unit at reduced pressure.

The sense port opens from the outlet cavity into the sensing cavity under the diaphragm. Thus, the sensing cavity contains outlet pressure forcing the diaphragm upwards. The diaphragm movement controls the regulating valve. The action is controlled by the preset pressure on the upper side of the diaphragm in the power chamber. This chamber is sealed from the inlet line and opens into the main gas dome through a restricting orifice. The dome pressure is preset to balance the outlet pressure, including the small resistive valve spring forces and the inlet pressure acting on the valve.

Initially, the diaphragm forces the inlet valve open, allowing limited gas flow until equilibrium is reached and the line flow and pressures are stabilized. The power chamber and dome pressures are equal at approximately outlet pressure. In the event of sudden change in outlet pressure, the restricting orifice and the small-volume power chamber act as dampers to prevent valve chatter.

Dome and inlet needle valves are used during regulator adjustment.

Shuttle Valve Unit. The shuttle valve unit provides manual selection of gas from either of two sources during semi-closed breathing modes of operation. In the normal mode, the diver uses gas supplied through the umbilical from the PTC. In the emergency mode, gas from the backpack cylinder supply is used.

In the normal mode the ball valve is closed, reducing pressure from the cylinder side of the shuttle. In this state, umbilical gas pressure forces the shuttle and its O-ring against the cylinder gas inlet allowing umbilical gas to flow through the shuttle ports into the common outlet. Pressure differential of less then 1 psi will actuate the shuttle. Dust or dirt is removed from the gas by a 10-micron, in-line filter.

In emergency mode caused by disconnection or pressure loss in the umbilical, the diver manually opens the ball valve, permitting cylinder gas to flow through shuttle ports into the common outlet. The manual ball valve serves the safety function of preventing inadvertent depletion of cylinder gas, which could occur without the diver's knowledge should umbilical pressure fall below the cylinder gas regulated pressure.

Flow Control Valve Unit. The flow control valve unit provides a constant mass flow of gas from the pressure source as determined by the shuttle valve. The gas from the flow control valve is mixed with recirculated gas from the canister which then enters the inhalation bag.

The flow control valve was designed to allow selection of one of three available orifices by a selector which blocked the unused ports. The present concept is to cap the selector port and plug the two unused orifices, and to insert the appropriate orifice for the scheduled dive.

When the by-pass lever is not depressed, gas enters the inlet port, passes into the by-pass chamber, through the filter and through the main duct to the selection chamber. From the selector port, gas passes through the single uncapped orifice and through the outlet port to the gas injection hose.

With the by-pass lever in normal position, the valve stem is seated against the gland nut by-pass spring, and the by-pass chamber is sealed by O-rings. By-pass gas flow is accomplished by depressing the lever, and raising the valve stem. Gas is then directed from the inlet port through the gland nut, around the groove and through the by-pass duct directly to the outlet.

Hoses. Three hoses with quick-disconnect fittings are used in the MK 8 SCUBA. All hoses are made from Synflex hose, 3/16 inches ID x 13/32 inches OD. The hose is multilayered, consisting of a seamless nylon core, double layer braided nylon reinforcement, and a polyurethane cover. Each hose can tolerate a minimum bend radius of 2½ inches without affecting service life. Working pressure is 2,600 psig.

Two hoses are equipped with female Snap-Tite disconnects on both ends and differ only in length. The 41-inch hose connects to the shuttle valve outlet, passes around the neck of the vest, under the breathing bags and connects to the flow control valve inlet. The 24-inch hose connects the regulator outlet ot the shuttle valve inlet. The disconnects are keyed such that the knurled sleeve notch must be rotated into alignment to connect or disconnect. The sleeve notch is rotated into out-of-align when connected to provide a lock.

The gas injection hose is of similar construction, but is equipped with Hansen disconnects. The male end mates to the canister outlet hose, and the female end joins the flow control valve outlet.

Umbilical Cord Set. The umbilical cord set is used for diver life support in the normal mode of the breathing apparatus, supplying gas and electrical connections from the PTC. Both the electrical wire bundle and gas hose are lashed together every 6 feet. Two cord lengths, 200 feet and 600 feet, are provided, terminating at both ends in electrical connectors and gas connectors.

The umbilical electrical cable interconnects the PTC to the diver's earphones, micorphone and the oxygen sensor. The umbilical gas hose carries

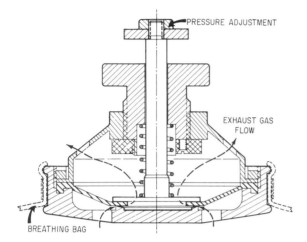

Figure AII-6. Cross section, exhaust valve, Mk 8 SCUBA.

mixed gas from the PTC to the diver. The diver's end is terminated in a quick-disconnect fitting that mates to the shuttle valve mounted on the right-hand side at the diver's waist.

Vest, Breathing Bags and Exhaust Valve. The vest assembly consists of an inhalation bag and an exhalation bag, which are attached to a vest with nut and bolt fasteners. The vest is divided vertically in front midline and is provided with a zipper closure for ease of donning and removal. Each bag extends from behind the neck, over the shoulders and down over the upper portion of the chest. When inflated, the volume of each bag is approximately 4 liters and is fitted with securing bands to hold the bag close to the diver's body. The twist-lock fasteners used for closing the weight pouches secure the shuttle valve unit and flow control valve units to the vest.

The forward set of hose connectors on each bag mates with the mask hoses; the aft set of hose connectors mates with the canister hoses. Located in the bottom of each bag are capped drain fittings. The exhalation bag is provided with a rubber seal opening for exhaust valve installation. A removable band-clamp is installed around the rubber seal to retain and seal the exhaust valve in the bag.

The exhaust valve (Fig. AII-6) is mounted on the exhalation bag. The major components of the valve are the valve body, diaphragm assembly, spring, cover assembly, stem assembly, pull grip, adjusting screw and retaining ring.

In operation, the adjusting screw compresses the valve spring which applies pressure on the valve stem. The valve stem covers and seats the diaphragm. Gas pressure in the system increases as both fresh gas is supplied from the flow control valve and gas is exhaled by the diver. When the system pressure exceeds the spring pressure, the diaphragm unseats and allows gas to escape until spring pressure can reseat the stem and diaphragm. The exhaust valve may be adjusted to unseat over the range of ¼ to 1 psig by adjusting the screw.

Demand Regulator Assembly. The demand regulator assembly provides an emergency open circuit breathing mode for the diver or for a buddy. A first stage regulator connected by a hose to a second stage regulator-mouthpiece constitutes the demand regulator assembly. The first stage regulator is coupled by means of a yoke to the manifold outlet closest to the cylinder to permit flow of cylinder gas to the demand regulator assembly. The semi-closed breathing mode regulator unit connects to the manifold outlet farthest from the cylinder.

The first stage regulator provides gas to the second stage regulator mouthpiece and incorporates an adjustment for ease of breathing. The second stage regulator provides demand breathing by means of a diaphragm actuated valve and incorporates a flush purge button. The interconnecting hose is non-kinking and of sufficient length to stow the second stage regulator mouthpiece at waist level when not in use.

The first stage regulator consists of a main body with high and low pressure chambers and a means of regulating and metering the output flow to the interconnecting hose. A screw plug opposite the hose outlet may be replaced by a gage to monitor existing tank pressure. A setscrew holds the manifold-engaging yoke in place at the high pressure chamber inlet port.

High pressure gas from the cylinder manifold flows directly into the high pressure chamber through a sintered metal filter. Retained in the cylindrical chamber by a lock ring, a force-rebalance valve seats over an orifice in the base of the chamber. This valve is raised by a pin protruding through the orifice whenever the support disc is actuated by the diaphragm. The diaphragm seals the low pressure chamber from the environment.

The hose port, the orifice duct and the internal chamber of the diaphragm make up the low pressure chamber. The action of the diaphragm upon the support disc and pin causes the valve in the high pressure chamber to open when pressure in the low pressure chamber falls below preset pressure. This falling may be due to inhalation (via the hose) or due to increasing depth effect on the external side of the diaphragm. When gas flow through the orifice is sufficient to raise the low pressure chamber to nominal, the diaphragm recedes and the metering valve closes. The nominal pressure setting is accomplished by adjusting a spring-loaded pad against the external side of the diaphragm; further adjustments to suit the individual diver also may be made. A 3/8-inch hex hole in the adjusting screw permits access of ambient pressure to the external face of the diaphragm.

The second stage regulator-mouthpiece consists of a two-piece metal body held together by a clamp ring, with the rubber mouthpiece and exhaust tube fastened to the main body. The interconnecting hose couples to the second stage regulator valve in the main body. The cover section contains the flush purge push button, and the diaphragm is sealed between the two metal body sections by the clamp ring.

Inhalation by the diver causes the diaphragm to deflect inward, operating the lever that controls flow through the second stage regulator valve. The shaped lever acts as a cam on the spring-loaded valve stem, withdrawing the seat from the inlet orifice and permitting gas flow into the breathing chamber. Upon exhalation the diaphragm deflects outward, releasing the valve lever and permitting excess pressure to exhaust through a soft, flat rubber disc valve. The external side of the diaphragm is open to the water through ports in the metal cover section, providing depth compensation.

Safety Relief Valve and Quick-Connect Coupling. The safety relief valve and quick-connect coupling are provided in the hose between the first and second stage regulators of the emergency breathing system. When the diver is using a mouthpiece unit, gas is provided through the quick-connect coupling to the second stage regulator in the demand regulator assembly.

A slide valve controls the open or closed condition of the quick-connect coupling. When the slide valve is positioned toward the sleeve, the coupling is open. When the slide valve is positioned away from the sleeve, the coupling is closed. When buddy breathing is required, the coupling is opened at the sleeve and the nipple. This provides quick access to the diver requiring emergency gas.

The safety relief valve is provided to prevent the regulator hoses from bursting if a malfunction occurs in the first stage regulator. The safety relief valve is preset to open at approximately 200 psig over ambient pressure. When the pressure is reduced in the regulator hose below the set pressure, the safety relief valve closes automatically. A manual relief knob is provided to manually operate the relief valve for checking proper operation.

Habitat Equipment

Peripheral equipment used in underwater living experiments, especially those involving extensive physiological testing programs, accounts for a major portion of project expense. The Sealab III instrumentation was the most comprehensive ever systemized for an underwater experiment. Tables AII-1, 2 and 3 list equipment and its functional use in the operation plan, indicating complexity of long-term, deep experiments.

(Tables AII-1, 2, and 3 begin on p. 168.)

Table AII-1
Sealab III Communication Stations

	Command Van	Medical Van	Sea-floor Habitat	Support Ship	Shore Support Ship
Electrowriter	2		2		
Bogen Intercom	2		4	1	
Open Mikes	M		5		M
Divers Intercom	2		2 + 6 for Aquanauts		
Closed Circuit Television	M	M	7 cameras		
Sound-Powered Telephone			3	1	1
UQC/BQC Under-water Telephone			1		
FM Speaker			2		
TV Monitor		1	1		

M = monitor; UQC/BQC receiver in support ship was part of diving system; FM speakers and entertainment TV monitor in sea-floor habitat were switched through the command van.

Table AII-2
Command Van Equipment List

Equipment & Manufacturer	Purpose
2 Tape Recorders (Magnecord)	For audio communication recording for future analysis and records
1 Intercom/Video Control (Nortronics)	To select remote cameras for viewing on the two monitors; also contained a 12-station, Talk-A-Phone intercom unit and a port-and-starboard diving system intercom-access unit
1 Intercom/Video Control (Nortronics)	Same as above except for additional selection of remote cameras for viewing on the medical van monitor
1 Speaker Panel (Nortronics)	Served as audio output for station monitoring, included FM and vidio audio; the short-and-open-alarm circuit of the habitat umbilical cable was also mounted on this panel
1 Speaker Panel (Nortronics)	Served as audio output for station monitoring
2 TV Monitors (Conrac)	Dual monitors for video display of selected remote-camera output to monitor aquanauts in habitat, DDCs and PTCs
1 Speech Unscrambler Control Panel (Nortronics)	Controlled primary station conversation switching of inter-stations for unscrambler input and output, tape recorder audio input, and station monitoring and output connections
1 Speech Unscrambler Control Panel (Nortronics)	Controlled secondary station conversation switching (same as above, except no video recorder audio access)
1 FM Entertainment Panel (Nortronics)	Contained a complete AM/FM receiver with stereo speakers used to monitor either FM output or TV video recorder audio output as selected
1 AM/FM Receiver (Fisher)	Served as part of entertainment systems for AM and FM commercial reception
1 TV Receiver (Conrac)	Served as part of entertainment system for commercial TV broadcast reception
1 TV Rotor Control (Nortronics Cornell-Dubilier)	Used to position TV antenna for directional reception

(Table AII-2 continues on p. 170.)

Table AII-2 *(continued)*

Equipment & Manufacturer	Purpose
1 Speech Unscrambler (Naval Applied Science Laboratory)	Used to unscramble speech affected by characteristics of helium atmosphere
1 Speech Unscrambler (Singer)	Used to unscramble speech affected by characteristics of helium atmosphere
2 Electrowriter Tranceivers (Victor)	Used for facsimile transmission between command van and habitat/PTC/DDC complex
2 Intercoms (Bogen)	Used as direct communication access. Primary communication station to lab; secondary station to lab, observation room, and diving station
4 Audio Amplifiers (Dynaco)	Two each at primary and secondary communication stations
5 Audio Amplifiers (Dynaco)	Used at primary communication station
1 Summation Amplifier (Nortronics)	Summation and isolation buffer for unscrambler control panel output
2 Mixer Amplifiers (Nortronics)	Composite mixing and amplification of multiple station input
2 Intercoms (12-station) (Talk-A-Phone)	Used for direct communication network between 12 stations located at primary and secondary stations, medical van, habitat, DDCs and shipboard stations
Voltage Regulator (Sorensen)	Provided 110v AC regulated line voltage for equipment power at primary and secondary communication stations, instrumentation recording station and medical van instruments
1 Circuit Breaker Panel, Voltage Regulator (Nortronics)	Contained circuit breakers to control 110v AC regulated power output
1 Power Supply, 28v DC (Transpac)	
1 Video Tape Recorder (Ampex)	Provided recording of video images received from the remote cameras for record and future analysis

Table AII-2 *(continued)*

Equipment & Manufacturer	Purpose
3 Multipoint Recorders	Used to provide a permanent record of habitat environmental, oceanographic and engineering evaluation sensor outputs
1 Galvanometer Input Control (Nortronics)	Provided a calibrating function for habitat data signals prior to recording on the oscillograph flow metering and power sensors
1 Oscillograph (Midwestern Instruments)	Provided a method of recording the outputs of the habitat flow-metering systems and power sensors
1 Integrator Calibrate Panel (Nortronics)	Provided calibrating and signal conditioning of habitat data sensor outputs for input to the scanning system
1 Magnetic Tape Recorder (Kennedy)	Provided a means of recording data sensor outputs from the habitat, controlled by the scanning system
1 Digital Voltmeter (Dana)	Provided digital readout of scanned signals from the habitat
1 Coupler and Clock (Ward/Davis)	Provided coupling and clock control for the scanning system reading habitat data signals
1 Scanner Control (Dana)	Controlled output of scanners to read habitat monitoring data on DVM or to record data on magnetic tape recorder
2 Scanners (Dana)	Provided continuous scanning of data sensor outputs from the habitat (used in conjunction with scanner control)
1 Time Totalizing Panel (Nortronics)	Provided time totalizing monitor for key functions in the habitat (refrigerator, freezer, washer-dryer, D/H pump motor and water heater)
1 Gas Pressure Sequencer Panel (Nortronics)	Controlled stepper switch in habitat to monitor gas storage pressures
1 Air Conditioner (Ellis and Watts)	Provided heat, cooling and humidity control of the command van atmosphere

Table AII-3
Medical Van Equipment List

Equipment & Manufacturer	Purpose
1 Gas Chromatograph (Loenco)	Gas analysis
1 Gas Chromatograph (Hewlett-Packard)	Gas analysis
1 Infrared Analyzer (Lira)	Carbon monoxide monitoring
2 Recorders (Texas Instruments)	Gas chromatograph output recording
1 Spirometer (Wedge)	Respiratory measurements
1 Oscilloscope (Tektronic)	Respiratory measurements
1 Camera System (Tektronic)	Respiratory measurement photography
1 EKG (Hewlett-Packard)	Electrocardiographic studies
1 Flame Photometer (Baird Atomic)	Blood, urine, sodium and potassium determination
1 Coenzometer (MacAlaster Scientific)	Enzymology
1 Osmometer (Advanced Instruments)	Osmometry
1 Colorimeter/Spectro-photometer (Bausch & Lomb)	General blood and urine analysis
1 pH/Blood Gas Analyzer (Instrument Labs)	pH and blood gas analysis
1 Air Conditioning Unit (Ellis and Watts)	Medical van cooling and heating
1 Autocytometer (Fisher)	Blood cell count
1 pH Meter (Colemen)	Urine pH measurements
1 Typewriter (Sears Roebuck)	Data logging
1 Barometer (Precision Theometer and Instrument)	Barometric pressure

Table AII-3 *(continued)*

Equipment & Manufacturer	Purpose
2 Microscopes (American Optics)	Microscopic studies
1 Electrophoresis (Hyland-Photovolt)	Isozymes determination
1 Spectrocolorimeter (Beckman)	Blood/urine, calcium and chloride determination
3 Water Baths (Thelco)	Temperature stabilization determination
1 Still (Corning)	Water distillation
1 Fume Hood (Lab Line)	Fume removal
1 Oven (Thelco)	General drying
1 Clinical Centrifuge (International Equipment)	General centrifuging
1 Incubator (Thelco)	Culturing
2 Viscosimeters (Hess)	Blood viscosity
1 Sphygnomanometer (Baumanometer)	Blood pressure
1 Metric Beam Balance (Detecto)	Body weight
1 Spirometer, Dry Gas (Instruments Associated)	Pulmonary studies
2 Refrigerators (Stoneite)	Reagent storage
1 Freezer (Stoneite)	Sample freezing
1 Air Compressor (ITT-Bell Gossett)	Provided compressed air for flame photometer

Table AII-3 *(continued)*

Equipment & Manufacturer	Purpose
1 Water Heater (Sears Roebuck)	General cleaning
1 Television Receiver (Conrac)	
1 Calculator (Wang)	Physiological calculations

Bibliography to Part II

Beckman, E.L. 1966. *Thermal protective suits for underwater swimmers.* U.S. Naval Medical Research Institute Research Report.

Cayford, J.E. 1959. Underwater work. In *A manual of scuba, commercial, salvage and construction operations.* Cambridge, Md.: Cornell Maritime Press.

Conference for National Cooperation in Aquatics. 1962. *New science of skin and scuba diving.* New York: C.N.C.A.

Coyle, A. J. and Eibling, J.A. 1969. The diver, the key to ocean today. *Battelle Research Outlook* vol. 1, no. 1, pp. 8-12.

Davis, R.H. 1962. *Deep diving and submarine operations,* 7th ed. London: Saint Catherine Press Ltd.

Dragerwerk Lubeck. 1969. *Diving techniques information,* vol. 1 and 2. Lubeck, Germany: Dragerwerk Lubeck.

―――. 1969b *Diving technics.* Prospectus 200 e Lubeck, Germany: Dragerwerk Lubeck.

Goldman, R.F., Breckenridge, J. R., Reeves, E. and Beckman, E.L. 1966. "Wet" versus "dry" suit approach to water immersion protection clothing. *Aerospace Medicine* vol. 37, no. 5.

Hackman, D. J. 1969. Power tools for divers. *Battelle Research Outlook* vol. 1, no. 1, pp. 13-16.

Hahn, W.A. and Lambertsen, C.J. 1953. *On using self contained underwater breathing apparatus.* N.R.C. #274 (taken from *Men Under Water*). Washington: U.S. Government Printing Office.

Hallanger, L.H. 1968. *Diver Con 1: a diver construction experiment, development problems and solutions.* Undersea Technology Conference. ASME pub. 69-UNT-10.

Hamilton. F.T., Bacon, R. H., Haldane, J.S. and Lees, E. 1907. *Report to the admiralty of the deep-water diving committee.* London: Stationery Office.

Harter, J.V. 1966. Fire at high pressure. *Proceedings, third symposium on underwater physiology.* Lambertsen, C. J., ed. Baltimore: Williams & Wilkins.

Haux, G. 1968. *Diving technics.* Lubeck, Germany: Dragerwerk Lubeck.

Krasberg, A.R. 1966. Saturation diving techniques. *4th International Congress of Biometerology.* New Brunswick, N.J.: Rutgers University.

Los Angeles County. 1964. *Underwater Recreation.* Los Angeles: Dept. of Parks and Recreation.

Mishler, H.W. and Randall, M.D. 1969. Underwater joining and cutting. *Battelle Research Outlook.* vol. 1, no. 1, pp. 17-22.

Northrop Corporation. 1969. *Semi-closed underwater breathing apparatus (SCUBA) Mark 8 Mod. 1.* Service manual for Dept. of the Navy, Deep Submergence Systems Project Office. Chevy Chase, Md.: Dept. of the Navy.

Office of Naval Research 1967. *Project Sealab Report, An experimental 45-day undersea saturation dive at 205 feet.* ONR Report ACR-124. Washington: Government Printing Office.

Searle, W. F. Jr. 1969. How deep sea work capability is growing. *Ocean Industry* vol. 4, no. 4.

Thompson, F.E. Jr. 1944. *Diving, cutting and welding in underwater salvage operations.* Cambridge, Md.: Cornell Maritime Press.

U.S. Navy. 1968. *Priniciples and applications of underwater sound.* NAVMAT P-9674. Washinton: U.S. Government Printing Office.

―――. 1970. *United States Navy diving manual.* part 1, NAVSHIPS 250-538. Washington: U.S. Government Printing Office.

Part 3
Diving Activity

7. Commercial Diving

Diving has grown from a small migratory employment field to a major business. It has undergone a series of changes that make it complicated to define. Is the scientific diving technician a commercial diver? Is the sport diver who spends his weekends scrubbing the bottom of yacht hulls a commercial diver? Is the photographer with his camera in a waterproof housing, using the techniques learned in his sport diving course, a commercial diver? In many instances, no clear delineation exists and the answers will vary according to the reason for asking.

For our purposes, we will consider the commercial diver as either a self-employed diver who makes all his livelihood from diving or as a man who is principally engaged in providing diving services for the company that employs him. Without a doubt, this excludes more than 75% of those who consider themselves "commercial divers"—but we will give them just due in other sections.

One question always raised by people looking at the business of diving, either for investment, career opportunity or just plain curiosity is, "How large a business is it?" In dollars and cents, the business is about 35 million for 1971, with reasonable assurance of being a 200 million dollar a year business in 1980. With regard to the number of people involved, figures are transitory. When the diving season is in full swing, the number may reach as high as 10,000—realistically, averaging the number of people who qualify according to our definition, the number is probably closer to 5,000 on a world-wide basis. The second question always raised concerns the fantastically high salary reportedly paid to a competent working diver. True, under unique circumstances, divers have made over $70,000 in a single year. The median salary for a working diver, averaged over all commercial activity, is somewhat under $10,000. A particularly effective diver with a good reputation in the field and operating within the constraints of good sense will make up to $30,000 per year. Divers who are willing to go to foreign locations and remain there for protracted periods can earn up to $50,000. The author has never met a commercial diver who honestly made $70,000 in a single year.

Earnings, however, are not a meaningful criterion for analyzing the commercial diving field. In practice, commercial divers can be better divided into two general classifications: the offshore diver and the harbor or mud diver. Regardless of which classification fits a diver, his general

Figure 7-1. Commercial diving, especially in harbors, is simply plain hard work. The beginning diver soon learns that the glamour of diving is in the telling, not in the doing.

job is to perform work in the underwater environment that is performed on land with less or no fanfare. There is little mystery about it—divers inspect things, bolt or weld things together, rig things for lifting, or cut and burn things loose. They carry on a myriad of other journeymen activities that are anything but glamourous. One of the first things a commercial diver learns is that the glamour of diving is in the telling, not in the doing.

People who are not familiar with working diving find the idea of diving not only adventurous, but similar to either movies in which the hero snatches the treasure from the hold of a sunken ship just seconds before it falls over the precipitous underwater cliff,—or more recently, similar to undersea movies photographed on beautiful sunny days with the water visibility of a swimming pool. This is not the harbor diver's environment, nor is it often the work site for the offshore diver.

The Harbor Diver

To put the diver in his proper perspective, we must look at him with reference to the work that requires his presence. For the harbor diver (Figure 7-1), this may range from common underwater labor to carpentry or plumbing. His reputation among his land-bound counterparts is anything but good, for several obvious reasons. First, he takes at least two or three times as long to perform a task underwater as they would at the surface. Further, while he is in their opinion carrying out this underwater malingering, he is making at least twice their salary and has a squire or tender to serve his needs. To make things even worse, he conducts his underwater activity with absolutely no supervision. From the standpoint of job comparison, the diver is definitely the most pampered, well-paid laborer in the construction business. A closer look at the conditions under which the diver works will show that whatever apparent advantages he may have are more than ameliorated by the conditions under which he works.

The harbor diver is not called a "mud diver" without good reason. Our growing civilization has always considered the oceans a general dumping ground for refuse and the last point of sewage concern. Concomitantly, the harbors have served as metropolitan garbage cans. All it takes to confirm this is a look into the water of any of our major harbors. These oil-slicked scum pots are the working environment of the harbor diver. At a depth of 20 feet in the average seaport, visibility is absolutely zero—all work is performed by touch. The problem is further complicated by cold. Even under summer working conditions, water at the harbor bottom is seldom warmer than 60°F; in winter it may be as low as 40°F. The diver's tactile sense does not function well for very long under these conditions.

However, lack of visibility is only one of the environmental problems. In addition, there are currents that run at speeds capable of sweeping the diver off his feet even when he is dressed out in the two hundred pound standard diving dress. And when fast currents are not present to scour the bottom, the diver must contend with the "bottomless bottom." This is not intended to be a paradox but is a valid description of a standard bottom condition where years of accumulated ooze forms layers of mud that get thicker as the diver descends; he finally reaches a point where the mud is dense enough to support his weight, but his whole body is immersed in mud only a little less viscous. This is a working condition that culls many would-be divers from the ranks. Working in complete isolation in the dark and with all movement restricted is often enough to bring out the most latent claustrophobia in the neophyte.

Although currents capable of sweeping the diver off his feet and the bottomless bottom are extreme conditions, they are by no means uncommon. At the other end of the spectrum, the best harbor diving conditions would not meet the minimum standards for most other types of diving.

One point mentioned in other parts of the book is particularly applicable to the commercial diver, so it is reiterated: *a diver does not get paid for diving; he is paid for what he does when he gets to the work site—diving is merely his means of transportation.* His hourly rate of pay is, of course, considerably higher than his surface-bound equivalent, but this is to compensate for his added skill in getting to the work site and the hazards involved in his "means of transportation."

In harbor diving, work is seldom performed deeper than 75 feet and is most commonly done at 40 feet or less. Although it is by far the most mundane of all diving work, harbor diving is the most consistent and employs the greatest number of divers throughout the world. The work, regardless of whether it is performed in New York harbor or Singapore, is characteristically similar, although the wages paid for the same work will show a great disparity. American divers employed in the U.S. or foreign locations are paid standard union wages. Local divers in overseas ports receive salaries ranging from 10-50% of those paid to imported Americans. It goes without saying that the situation has not resulted in a happy international commercial diving community, but this is gradually changing.

For the American diver, the depth at which work is performed is the key to his earnings. Because hazards once increased as depth increased, a compensation scale was designed to provide extra pay for the diver for each foot of depth beyond 50 feet. Today, with "depth" pay still in effect, a 350-foot dive yields the diver as much as a month of performing the same job on the surface. Harbor divers seldom dive deep enough to receive these extra pay benefits.

Most harbor work is directly related to waterfront construction—installing, maintaining and repairing shipping piers and industrial sites. The diver's function in these jobs combines underwater inspection, rough carpentry, matting and connecting large-diameter piping, cutting off decayed pilings at the bottom, and plain manual labor. Many harbor firms specialize in other harbor activities, principally ship repair. With proper equipment and techniques, a diver can perform much routine ship maintenance and repair, such as temporarily sealing up various ship machinery water intakes. If the job is not performed by divers, the ship must be drydocked.

Although sealing water intakes sounds like an exceptionally simple operation, only a few diving companies can handle the job safely and economically. "Few" is used in terms of percentage, because more diving companies are probably listed in every phone book than there are bonafide commercial divers in the area. The reason for the large number of "companies" is that shallow water diving is the easiest to outfit. For less than $2,000 a diver can outfit himself, put an ad in the Yellow Pages, and wait for a call. The attrition rate of these enterprises is high. Ten years ago, it was feasible to start in the business with a minimum of equipment and experience; today, the competitive environment of the harbor diving business, combined with requirements for a high degree of financial responsibility, nearly precludes the possibility of success for underfinanced diving operations.

The better diving companies in the harbor business operate under the auspices of a journeymen union—either the Piledrivers or the Carpenters. According to most operators, this is out of sheer necessity. The waterfront in all ports is tightly unionized and there is no possibility of operating on a union pier with a nonunion crew.

Divers' wages are set by union contract. The scale varies according to location but is approximately $100.00 per day. There are other wage scales for the diver who provides his own equipment, and still a third basic daily wage scale for the diver-entrepreneur who provides himself, his equipment and his own tender.

Computed on a five-day work week, a union diver on the West Coast should make between $20,000 and $30,000 per year. An analysis of diving companies shows that for harbor divers working for the larger companies, $12,000 is a high median figure, with top hands drawing down $19,000. The average over all companies, including nonunion companies that perform most of their work away from unionized shipping, is closer to $9,500. There are two reasons for this low average. First, there is not sufficient work for union divers to work a five-day week throughout the year, and secondly, nonunion diving companies pay their divers approximately $75 per day.

Despite the abominable working conditions and the uncertain pay, harbor diving continues to attract new divers. It is by all standards the least attractive diving work. Because of this, a continual job vacuum exists since the top hands move to the better and more lucrative offshore diving working conditions.

Offshore Diving

The last to feel the impact of business shift and technology has been the smaller company operating strictly in harbor business or inland waterways. The operator in offshore diving now lives in an entirely different business environment.

Diving companies are changing from small businesses to large corporate structures because of the increasing technological demands of the offshore oil and pipeline industries. At one time, these operators employed the single diver contractor to perform routine underwater maintenance and installation chores. The ability of these divers to fill the requirements in the early 1960s was limited and, in fact, led to a situation where the major offshore oil producers were willing to spend millions of dollars to "engineer-out" the need for diver services. Some, such as Shell Oil, went so far as to invest several million dollars in remotely controlled underwater robots to replace the diver. At this juncture, the future of the diver in the offshore oil patch was at best tenuous. The oil producers wanted better diver performance, a demonstration of ability to perform assigned tasks in a reasonable time frame, and more work for their dollar.

Fortunately, several companies emerged that were capable of accepting the challenge of mating available technology to problem solutions. The period 1960-1965 was formative for the major diving companies and, in fact, the healthy prognosis for future development of the diving business is directly attributable to action of such companies as Taylor Diving, J&J Marine, Dick Evans Divers, Ocean Systems and World Wide Divers during that period. All of these organizations have prospered from their efforts and grown at a rate that clearly shows they have been successful in solving the technological problems of deeper diving.

The major offshore diving companies in 1971 are hardly recognizable when compared to their status ten years ago. The equipment required to support a diver on a saturation dive was not even in existence in the early 1960s. Specialized systems such as underwater welding habitats are now a standard part of the company's inventory. What was once the larger portion of the diving business equipment, the diver's life-support gear, is now only a small portion of the heavy apparatus that constitutes the diving system. Most of the equipment and techniques used by the major offshore diving concerns have been developed recently to meet the expanding technology involved in ocean resource recovery. The changes in methods and machinery are also reflected in the men who perform the new tasks and operate the new machines.

The principal difference between the offshore diver of today and his counterpart of ten years ago is in basic capability. The problem of diver capability, which almost led to the demise of the industry, was that the offshore diver was merely a harbor diver working in deeper water. His knowledge, on the average, did not include a broad technical background in offshore technology; his welding capability was limited to wet welding and burning techniques, not necessarily of high quality; his knowledge of the broader aspects of

Figure 7-2. Diving is merely a means of transportation for the working commercial diver. This French diver is a career technician, expert in the work he is to perform on a wellhead and generally a tough competitor for the lucrative business of diving. (Photo: COMEX.)

offshore construction was meager; his attitude toward the operator was poor and his desire to acquire new technical knowledge to equip him for better performance was not good. These comments refer to divers as a whole—the exceptions to the rule are still in the business today. Those who fit the comments are either completely out of the field or are back working in the rivers, harbors and bayous.

The offshore diver today is a well trained, experienced technician, competent in basic mechanical engineering, possessing both learned technology and a natural ability to handle tools. He understands basic physics and physiology and is competent in the operation of both air and mixed gas diving equipment; he is an expert rigger, welder and underwater observer. In the context of the term "diver," he represents an entirely new and different breed and, in reality, should not even be called by that name; "underwater marine technician" more closely describes his function. He is the first of the commercial divers to use diving techniques purely as a means of transportation (Figure 7-2).

The rapid change in diving technology and business atmosphere has caused side effects on the offshore diver's career opportunities and future. Few if any training courses are available that encompass all the technology required by the diver to do an effective job underwater. Recent changes in training, particularly the creation of two-year marine technician training courses at local junior colleges, have raised the proficiency of the beginning diver, but even this will not provide the technician with enough specific knowledge to let him step into an offshore diver's slot.

To enhance their own competitive position, the larger diving companies have instituted in-house training to assure a basic standard of quality control on work and to prevent diving accidents caused by inadequate knowledge of the company's diving systems. Nearly all companies have built diving systems designed specifically to meet the needs of their particular field of expertise in the oil patch. Although most operate on the same fundamental principles, knowledge of their idiosyncrasies and operational nuances must be part of the diver's repertoire. In-house training, especially in a migratory business, is an expensive overhead proposition. For example, training a diver to use a new welding technique and to produce high-quality results in an underwater habitat requires months of work. For this particular skill, a conservative estimate of the investment in each trainee is close to $5,000. An operational weld performed on a pipeline at a depth over 200 feet costs the oil or pipeline company more then $250,000. For that kind of money, they want quality work. The training investment, with regard to quality work, has begun to pay off but has created other problems. If a diving company must invest in a man, they want to keep him available. Itinerant divers simply do not provide the continuum of expertise required to keep the company in a competitive position.

To meet company needs, many divers have been shifted from "hourly" or contractor status to that of a salaried employee. All large diving companies now maintain a cadre of highly trained and skilled divers on a yearly basis. Although the salary may appear to be less than what the diver can make by jobbing himself out at standard union diving wages, the average salary over several years is in most cases more than the diver would earn otherwise. Added to the other benefits he enjoys in an employee position and the opportunity to move through the corporate structure to higher responsibilities, the salary idea is being rapidly accepted within the field.

Maintenance of technical expertise, although it was a compelling reason for initiating the concept of the salaried commercial diver, was not the only reason for the change. Competition from European diving companies was also a major force. European diving companies have always been structured with the idea that their employees were principally engineers or technicians with the added skill of diving. Because European diving companies were prevented from operating in the continental United States, American diving companies felt little pressure in their own back yard. However, recent developments in oil exploration have changed the situation drastically. American companies are forced to go to foreign locations to continue their expansion and to keep pace with the business. Diving outside the United States is now more of a market than that available within its protected boundaries. When the American and foreign (particularly the French) diving companies met in head-on competition for work in places like the North Sea, Africa, South America, Australia, Indonesia and the Near East, the Americans, because of their higher wage scales, were coming out second best. The concept of salaried divers has helped this situation, and the U.S. companies can now be more competitive.

While many American companies started as associations of divers, with a titular head for negotiation, the Europeans have always been organized on a career basis. A good example of the larger and extremely effective European diving companies is the French firm Compagnie Maritime d' Expertises (COMEX) of Marseille, France. Organized in the 1960s, the company employs nearly 100 people, 60 of whom are qualified divers. Employment with this firm is a permanent situation. Benefits such as retirement payments and health and disability insurance are all part of the package. Each man is trained in a specific non-diving capacity so that when diving slacks off

Figure 7-3. The offshore diving world is now big business. Efficient work in deep water requires heavy capital investment in equipment such as this 166-foot diving workboat operated by COMEX of France. The ship is equipped with a complete saturation diving system located midship on the starboard side. (Photo: COMEX.)

during the winter months, all are gainfully employed within the company. To carry off this plan, salaries of divers in the company are quite different than those enjoyed by the American diver. The maximum salary paid is in the vicinity of $1,200 per month. The divers are continually encouraged to increase their underwater capability, even to the extent of mandatory physical training on company time. COMEX designs and builds most of its own equipment (principally during the off-season months), which has resulted in diving systems and techniques that are among the most advanced in the field. In addition to their equipment research and development activity, they operate one of the leading hyperbaric research installations in the world. In business matters, they epitomize the lean, efficient companies that will be the hard

competitors for the world-wide business of diving. From a manpower standpoint, their progressive policies of changing diving from an itinerant career to a permanent one is building a group of dedicated, effective underwater marine technicians and engineers, representing some of the best deep diving capability in the world.

Deep diving capability is the most publicized of all commercial diving activity. In the 1960s saturation diving techniques were developed, allowing divers to remain pressurized at near diving depth for the extent of the operation. This made it possible to perform jobs that could never be done with standard "bounce" diving procedures. All major companies, both American and European, now have this capability (Figure 7-3). From the commercial diver's standpoint this new procedure increases the margin of safety of deep dives by an order of magnitude, and also increases his earning power by extending the amount of time he can spend "on the bottom" and, simultaneously, his depth pay earnings.

While deep diving is one credential used by a diving company to demonstrate capability, it does not make up a major portion of the work at this time and probably will not in the near future. To see where the bulk of work will be performed in the next five years it is only necessary to look at the proposed offshore oil lease fields. The average maximum depth throughout the areas that will actually be produced is slightly over 300 feet. There are, of course, many fields in the range of 300-1,600 feet, but these are primarily the subjects of experimental drilling and completion techniques. The commerical diver's job site is dependent upon the economics of resource recovery. Knowledge of oil existence in a particular location is not the controlling factor in a recovery attempt; the controlling factor is simply whether the resource can be brought ashore at a competitive price. In terms of continuing employment of divers in the near future, the question is academic; there is sufficient work available in the nearshore waters of the world to absorb the growth of all diving companies for the foreseeable future.

The problems that control the growth of individual companies and therefore effect the career opportunities of commercial diving are found in the companies and the peripheral support activities that have grown up with them. Even though commercial diving has enjoyed an explosive expansion in the past ten years, many of the entrepreneurial management characteristics have survived. Some of these, such as lack of standardized safety procedures and minimum standards for equipment performance, have ramifications extending well beyond the actual diving process.

In an effort to enhance its competitive posture, each company considers its equipment and procedures highly proprietary, and anything related to these areas is treated with a "company confidential" attitude. This attitude of "playing their cards close to the vest" can in some cases be justified. A good example is the development of deep diving decompression tables. Speaking to any company, American or European, about the subject of decompression tables will yield a reply that clearly indicates that they have found a "window" in the physiological process of decompression that permits them to operate more efficiently than anyone else at depth. There is no doubt that the American companies, particularly Taylor Diving and Ocean Systems, and European companies such as COMEX and S.S.O.S., spent large sums of money to develop proprietary tables and consider them as competitive tools. Looking at the field of diving as a whole, however, tools such as these are factors in diving safety, and diving as a business will never enjoy its full growth potential until a means is found that will provide a return on a company's investment but will make the process available to all.

This same competitive attitude adversely affects growth in other areas within the field, among them diving safety and equipment. Aside from its effect on the well-being of divers, diving safety has a secondary importance in relation to insurance.

Insurance has always been a major expense in diving. For the diving contractor, it is a charge to be passed on to the customer. The diver as an individual has had a problem obtaining personal life insurance. Most insurance companies will simply state that not enough statistical data is available to rate the risk involved. The primary reason for the scarcity of data is again the "close to the vest" attitude of diving companies. Specific questioning into diving accidents will elicit extremely vague replies, usually containing little, if any, information. Pressed for an answer, the diving company's classic reply is usually "We can't answer that question because we feel that it will adversely affect our insurance rating."

By nature of their work divers are nonconformists and not joiners. Any attempt to standardize their operations is normally met with almost complete rejection. Being in a business that requires an exceptionally high degree of self-reliance and constantly being in a position where life depends on their capability and self-control, divers are not easily convinced that the manner in which they perform their work should be regulated from the outside. The premise is easily understood, but, as in many other highly specialized occupations, the diving industry is faced with a decision either to regulate itself or have it done by nonprofessionals from the outside.

Under the auspices of the Marine Technology Society, the principal diving companies have buried their competitive hatchets and are participating in the Man's Underwater Activity Committee which has made headway towards creating a set of minimum commercial diving safety standards and establishing operational criteria for diving equipment. If the major companies are willing to accept the final products of this work, there are few other impediments to continuing business growth and acceptance of offshore diving as an attractive professional opportunity.

Breaking In

The normal way for a young diver candidate to break into the field is to serve an apprenticeship as a tender (Figure 7-4). The term tender is entirely descriptive of the function. Literally, the tender is the squire to the diver, who often sees himself as some form of Knight of the Deep. The tender is responsible for keeping the diver's equipment in reasonably good condition, for laying out his diving dress, arranging all gear and equipment so that the diver can dress out with a minimum of trouble, inspecting, starting and assuring the proper running condition of the compressor, flaking out the diver's hose to assure free running during descent, and generally doing all the things a good diver should do for himself.

Once the diver is dressed and his system checked out, the tender helps him over the side of the barge. A careful diver will make a descent to the water using a ladder. If, however, he really believes that he is a Knight of the Deep, this entry process may be theatrically enhanced by a leap into the water calculated to astound all landbound onlookers. Accident reports indicate that it is usually the errant Knight of the Deep who is astounded by a good head rapping from the helmet.

Once the diver is in the water, the tender's responsibility is to keep his air/communication

Figure 7-4. One or two years as a tender is a prerequisite for diver candidates. In harbor diving, the tender is the diver's link with the surface. His responsibilities include dressing and undressing the diver, tending to umbilical lines, and a myriad of other chores to insure a safe, effective dive.

umbilical free and to continually "fish" the diver to avoid snagable slack in the umbilical. He records the diver's bottom time for decompression calculation and performs other miscellaneous tasks such as sending down tools the diver did not think he would need and other hardware the diver has lost in the muck. If the dive requires it, the tender sets up the deck decompression chamber in preparation for the diver's arrival at the surface. When the diver surfaces, the tender helps him up the ladder, undresses him, puts him in the decompression chamber if necessary, and finishes the day by cleaning and stowing the gear. This is, of course, what a good tender will do. Tender quality varies nearly as much as diver quality. The best tender is usually the young diver candidate with enough knowledge and experience to understand what he is supposed to do and an active desire to get at the "big money" diving.

After serving his apprenticeship as tender, and assuming that he has all the other credentials such as a graduation certificate from a diving school, the neophyte is ready to try his hand at diving. His first opportunity is usually unplanned—a job is just too long for a single diver or, not unlike the theater, an unplanned sickness of the star performer may occur. The time span required for a bright young candidate in the tender job to reach diver status is usually one year; in the larger diving companies it may be longer. Union scale pay for a tender is just under $7.00 per hour.

Certification

One of the continuing problems in commercial diver capability has been the lack of standard certification. Even with the growth of larger companies, there has been a concerted resistance of management to submit to some basic standardization of qualifications. Each company con-

siders itself unique within the business and establishes its own, usually unwritten, job standards. In the explosive expansion years of commercial diving (1965-1968) when there was a continual shortage of qualified divers, this practice, combined with the lack of adequate operational safety standards, led to some of the most freakish diving accidents ever recorded. The situation has improved somewhat in the last few years, principally because the people who buy diving services are more demanding. The slight squeeze that the diving business felt from mid-1968 to 1970 has increased competition and further pressured the diving contractor to improve his performance in order to get the work.

Most harbor diving is still performed by entrepreneurial one- or two-man outfits. How qualified a diver may or may not be is still transmitted by word of mouth. The final control of qualifications, however, rests with the service purchaser. The market is becoming more sophisticated, the business is now big business, and the qualified diver created through word manipulation is becoming a thing of the past.

8. Military Diving

Military divers comprise the largest segment of the diving community. not only does military diving involve the greatest number of divers, it covers the full spectrum of underwater activity including salvage, construction, maintenance, scientific and many other facets indigenous only to military operations. Military divers are trained to operate in all environments (Figure 8-1).

Every major navy of the world maintains a cadre of qualified divers in their operating forces. Each navy's program is organized to best suit the support functions necessary for their primary mission. As a result, there are as many different structures for military diving activity as there are navies. However, a certain degree of uniformity exists among all naval diving, and in this chapter we will consider the diving activity of the Royal Navy, the Royal Canadian Navy, the Federal German Navy and the United States Navy.

There has always been a free exchange of technical diving information among the navies of the free world. Which navies constitute the "navies of the free world" is of course variable. But in late 1968, the United States sponsored the First Annual International Naval Diving Conference at San Diego, California, with attendees from the naval forces of Australia, Canada, Great Britain, Germany, France, Sweden, Switzerland and, of course, the United States. Each of these countries described their diving activity during this conference, and much of the following information comes directly from that meeting.

Royal Navy

Diving activity of the Royal Navy was outlined by Comdr. Philip White, Superintendent of Diving, Royal Navy. According to Comdr. White, the Royal Navy is responsible for all types and methods of British military diving and for the design and development of diving apparatus. To fulfill their diving requirements, two general classifications of divers, the *Clearance Diver* and the *Ship's Diver*, are used.

The *Clearance Diver* is a full-time specialist, qualified in all types of self-contained equipment. He is also a demolition and ordnance disposal expert. One of his most important tasks is rendering safe and disposing of explosive ordnance found below the high water mark. Even though World

Figure 8-1. Navy combat swimmers operate in all environments and are trained for survival in all water, from the tropics to the arctic. These two UDT divers have just completed a training exercise in less than desirable diving conditions. (Photo: U.S. Navy.)

War II has been over for twenty-five years, three Royal Navy Clearance Teams are continually busy with this task. As succinctly stated by Comdr. White, "I don't have to tell you that the stories of these particular exploits are long and varied, and rather like fishermen's tales. But, invariably the task boils down to dealing with a piece of explosive fitted with a means of exploding which, for some reason or other, has failed. With last-war ordnance, it would seem reasonable to expect that, through age and years of submersion, these pieces of ordnance could be regarded as safe. Unfortunately, this is not true."

The *Ship's Diver* of the Royal Navy is trained to operate self-contained breathing apparatus to 120 feet. Unlike the Clearance Diver, he does not carry a rating as a full-time diver, but rather diving is a qualification in addition to his rating or rank. With the reduction of British overseas bases, this

diver plays a role of growing importance in ship maintenance and repair. Free-swimming Ship's Divers now routinely practice changing mine-sweeper and submarine propellers, underwater sonar domes and telephone transducers. It is anticipated that in the near future Ship's Divers will also be called upon to clean and paint ship hulls under water.

Hard-hat divers have been gradually discontinued in the uniformed service of the Royal Navy. With the exception of some possible future developments for deep diving systems, all hard-hat and tethered diving is performed by civilians employed by the Boom Defense and Salvage Departments.

Royal Canadian Navy

Royal Canadian Navy diving activity is structured in a manner similar to the Royal Navy, with some minor modifications. They use both the Clearance Diver and Ship's Diver classifications, the clearance rating being a primary function and the ship rating being an auxiliary capability.

In the Canadian Navy the Ship's Diver is only trained to use open circuit SCUBA. His training is designed to qualify him for diving to a maximum depth of 50 feet. His primary diving function is inspecting ship bottoms for unfriendly ordnance that may have been attached. Light repair and damage inspection have recently been added to his responsibilities.

One unique use of Ship's Divers was initiated by the RCN. They are now used exclusively for man-overboard rescue operation. The normal procedure of lowering a small boat to pick up a man overboard has been discontinued. Now, once the ship is in position, the Ship's Diver, with line attached, jumps overboard to effect the rescue. Instead of a small boat, they have also recently begun using a diver to perform buoy mooring procedures.

The function of the Canadian Clearance Diver is identical to that of his Royal Navy counterpart. The rating is a career situation, and the CDs are organized into teams. However, the Royal Canadian Navy Clearance Divers receive their explosive ordnance training with the United States Navy Explosive Ordnance Disposal Training Unit at Indianhead, Maryland, rather than with the Royal Navy.

Federal German Navy

According to Comdr. Wolfgang Brinckmann, divers in the Federal German Navy come under one of two general classifications: *Technical Divers* and *Weapons Divers*. Their responsibilities are comparable to those of divers in the U.S. Navy.

Technical Divers, supervised by the Naval Ships Technical Command, may qualify as either *Helmet Divers*, equivalent to the U.S.N. Deep Sea Diver, or as *Swim Divers*, equivalent to the U.S.N. SCUBA Diver. *Helmet Divers* are responsible for salvaging ships, torpedos and other heavy objects. They are also directly involved in heavy underwater construction work on both ships and harbor installations. In these capacities they are trained in underwater rigging, demolition, welding and burning.

A technical diver of lesser grade, trained only in air SCUBA, is in general use in the fleet. The basic mission of the *Swim Divers* is the inspection of ship hulls to detect underwater ordnance and performance of minor repair and maintenance tasks.

Weapons Divers in the Federal German Navy are under the direction of the Naval Weapons Command and are divided into two classifications: *Clearance Divers* and *Battle Swimmers*.

The *Clearance Diver* is again primarily engaged in the disposal of underwater ordnance. His mission is to search, locate, classify, identify and neutralize unfriendly mines and other underwater explosives.

Battle Swimmers are organized and have the same mission as the U.S. Navy UDT Teams. They perform covert underwater and beach reconnaissance in support of amphibious landing and can also be assigned long-range underwater or land assault missions.

United States Navy

For many years, the U.S. Navy was the prime mover in diving activity, the leader in underwater research and the source of personnel for most diving enterprises. In 1971, it no longer maintains this position, partially from intent and partially because of the system from which it operates. The fact that it has not maintained its leadership in diving technology is irrelevant to naval diving. Unlike commercial diving, where each advance in diving capability represents a measurable step in exploitation of the ocean resource pool, naval diving is tuned for response to military operations and must be considered in that light.

In appraising the effectiveness of naval diving, major consideration must be given to its role within the naval system. The primary mission of the Navy is operational support of United States's policy through sea power. To prosecute this mission requires the operation of surface ships, submarines and aircraft. With the exception of mine clearance divers and combat swimmers, the tactical importance of diving in these operations is minimal and the strategic importance is nil. Since tactical and strategic importance are the controlling factors in budgetary considerations, research for the technical advance of diving has always, except for the Sealab program, been on the tail end of the money channel.

At the First International Naval Diving Conference in 1968, Captain Willard Searle, then Supervisor of Salvage for the U.S. Navy, addressed the gathering with closing remarks which give an excellent picture of previous advancements within the context of U.S. Navy diving:

In 1917 we lost a submarine, the F4 in Honolulu, Hawaii. It went down in 320 feet of water. At that time we had a small Experimental Diving Unit operating in New York City based at the Brooklyn Navy Yard. These fellows, under a gunner by the name of Stillson, were really trying to introduce Haldane's theories of diving and the Siebe Gorman equipment into the United States. They had gone around the world and bought one or two copies of each piece of diving equipment they could find, had built a "pot" and were diving. This was the first bit of diving research ever done in this country. (Authors's note: Captain Searle was undoubtably referring to diving research under U.S. Navy auspices.) The submarine went down, was found, they needed divers. Gunner Stillson with four of his men and one doctor were hastily taken across the country on a train, put on a cruiser and steamed at forced draft to Honolulu where they dived 318 feet, which was a world record on air. It's a dive that we don't want to do again today. They did a magnificent job.

In the mid '30s we were doing research in helium. SQUALUS went down in 1939, and we hastily threw the helium diving system and the helium diving teams from the Experimental Diving Unit in Washington, DC into the gap, and those people used helium diving for the first time in an operational sense on the SQUALUS. In May of this year (1968) SCORPION went down; we didn't know where it was. We had early reports of debris in

an area below the Azores, not in the area in which it was eventually found. There were sea-mounts in the area. Depths of the sea-mounts from the surface were in the neighborhood of 900 to 1000 feet. We hastily looked around on the assumption that we might find the submarine parked on top of a sea-mount, put together diving systems, ad hoc diving teams, diving medical officers and flew them to the Azores, and we were in fact prepared to dive to 1000 feet if necessary.

The point is that in each major advancement: deep diving, deep helium diving and now real deep diving, we have been pushed. We found ourselves by some emergency pushed to do a job before we were ready.

Captain Searle's remarks basically describe the character of naval diving—reaction. Perhaps the problem is the mental attitude of planners toward the role of diving in naval activity for who would really like to think (and believe) that a $100 million vessel now on the drawing board would ever be resting on the bottom awaiting the services of a salvage crew to bring her back to the surface. Regardless of the reasons—mental attitude, strategic or tactical importance—it was not until the middle 1960s that naval diving activity began to assume importance.

This change in attitude was not motivated by a new awareness of the relative importance of men in the sea, but rather was an extension of the world-wide interest of all government activity in exploitation and use of the sea. Many programs dreamed of by system-bound navy men showed signs of becoming realities: Large Object Salvage System (LOSS), the Sealab experiments and many others rapidly went from the paper proposal stage to prototype equipment. Within the 4-year period from 1964 to 1968, the U.S. Navy expended more research and development funds to advance man's diving capability than had been spent in the preceding 40 years. From this effort, new equipment, new information and new diving techniques emerged that will affect the course of diving technology for the next decade. The vagaries of government spending, however, always prevail and during the closing years of the decade, clenching of the fiscal fist cut the flow of diving funds to a level commensurate only with maintaining minimal program life.

In 1970 Captain Eugene Mitchell, Supervisor of Salvage and Diving, accurately summed up the U.S. Navy's position in the business of diving when he said in an interview with *DATA* magazine,

> The diving equipment which is being used by the Navy today has suffered from many years of neglect. We used to be the leader in the diving field, but at some point this initiative was lost, and we found some years ago that commercial divers had begun to outspace the Navy both in equipment and in ability to operate it. We have a good start toward catching up.

Diving Structure

Part of this catching up has been the development of new deep diving systems, new techniques and the development of a new diving activity structure within the Navy. Diving activity within naval cognizance is varied and because of its wide range, the structure is both loose and complex. Since 1969 all diving activity has been in some way under the Supervisor of Salvage and Diving. The "and Diving" portion of the title, added in that year, at least gave diving activity more visibility, even if it did not make a measurable improvement in funding.

The naval diving structure can be divided either horizontally, proportional to the skill required for each assignment, or vertically, according to the nature of the specific diving activity. The vertical division is more discrete and conforms closely to the description of activity in other navies. Irrespective of divisions used for naval activities, divers in the U.S. Navy are most logically classified as either *working* or *tactical* divers. The *working* diver classification includes salvage, inspection and construction, while the *tactical* classification includes underwater demolition (UDT), explosive ordnance disposal (EOD) and Sea-Air-Land assault diving (SEAL).

There are no specialty "diver" ratings in the U.S. Navy personnel structure, so all divers carry diver qualification in addition to a basic technical rating. At first glance, this seems disadvantageous to the career diver. If he is serving in a diving billet for many years, the logical assumption would be that he is far removed from the study and practice of his journeyman skill—the skill with which he advances both in pay and rate. Advancement records, however, indicate the contrary. Analysis of career progress of UDT enlisted personnel shows that advancement rate within the teams is nearly 20% faster than in the fleet. The Navy also feels that from the enlisted career security standpoint, this system has another major advantage. The rigorous physical requirements for diving, especially with the tactical units, present a continuing threat of disqualification during a 20-year career. In the event of physical disqualification or disenchantment with underwater work, a man need only drop his diving qualification, not his navy career.

Working Divers in the U.S. Navy are either Deep Sea Divers or SCUBA Divers. Both undergo the same intitial training. SCUBA Diver training is, however, considerably shorter and after its completion, the diver is returned to a fleet assignment in his technical rating. His diving function in fleet operations is similar to that of his counterparts in other navies of the free world; that is, he is principally concerned with underwater hull inspection and performs miscellaneous underwater activity consisting of clearing fouled screws and routine maintenance.

The *Deep Sea Diver* is trained with the idea of making him a competent underwater workman—a professional diver. Throughout his career, opportunities are provided for advancement through further training in the diver qualification. The levels of skill currently recognized are *Second Class Diver, First Class Diver,* and *Master Diver*. Each requires a specific training course at a Navy Diving School and an intervening period of experience and proficient performance. The basic skills taught in these schools, and most work in the fleet, is related to salvage—not the movie scenario brand of salvage where the hero recovers chests of golden doubloons, but the real world of salvage—the recovery of sunken or stranded vessels. It is difficult to give a meaningful overall description of this work. A typical operation where divers played an important role in the salvage work will, however, provide an overview of salvage diving tasks.

On July 18, 1965, a radar picket ship of the U.S. Navy, the USS *Frank Knox* (DDR 742), was steaming at 16 knots in the South China Sea about 180 miles south of Hong Kong. At 0235 she grounded of Pratas Reef, a circular coral atoll with a small island on its western side. The hull was well aground on the shallow reef and by 0300 the captain sent out a call for assistance. First to arrive at the scene was a cadre of Nationalist Chinese UDT swimmers from nearby Pratas Island. They made an inspection dive and reported that the ship was aground from the bow to a point just aft of midship, in addition, all propeller blades were either badly chipped or bent. The ship's engines were unable to free it from the reef and by 0400 a full-scale salvage operation was ordered from the closest naval base, Subic Bay.

None of the crew were injured in the grounding, but because of the exposure to high surf on the reef, both vessel and crew were considered to be in jeopardy. All but a skeleton crew were evacuated. The ship, built in 1945, was equipped with the latest in electronic warfare and radar equipment. The value of the Combat Information Center alone, if it could be salvaged intact, was over $15 million. It was assumed that with proper towing the *Knox* could simply be pulled from the reef. The arrival of the first salvage vessel, the USS *Grapple,* got operations started within 24 hours. An inspection by *Grapple* divers confirmed the earlier findings: the ship was solidly aground on a bottom composed of hard, smooth coral with sharp-sided crevices running perpendicular through the reef. During the next several days, standard unbeaching techniques were used to extricate the *Knox* from her predicament, all to no avail. More ships arrived at the scene and new procedures using their combined pulling power were tried. Again these turned out to be exercises in futility. The weather was worsening and reports indicated that the area might be hit with a typhoon.

High surf made diving external to the ship an impracticability. Repeated efforts to dislodge the vessel by towing met with little or no success. The expected storm did hit, causing a temporary cessation of salvage attempts and driving the salvors away from the site. Swells as high as 12 feet were buffeting the atoll and the stranded vessel.

By July 22 the storm had abated, but continuing swells and surf still made external diving unsafe. Divers were sent into the flooded compartments of the ship for further inspection and to attempt shoring up the holes in the engine room compartments. They did manage to shut off all the leaking sea water lines and made some progress at plugging the small holes left in the hull by rivets that popped during grounding and subsequent surf beating. Continued attempts to pull the *Knox* free, even using as many as five salvage vessels, proved to be unsuccessful.

The water conditions by the end of July had improved to a point where external diving could safely be resumed. Underwater operations to seal off the ship's hull were begun using pads and shoring material, and the divers attempted blasting the coral under the ship with the hope of reducing drag for the next pulling attempt. Using plastic explosive charges, they managed to clear large areas of coral that were either holding the vessel or presenting a threat of impaling the hull when it started to move. Attempts to seal the hull with external storing and internal cementing were unsuccessful so a decision was made to use foamed-in-place polyurethane as a sealer and buoyancy material for the damaged compartments. Divers were sent into the compartments to begin filling the hull with the plastic material. This was the first major use of self-foaming plastic by Navy salvage crews and it turned out to be an educational process.

The combination of foaming, coral blasting and the addition of more pulling gear finally turned the trick. Early in the morning of August 24 the USS *Frank Knox* was freed and successfully towed to Yokosuka, Japan.

Although U.S. Navy Deep Sea Divers are usually employed on more mundane tasks than saving a line ship from a potentially disasterous ending, the operations performed on this job are typical of naval salvage activity. The diver must be well trained in underwater rigging, use of explosives, underwater welding and burning. Above all,

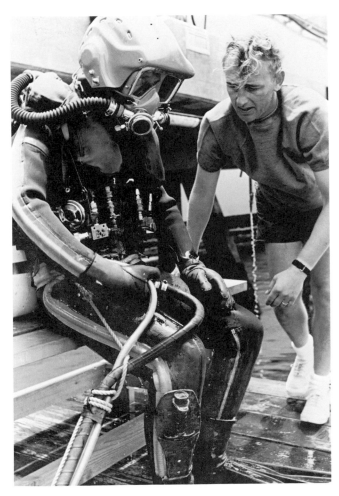

Figure 8-2. A primary mission of the U.S. Navy Salvage and Diving Office is increasing the Navy's salvage diving capability to 1,000 feet. This aquanaut is preparing to test a diving system designed for 850 feet. (Photo: U.S. Navy.)

he must be able to improvise in new and unique underwater problems.

The present short-term goal in this area of U.S. Navy diving is to extend capability to a working depth of 1,000 feet (Figure 8-2). Beyond that depth, the practicability of using divers with in the next 5 years is questionable.

Extending the capability to 1,000 feet has required initiation of new diver training activity and development of new operational techniques. Recognition of specialized forms of diving has broadened the classification of diver rating to include two new levels of expertise: *Saturation Diver* and *Aquanaut*. These classification are currently used only in the research and evaluation activity of naval diving, but as saturation diving techniques move into wide fleet usage, the *Satura-*

tion Diver will become a standard operational rating. The *Aquanaut* rating was primarily established to provide some differentiation for diver engaged in high risk, experimental diving programs such as Sealab I, II and III, and will probably never by widely used in fleet operation.

Combat Diving

Tactical Divers in naval activities perform myriad underwater operations, some with counterpart activities in other navies, some unique to U.S. Navy operations. The mission of these men can be either defensive or offensive: the *Explosive Ordnance Disposal Diver* performs purely defensive functions while the combat swimmers, the *Underwater Demolition Team* members and the *SEAL* are primarily engaged in offensive actions.

The *U.S. Navy Explosive Ordnance Diver* (EOD) is equivalent to the Clearance Diver used by other navies. His primary function is to locate, identify and render harmless all unfriendly ordnance found in naval operating areas. To the EOD man, diving is only a means of transportation. His equipment is specially designed SCUBA and his most common assignment is the disarming of mines. The nature of modern mine techniques makes the use of specially designed equipment mandatory. Nearly every type of standard underwater breathing apparatus has both a magnetic and acoustic signature; that is, the noise generated by its operation and the ambient magnetic field changes caused by its presence are definable qualities. It is quite possible, therefore, for these signatures to be used to detonate the ordnance by simply tuning the detonator sensors to these characteristics. EOD equipment is designed so that its noise output is of the same order of magnitude as normal sea noise, and all metal parts are nonferrous to produce minimal magnetic anomalies.

The basic tools of the EOD diver are closed or semi-closed life-support equipment, hand-held diver sonar equipment, a high degree of technical ordnance knowledge and an inordinate supply of "cool."

Operational use of EOD divers varies with the type of mission, most of which are highly classified, but a typical example of their activity would be the destruction of a bottom-secured mine. The EOD team is temporarily billeted on a minesweeper-type vessel. Ship sonar is used to locate underwater objects. When a suspected contact is found, three EOD men leave the ship in a small boat, usually an inflatable with a low noise outboard motor. The small craft is brought to the general vicinity of the contact by guidance from the minesweeper. The diving team then swims in to the ordnance for specific identification and classification. Depending on its type, the disposal method is selected. Any number of methods are available, among them underwater disarming (seldom done) and spiking with a delayed explosive charge to detonate the device after the EOD personnel have cleared the area. Regardless of the methods used, a constant battle of intelligence

runs between the disposal unit and the unfriendlies: the disposal people want to quickly classify and find a means of destruction—the unfriendlies want to work out new methods of making the devise lethal to disposal.

Most ancient literature references to divers mention divers as assault swimmers. In early sea battles, they often proved to be the deciding factor in victory. Somewhere along the line, however, naval tacticians began to feel that although covert swimmers had been effective weapons against ancient Greek galleys, their tactics could never work against the sleek, fast and powerful warships of the day. This trend of thought managed to survive while warships progressed from powerful square rigged frigates to invincible battleships boasting 16-inch cannons.

Today, when a single ship carries the offensive firepower of entire fleets of bygone eras, the combat swimmer has come into his own. The covert destruction of enemy shipping is no longer his principal mission. However, all modern navies of 1971 recognize that not only is the combat swimmer a formidable weapon, but water is the ideal medium for delivery of small, clandestine assault forces.

The modern use of combat swimmers must be credited to the Italian Navy. In several stunning and completely surprise attacks during World War II, they wrecked havoc with the British Mediterranean Fleet. Many Italian tactics, including swimmer delivery vehicles to extend swimmer range, are still in use today. Combat swimmers in the U.S. Navy and the Royal Navy made their appearance after intelligence finally figured out what had happened. The U.S. Navy's first use of combat swimmers came in the latter part of World War II. The early exploits of these men are carefully detailed by one of the participants, Comdr. Fane, in his book, *The Naked Warriors*.

All U.S. Navy Combat Swimmers now operate under a single command, the Naval Special Warfare Group. In addition to providing administration, NSWG also tests, evaluates and develops new equipment and tactics for swimmers. Of the four groups under NSWG, two—the *Underwater Demolition Teams* and the SEALS—are combat swimmers. The other two—the Beach Jumpers and Boat Support Groups—are directly concerned with near-shore and amphibious warfare.

The *Underwater Demolition Teams* are the outgrowth of activities started during World War II, but their organization, equipment and tactics bear little resemblance to the surface swimming demolition experts used during the invasion of the Normandy beaches. Five active UDT groups are authorized: Teams 21 and 22, based at Little Creek, Virginia, and Teams 11, 12 and 13, based at Coronado, California. Each team has billets for 100 enlisted personnel and 15 officers and is sub-structured into four platoons of 25 enlisted men and officers. The teams based at Little Creek support the Atlantic Fleet, the Coronado teams support operations in the Pacific.

The primary function of the Underwater Demolition Team is not assault, but reconnais-sance (Figure 8-3). They are closely integrated into the Navy's amphibious landing capability and when deployed, they are under the command of the respective amphibious commander. Basically, the UDTs are responsible for determining whether or not specific beaches are suitable for landing of troops, for providing hydrographic information about the area and for clearing landing sites of natural or man-made obstructions.

To those who are not aware of the paucity of information about the shallow portions of the continental shelves throughout the world, this might seem like a rather simple task, one which might be simply solved by referring to the proper charts. Unfortunately even the best of hydro-graphic charts are relatively inaccurate at water depths needed for amphibious operations—the depths from 20 feet to the shoreline. The basic assumption is that a detailed survey of the beach must be made if an amphibious force is to be landed. Since this beach is usually unfriendly, the survey operation must be carried out in a covert fashion.

Once an area has been selected as a possible operational site, the UDT's mission is to determine if such operations are possible. The first task is called a beach feasibility study, which is simply a gross survey of the areas to see whether or not the

Figure 8-3. Underwater Demolition Team swimmers perform reconnaissance missions and clear beaches for amphibious landings. (Photo: U.S. Navy.)

approach might be suitable for the amphibious forces. Usually, this is performed by making several runs parallel to the beach with an outboard powered rubber boat, or clamshell. Fathometer recordings are made, and these data plus sight observations made by the patrol are used as a decision basis for further investigation. If the decision is made to carry out a detail survey, the mission is assigned to one or more platoons. In the light of modern technology, the methods used to perform the surveys are crude. But as with all work in the ocean, it is the rather crude, unsophis-ticated systems and methods that are the most consistently successful.

The survey platoon is deployed along a line at the 20-foot contour. The swimmers are spaced 25 yards apart and each is equipped with a marking slate and sounding line. The surveyors then swim towards the beach, recording all major bottom features and periodic depths. The result is a highly accurate series of profiles 25 yards apart from the 20 ft. contour to the shore. This information can then be correlated to tide data for detail wave forecasting and selection of approach channels. If

feasible, the team will go ashore and record data about the hinterlands and shoreline area.

The simplest and most common way to make these surveys is during daylight hours, using surface swimmers. Of course, this is not always possible, and alternate methods have been developed for making surveys while completely submerged and using SCUBA. In their operations, the UDT uses three basic types of SCUBA equipment: standard open circuit air, mixed gas semi-closed MARK VI circuit and the closed oxygen system. The nature of the mission and the method of swimmer delivery determine the type of equipment employed.

All combat swimmers are trained and qualified for several different methods of delivery to the mission area. The most common delivery method is by ship and then small craft. Where more covert methods are required, the swimmers can be delivered by fleet submarine. Few sub-

marine commanders will bring their boats in beyond the 120-foot contour, so in these deliveries the swimmers must lock out at depth and usually have a long swim to the site (Figure 8-4). The showiest and least used delivery method is the parachute; however, all UDT personnel are trained to make water entries in full swimming gear from both low and high altitude parachute drops.

A more recent and practical addition to the operational repertoire is helicopter delivery. Almost everyone in the free world has seen the recovery on American space shots. In test operations the helicopter hovers at a low altitude and the swimmer jumps into the water. (The swimmers used in the recovery of the Apollo space capsules were UDT personnel.)

In the Korean conflict, UDT was concerned with reconnaissance of beaches, the demolition of obstacles to safe amphibious landings and demolition problems in areas close to the shore line.

Figure 8-4 (left). These combat swimmers are preparing to lock out of a fleet submarine. Figure 8-5 (above). Probably the most efficient fighting machine ever developed, the U.S. Navy SEAL Team has changed concepts of marine special warfare. Trained first as combat swimmers, then as long-range assault troops, SEAL teams are suited for antiguerilla warfare. (Photos: U.S. Navy.)

Operations in Viet Nam have added to the requirements for more combat capability.

Special Warfare

Developing tactics to meet new threats is what special warfare is all about. The term *special warfare* is relatively new. It basically refers to naval warfare that is not classical—that type of naval warfare not conducted by textbook maneuvering of ships and aircraft—that type of warfare where sheer power is impotent. Several years ago it was called *unconventional warfare*. This unconventional warfare has become so conventional that it has caused major changes in nearshore naval tactics.

In the naval warfare associated with Viet Nam, the effective naval forces have been those units able to change from "blue water navy" to "brown water navy." Riverine warfare has been the important consideration in the past few years and the extensive waterway system throughout Southeast Asia has been one of the primary war zone battlegrounds. This new warfare has altered many of the UDT tactics and led to the development of new mission assignments.

The primary work of UDT, reconnaissance and clearance of obstructions, is still valid, but when this activity moved from the nearshore environment to inland waterways, the tactics necessary for survival changed drastically. More effort now goes into such tasks as overt assaults on riverside bunkers and fortifications. Constant patrolling is required to keep rivers and canals free of obstructions and constructions that would make the waterway unnavigable or tactically disadvantageous. In short, changing special warfare operations are making the UDT swimmer a shoreline assault specialist. The change in naval warfare tactics has not only been answered by change in operational missions of existing units, but entirely new groups have been organized to conduct these operations.

Sea-Air-Land Units

On January 1, 1962 President John F. Kennedy directed all U.S. military services to develop a limited "unconventional" warfare capability. The Navy's response to this order was the organization of Sea-Air-Land (SEAL) units (Figure 8-5). Their specific charter was to conduct special warfare operations in the "maritime atmosphere." The maritime atmosphere is subject to wide interpretation and today is taken to mean rivers, deltas, harbors and shorelines. Operations are sometimes overt, sometimes covert, but primarily clandestine.

In the short time that the SEAL teams have been in existence, they have acquired the justified reputation of being the elite naval special warfare corps, although bad publicity in the pulp magazines has created unwarranted feelings that they are "beady-eyed killers." Actually, they are UDT-trained Navy men with additional expertise in land warfare tactics. They have proven to be the most economical and effective weapon against guerilla-type warfare. According to Lt. Comdr. E. L. Schaible, Commanding Officer of SEAL Team #1, their disproportionate success in recent operations is due to extensive training that enables them to maneuver from the "marine atmosphere" into difficult terrain, using only small groups with highly original tactics—tactics that allow them to bring the war to the unfriendlies.

The SEAL team members are not only closely associated with the UDT, they are selected from the same personnel pool. When a candidate volunteers for special warfare service he undergoes Basic Underwater Demolition SEAL training (BUDS). At the conclusion of this 20-week training course, part of the class is assigned to the UDT units and part to the SEAL teams. Assignments are arbitrary and based upon personality and performance during the initial training. Those assigned to the UDT teams undergo an additional 3-week course in parachute techniques and are then assigned to an operational team. For those assigned to one of the SEAL teams, a long road of additional training lies ahead: parachute, weaponry, navigation, small unit tactics, scouting and patrolling—generally a complete training in land warfare tactics. But regardless of additional training, the SEAL team member is a tactical diver, the epitome of the assault diver.

Presently, there are two authorized SEAL teams in the U.S. Navy: SEAL Team #1, operating out of Coronado, California, and SEAL Team #2, headquartered at Little Creek, Virginia. The published, authorized level of these units is 33 officers and 195 enlisted personnel, a remarkably small number for their present mission activity.

Army Diving Activity

Using the term "military" and "diving" in contiguous positions brings to mind navy and water so military diving is assumed to be an interchangeable term for naval diving, but this is not true. The second largest contingent of divers in the United States is under control of the United States Army. Since the U.S. Army operates the third largest "navy" in the world, this is really not a surprising fact. The primary function of their diving corps is the support of the Army "fleet" and differs little from their navy counterparts. The description of naval salvage activity provides a close approximation of their tasks. In addition to these working divers, the Army and Marine Corps have trained and are operating special forces groups that use diving as a delivery method for performing their primary mission.

For the first half of the coming decade, military diving will continue to numerically dominate diving activity. Basic physiological research conducted by the military will continue to play an important role in man-in-the-sea advancements, but it is doubtful that the gap between commercial and military diving will ever again close.

9. Scientific Diving

Diving for scientific purposes has done more than any other single factor to shape the future of man-in-the-sea. Scientific divers are found as far back as we can trace diving history, but scientific diving as we know it today is only 20 years old.

In that 20 years, diving extended almost every scientific discipline connected with the earth system, including space exploration. For example, marine biology is no longer just the study of specimens and ecological hypotheses; marine geology in moderate-depth water is now akin to dry land geology; diving physiology has led to great strides in hyperbaric medical therapy; diving archaeologists have uncovered the untouched remains of archaic and unrecorded civilizations on the weatherless ocean floor and astronauts are trained for space travel in the zero-G diving state. The scientific diving community's growth has been exponential. In 1950 there were probably no more than twenty scientific divers in the world. In 1970 a conservative estimate is 2,000, with over 1,000 being trained each year. Considered as a group, rather than in the context of their individual disciplines, the community is a major entity in the business of diving. These men are having profound effects on the acquisition of knowledge about what man can do in the sea and, still more important, what the sea can do for man.

The Bio-Diver

Organisms living in the ocean have interested man for centuries. Of course, part of this interest has been self-effacing. The ancient mariners' tales of conquering dreadful sea monsters added to their stature as extraordinary adventurers. Darwin's work forged a main link in the man-sea creature relationship. The hypothesis that we are directly related to the inhabitants of the ocean depths heightened interest beyond mild curiosity. Modern diving technology is now providing apparatus to satiate this desire to know more about the controlling links in the earth's ecosystem.

At the beginning of the twentieth century, the marine biologist had very little opportunity to be an *in-situ* observer. Unless he was a highly skilled adventurer willing to risk life and limb to the vagaries of a 200-pound diving rig topped with a brass pot, with his air coming from a hand pump operator at the surface, he was restricted to wading in tidal pools. Deep diving submersibles were nonexistent, and the use of diving bells was limited by expense and technology.

However, pioneers did appear. In 1930 C. William Beebe and Otis Barton descended more than 1,200 feet in a bathysphere. The reports from their series of dives tell of a "memory of living scenes in a world as strange as Mars," and did much to spark marine biologists' interest in seeing the underwater world firsthand. Development of underwater photographic techniques had a similar effect. Photographic evidence showed that the underwater environment was not being thoroughly investigated by remote gathering methods and convinced marine ecologists that to study the "houses in which they live" they must see these creatures undisturbed.

The arrival of the Aqualung at Scripps Institution of Oceanography in 1950 had a dramatic effect not only on the marine biologists themselves, but on the entire field of marine biology. A whole new world of nature was opened for investigation.

According to marine biology divers like Dr. Andrew Rechnitzer of North American Rockwell Corporation, and James Stewart, Diving Officer at Scripps, the most astonishing thing about their first dives was the abundance of sea life and the number of species. Many "rare" fish were found to be prolific within a few hundred yards of the shore. As the early investigations proceeded, it became evident that "rareness" was related to the creatures' ability to avoid the previous catching methods, not their abundance in nature.

Reports of early diving marine biologists are filled with the same awed literary responses as those of Beebe on his tethered trip. After the novelty of being a free swimming creature in the sea finally wore off, the new science of bio-diving started about its objective scientific business.

The impact of diving on the staid science of marine biology is difficult to appraise, since the present activity bears little relationship to pre-diving marine biology. Knowledge gained in the early days has caused an exponential growth in the areas of interest, and the adventurous "last frontier" nature of the work has attracted an ever growing body of young scientists. The scope of

study now transverses every aspect of ocean science including aquaculture, fishing, conservation, pollution and geography, and is sponsored by diverse groups—private enterprise, academic research, government, military, as well as by individual curiosity.

One bio-diving pioneer was the late Conrad Limbaugh of Scripps Institution of Oceanography. He was an underwater naturalist who was largely responsible for setting up the scientific diving training program at that institution. Among his publications is an extremely interesting treatment of *Cleaning Symbiosis* in marine animals, which describes how a large number of sea creatures live or benefit by cleaning other organisms. His direct *in-situ* observations showed that the cleaners were not only important ecological instruments for maintaining a healthy population, but that they did business at fixed "cleaning locations."

To demonstrate the importance of symbiosis on local fish populations in a simple experiment, Limbaugh removed the known cleaning organisms from two isolated reefs in the Bahama Islands. Within a few days the number of fish inhabiting the reef lessened; several weeks later the reef was deserted except for a few territorial residents.

The significance of this type of work—the ability to observe such processes in the natural environment and to verify laboratory hypotheses in valid ecological situations—was not immediately accepted by practicing marine biologists. It took an "academic generation," about 6 years, for bio-diving to establish itself as a safe and useful scientific tool. At the close of the 1960s, bio-diving had become the largest single segment of the scientific diving field and was actually a bridge linking marine biology with other disciplines previously considered isolated studies.

Ecological Balance

The relationship between geography and biology may seem rather remote—unless you happen to reside on one of the Pacific's coral atolls. Bio-divers recently made the startling discovery that many of the islands are being threatened by a moving front of a sixteen-legged starfish, the *Acanasther planci*, commonly known as the

"Crown of Thorns." According to Dr. R. H. Chesher, marine biologist and bio-diver at the University of Guam, the starfish have destroyed over 100 square miles of the Great Barrier Reef off the coast of Australia.

Similar reef destruction was observed by Dr. Chesher and other bio-divers around Guam. Destruction of the coral reefs would be an economic and geologic disaster for the inhabitants of these islands. The reefs are the ecological foundation for marine life in these areas, and the protein supply of all Oceania is based on marine resources. The reefs also form an integral part of the islands' structure, and provide natural breakwaters from erosion by wave action and storm damage. Without them, a rapid change of the islands' geology and geography could be expected. On the smaller islands, the loss of protective reefs would perhaps cause their complete disappearance.

The Crown of Thorns is one of the "rare" species. The creature, which measures up to 2 feet in diameter, is found in a variety of colors and is capable of inflicting painful and sometimes toxic punctures on the unwary diver. It feeds on living coral polyps by extending its stomach through its mouth and covering them, then dissolving the reef builders in gastric juices and ingesting them. Left behind is the structure's skeletal shell which is easily eroded by water currents. Normally, the starfish are kept in check by the coral and other local gastropods. When the starfish larvae land on the living coral soon after their birth, they become food for the coral polyps. On a healthy reef, both competitors live in balance.

Unknown causes upset this sensitive balance in the late 1960s and the Westinghouse Ocean Research Laboratory in San Diego sent some forty-five bio-divers and diving technicians through all the U.S. Territory Islands to find out why. With this group supporting him, Dr. Chesher discovered that even advanced stages of local infestation could be brought under control by injecting a solution of formalin into the adult specimens. Using a special underwater syringe, the bio-diver or diver technician can cover a wide area in a short time.

However, using bio-divers to control this onslaught and bring the situation into balance is not the intention of the Westinghouse program. The reason for the unbalance is being sought, as well as a method of effectively controlling not only the symptoms but the cause. Most probable causes so far suggested point to man as the culprit: excessive dredging, pollution, and even atomic testing in the atolls. Other scientists suggest that it is perhaps just another natural population explosion that will right itself. The natural rebuilding of reefs, however, takes time—in the sense of geologic time, not calendar time. Concerned about the survival of their food supply and their land, the islanders of Oceania do not take such an objective view.

Shark Behavior Studies

Sharks have been the most dreaded ocean creatures since antiquity, for several valid reasons. First, they are the most effective predators in the world of nature; their efficiency is indicated by their eonic survival in the same basic form. Secondly, when man is in the water, he sometimes becomes their prey.

How and why the shark is so effective has long been a major question. Marine biologists learned the physiological structure of the shark and other elasmobranchs many years ago, but the recent acquisition of behavioral information by bio-divers sheds light on the habits of these sleek, voracious carnivores. Gathering true behavioral information about such animals must, of course, be done in the same manner as gathering information about the most timid sea creature—in the water with them. To the layman, placing oneself in such a situation is akin to madness. To Dr. Donald R. Nelson, marine biologist at California State College at Long Beach, it is a valid method of obtaining ecological information.

Dr. Nelson has been investigating sharks' responses to underwater noise and their use of specific sounds to locate potential prey. Although most divers disagree on the characteristic behavior of sharks, two points of agreement are always found: sharks respond to vibrations in the water given off by wounded or captured fish and, secondly, shark behavior is unpredictable. Conrad Limbaugh, Eugenia Clark, Jack Randal, Jim Stewart, Hans Hass and Jacques Cousteau have all recorded the sensitivity of sharks to certain sound frequencies and their ability to home in on the source. Nelson observed the response several times in his early skin diving experiences. In one instance, he witnessed a shark homing in on a speared fish under circumstances that would have prevented use of either sight or olfactory sense.

After taking part in a program to test shark hearing at the University of Miami, Nelson conducted a series of *in-situ* observations of the pelagic sharks' response to an acoustic sound source in the deep water of the Tongue of the Ocean, off the Florida coast. Using an acoustic source emitting pulsed sound frequencies of 50 to 200 Hz from a source mounted close to the surface in several thousand feet of water, Nelson began the tests. Silky sharks, *Carcharhinus falciformis*, came from every direction. To keep track of visitors, a number of sharks were tagged with brightly colored plastic streamers. The tags were attached by spearing the animal in a nonvital area. The sharks responded to the sounds a number of times until they *learned* that the noise was not made by a wounded fish. This testing is still underway in both the laboratory and the sea. Once shark behavior is determined, it may be possible to control them, or at least remove the "unpredictable" label from their species.

Flora and Fauna Investigations

Unlike the unpredictable behavior of the shark family, man's behavior towards the nearshore environment has been accurately predicted. However, what effect man's actions will have on the coastal ecosystems is another question. Much speculation is taking place about the perturbations caused by the ever increasing level of man/technology activity in the nearshore. The extent of ecological modifications caused by our "civilized" pollution is difficult to assess. The core of the problem is that determining change or long term effects is impossible unless the normal ecosystems of an area are known in great detail. It is imperative that the base be established on data gathered over a reasonable length of time so that normal cyclic modifications are not attributed to external causes.

Studies of the nearshore environment adjacent to California have been underway for several years

through programs sponsored by both private and public institutions. The information gathered from these areas provides a datum for evaluation of both short term, or explosive, pollution and long term pollution.

The much publicized oil leakage from a drilling accident off Santa Barbara in 1969 is a good example of the need for this data. In this case, an excellent basis for comparison was available. Among the hardest hit areas was Anacapa Island, one of the group of Southern California islands, about 30 miles from the shore.

Because these islands are relatively isolated and thought to reflect typical coastal oceanographic conditions, they were selected for detailed shallow water flora and fauna investigations. Since 1965 studies at Anacapa had been underway to determine distribution of organisms from the lower limits of the intertidal zone down to a depth of 150 feet. When the oil leakage took place, information about the effects of pollution on the flora and fauna communities was readily available for objective evaluation.

The Anacapa studies are typical of those conducted by bio-divers. These detailed surveys are usually made along a permanently installed survey line. Reports include accurate physical descriptions of geological features, measurement of currents, temperature and other pertinent oceanographic parameters, descriptions of species and abundance of flora, and identification of all visible organisms. In some instances, the profile line is extended laterally by marking off quadratures of several meters on the bottom. Within these confines, detailed ecological studies of territorial animals are conducted over a period of years. Figure 9-1 shows a portion of the results from a study at Anacapa Island conducted by Michael Neushul of the University of California at Santa Barbara, William D. Clarke of Westinghouse Research Laboratory and D. W. Brown of the University of California at Santa Barbara. The figure represents a transverse section, or transec, along a survey line with depth ranging from 72 feet to 115 feet.

In addition to the surveys, researchers collected a considerable amount of data concerning the effects of currents on the growth of vegetation, particularly the giant kelp. Using hand-held sonar, divers measured the position of kelp formations under changing current conditions. They found that with strong currents, the plants form overlapping layers which increase the shade on areas of the bottom; with little or no current, the plants float vertically to the surface, once again changing the light pattern on the bottom. The results show that not only were smaller flora growing under the kelp plants affected, but modification of the light causes new kelp plants to move to greater depths where less light is available. This and other natural causes can contribute to vegetation loss of up to 50% in a given area without any man-caused perturbations.

With the increasing awareness of the need to maintain ecological balance, the role of the biodiver will continue to become more important both in the scientific community and the community at large.

The Physio-Diver

The metamorphosis of free diving from a mere curiosity to a scientific method, a commercial tool and an effective weapon system has focused attention on the meagerness of available information concerning man's ability to survive and perform in the ocean environment. As diving interest grows, the problem becomes more acute. The roles of diving physiology and psychology, where most of the serious problems are seated, are now major segments of scientific diving activity.

Hyperbaric physiological research has been underway for generations. Studies of the human machine's responses to the primary and secondary effects of pressure moved in a linear effort until technology created the equipment and the need to push divers beyond the limits of available knowledge. Today, the ever increasing number of physiologists and psychologists working on man-in-the-sea problems are hard pressed to provide answers, or even reasonable hypotheses, in advance of operational activity.

Major contributions to physiological diving research have been provided by the military, prompted by the development of free diving as a sophisticated weapon system. Through Office of Naval Research funding, basic research is now underway in U.S. Navy Laboratories and private research institutions that will eventually enable

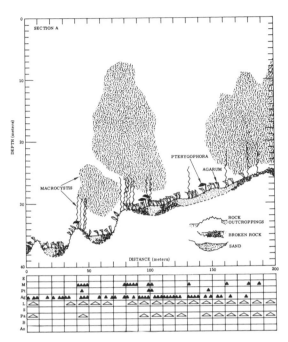

Figure 9-1. Transect off Anacapa Island, California, showing the types of bottom and indicating the distributions of the major plants. The occurrences of the different plants are indicated by symbols which are identified in the figure. The shaded areas above the bottom represent Macrocystis plants that lean over the survey line because of a current flowing from west to east. The table at the bottom of the figure indicates the occurrences per two and one-half meter section of four plants: E, Eisenia arborea; M, Macrocystis pyrifera; Pt, Pterygophora californica; Ag, Agarum Fimbriatum; and the presence or absence per ten meter section of six animals: L, Lytechinus anamesus; S, Strongylocentrotus franciscanus and S. purpuratus; Pa, Patiria miniata; B, Botruanthus benedini; An, Anthopleura xanthogrammica.

man to work safely and effectively anywhere on the continental shelf. The economic press of impending work at great depths, however, has caused a change in the rate of underwater research since 1960. Dissatisfied with the information flow from government research, commercial diving enterprises have accelerated their increasing investment in basic research within the past few years. The motivation is simple: there is work to be done, profit to be made, and the knowledge required to do the job is not available. To develop the technology, basic information is required; to maintain a suitable posture in the marketplace the individual diving companies must acquire the information on a proprietary basis.

The current surge in underwater physiological and psychological research has created a new group of divers within science: the physio-divers. The research being performed by these men subdivides into two major areas—problems concerned with the body functioning under the physical stress of the deeper ocean environment, and problems dealing with the efficiency of the diver as an underwater system.

Most medical research dealing with life-sustaining functions continues to be centered in the laboratories of the submarine medicine institutions of the world. Human tolerance to exotic breathing mixtures, the medical aspects of specific compression and decompression methodology, and the effects of immersion in extremely cold water under elevated pressure must first be determined in the laboratory under safe, controlled conditions. However, laboratory conditions do not provide information about human responses to the environment. Poor visibility, low ambient temperatures, a degree of external threat from the environment and the threat from the inner man present a set of variables that cannot reasonably be simulated in the laboratory. The diving physiologist and psychologist are providing the bridge between experimental medical data and the realities of purposeful man-underwater activity.

Actual sea conditions add a great deal of complexity to the investigator's task. In fact, they create an entirely new set of data-acquisition problems that do not exist in the laboratory test tank. Observing responses to situations and recording the information must often be performed *in situ*. The observation team and the principal investigator must be competent divers, capable of performing in the stress environment. The problem of man-rating underwater engineering so that tasks are within the diver's capability—determining the effects of stress on diver response to abnormal situations and quantitatively evaluating the effectiveness of underwater apparatus as part of the man/equipment system—are just a few of the specific considerations quantified by the physio-diver.

In a recent UDT/SEAL program, Dr. W. S. Vaughan of the Oceanautics Corporation of San Diego, California, was tasked with evaluating crew performance in vehicle depth control, heading control and acoustic homing capability using a new "wet" submersible. Operational requirements demanded performance for 4 to 6 hours and exposure to sea water temperatures of 60°F or less with near-zero visibility. To acquire a realistic feel for the nature of the task involved, Dr. Vaughan attended the U.S. Naval Diving School at San Diego before proceeding to the Naval Amphibious Base at Coronado, California, where a 28-week training and evaluation program for vehicle operators was to take place. Although only part of the data taken was through direct observation, it would have been impossible to develop the instrument recording system without practical experience in both the medium and the vehicle to define the specific problem areas. Testing results produced a variety of data that will contribute to future vehicle design and help define clearly the minimum vehicle instrumentation necessary for operational requirements. The data also quantified the maximum human performance that might be expected under operational conditions.

Data were obtained from determinations of depth-holding accuracy during operations and the effect of experience upon this capability. The sharp increase in expertise after several hours of operational training quickly levels off to a value of ±1.5 feet. After 10 hours of training, the capability improves slightly to ±1.0 foot, with the maximum accuracy for any run being ±0.5 foot. For long runs, divers studied the effects of thermal stress on depth-holding ability caused by exposure to water temperatures of 60°F or less. Operators were instrumented so that their body core temperatures were continuously recorded.

In addition to the basic performance data, a considerable amount of intelligence was acquired relating to human engineering concepts inherent in wet submersible design. Criteria were established for physically coded controls that are operable in reduced visibility with a simultaneous low operator tactile sensitivity, controls operable under zero gravity work conditions, and minimum creature-comfort seating arrangements for extended operations.

Underwater diver performance capabilities critically affect the design of swimmer equipment. The ability of a free swimming diver to carry weighted objects while swimming a prescribed compass course at a specific depth was the subject of a recent study program by Birger G. Andersen under sponsorship of the Office of Naval Research. Andersen's work quantified many answers arrived at by experienced divers in more empirical and subjective learning. During his field experiments, which were all conducted on a special underwater range at the Scripps Institution of Oceanography, only divers with more than 5 years experience were used as subjects.

To obtain data about the subjects' air consumption and depth-keeping abilities without encumbering them with telemetry wires, Andersen continuously sampled and telemetered rate of air consumption and depth over an Aquasonics acoustic link. The underwater test range was a measured compass course, 780 feet long, and a 250-foot error measurement array to determine the diver's course accuracy. Each subject swam the course a number of times in various weight conditions: neutral buoyancy, 3 pounds negative, 6 pounds negative and 9 pounds negative. The negative buoyancy test runs were done in two ways— with the extra weight attached to the diver's belt (near his CG), and with the diver carrying the weight by hand in a canvas field pack. The average compass course error for all subjects was 5.2°, regardless of weighting. Accuracy of all individuals increased with practice—in an almost identical manner to that shown in Vaughan's experiments.

Measurement of diver swimming speed indicated that the weight condition did not vary the diver's normal swimming speed. An average speed of 1.2 knots was maintained on all runs with the fastest at 1.5 knots, the slowest at 0.7 knots. Correlated to depth-keeping capability and air consumption, however, the weight and its location had marked effects. Air consumption increased from a nominal 1.0 cubic foot per minute for the neutral buoyancy runs to over 2.0 cubic feet per minute for weighted runs with the diver hand-carrying the weight. Similar effects were observed in the depth-holding data. The hand-held 9-pound weight caused an average depth deviation of approximately 5.0 feet. Performance worsened as the diver proceeded along the course. If the tests were conducted in deeper water where no bottom was visible for reference, and if the course were longer, further marked degradation of performance probably would have resulted.

The data from the initial phase of this program is finding immediate application in equipment design. Criteria for diver-held equipment now includes negative buoyancy limits in keeping with the observed data. In addition, the course-holding capability of a free swimming diver using dead reckoning and a simple compass is now quantified and should lead to development of a new and more accurate diver navigation system.

Underwater Communication

As divers' tasks become more complex and require deeper work situations, one of the peripheral problems of shallow water work has become a major concern—communication. Speaking and hearing in the water are listed among the primary research needs by almost every major recent underwater-man project. One cause for the paucity of basic information about man's ability to communicate through water is traceable to the assumptions made by the "Panel on Underwater Acoustics," which in 1948 discouraged government-supported research in this area by stating that man could react to underwater sounds only in terms of SONAR systems. Fortunately, contrary opinions developed, and we are beginning to get the basic information necessary for good underwater voice communication.

One of the most active groups working on the physiological and equipment considerations involved is the University of Florida Communication Sciences Laboratory. Under sponsorship of the Physiological Psychology Branch of ONR, the University program is working to answer two questions: (1)What are the capabilities of human speech and hearing systems under elevated pressure in the water? and (2)How well does the available underwater communication equipment work and what can be done to improve the systems?

According to Dr. Harry Hollien, principal investigator on the University of Florida project, inadequate communication among divers has been a major limitation in man's exploitation of the sea.

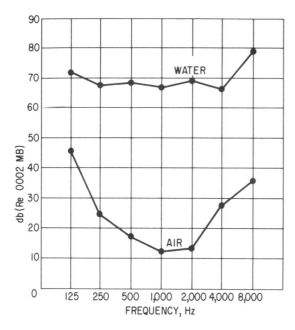

Figure 9-2. Typical of the data developed by physio-divers, this graph shows the difference between human hearing thresholds at various frequencies in air and water. This type of information makes the design of usable underwater communication equipment possible.

Although basic equipment for divers had made impressive gains, voice communication systems still remain relatively undeveloped. A number of underwater speech communication systems have recently become available; however, so little is known about man's basic ability to communicate orally underwater that the design of these systems is of necessity based solely on electronic considerations.

Using the DICORS (Diver Communication Research System) developed at the Communication Sciences Lab, Drs. Hollien and Brandt performed basic investigations into hearing activity and speech reception thresholds (SRT) for the auditory system in water. The results show that losses up to 60 db can be expected in the speech range. Figure 9-2 shows a graph averaged over a number of subjects that illustrates the degradation that occurs when the ear is occluded with water. The results of this portion of the investigation, comparing the air/underwater hearing capability of the human system, reveal that the normal hearing (sensory) processes cannot be used underwater,

and it is necessary to provide some compensatory mechanical substitutes such as bone conduction and amplification.

The problem of speaking underwater, especially while using SCUBA equipment, is also under study at the University of Florida Laboratory. Investigation to determine an ideal configuration for a mouth enclosure that is compatible with life-support requirements should lead to a mathematical model and eventually to a system that will permit intercommunication between swimming divers—an absolute necessity on complicated tasks requiring teamwork.

All underwater communication systems, both commercial and military, are being evaluated for the first time in a comprehensive program that will result in objective, quantitative data. In these tests a wide range of divers, including professional phoneticians and experienced military and scientific divers, read specially prepared word lists underwater. The speech is recorded in the DICORS system and then played to groups of listeners who independently identify the words for scoring. Initial results indicate that the divers' ability to communicate is not a matter of electronics. Rather, it appears more closely related to difficulties imposed by the environment, such as ambient pressure, and to problems like isolation of the nasal cavity by the face mask and back pressure, which are inherent in the life-support equipment.

With projects like these, the physio-diver is making major contributions to diving methodology, technology and equipment, and simultaneously lowering the level of "acceptable risk." The term acceptable risk has always been part of the diving business jargon. Translated into a more objective context it means, "we think we know enough about the physics, physiology and psychology involved to risk someone else's life in the situation—with the statistical probability being in favor of survival." Physio-diving promises to improve the current survival probability by an order of magnitude.

The Geo-Diver

The geology and geography of the ocean bottom have been studied for centuries. Until the 1960s these studies consisted mainly of profiling the bottom with sound. Seismic instrumentation was used for sediment studies, along with remote grab and coring devices to sample bottom composition. This is like making a geological survey of the earth's land surface from a helicopter—with an impenetrable cloud cover always below.

In the very deep portions of the ocean, remote methods are still the only practical ones. However, in the shallower water the situation has changed. Easily transportable and usable diving equipment, together with the scientific curiosity of geologists, has generated a new breed of scientific divers—the diving geologist or, more simply, the geo-diver.

According to Dr. Robert Dill, the scientist/diver is an effective, economical instrument system down to 150 feet. Dr. Dill was a pioneer in establishing SCUBA as a scientific tool, and he has made over 5,000 dives to obtain geological information about the seafloor. He states that the diving geologist is after the same information as his land counterpart. The most reliable geologic data is obtained from rock outcrops. Working on rock strata cropping out on or near the seafloor, Dr. Dill measures stratigraphic sections and structural attitudes and traces key beds and faults.

Unlike the dry geologist, the geo-diver usually works against time measured in minutes. His effectiveness depends on how much reliable data he can obtain and retain within the decompression limitations. For 10 minutes working time on the bottom at 150 feet, the diver must decompress for 3 minutes. If he extends his bottom time to a more realistic working period of 45 minutes, decompression time on ascent jumps to 86 minutes. Even at the moderate depth of 150 feet then, where decompression time can be two or three times the available working time, the economics depend almost entirely on how efficiently the geo-diver can perform his tasks.

Maximum efficiency underwater requires good tools. Where does the geo-diver get them? Until recently he had no choice—he made them. Ten years ago the underwater geologist, through necessity, had to be a combination scientist, diver, inventor and mechanic. His situation is a little better today.

One area that has received the geo-diver's attention is data recording. Most underwater observations were previously recorded by writing the data on a plastic board with grease pencil. In cold water, with limited visibility and light, this was an uncomfortable, tedious and time-consuming task. Today the geo-diver can strap an underwater tape recorder system onto his tank block and voice-record all his observations and measurements. A special diving regulator with a built-in microphone gives enough intelligibility to allow the record of the dive to be transcribed by a laboratory stenographer.

Geo-diving is not restricted to the continental shelf of our oceans. A group of diving scientists at the University of Michigan Great Lakes Research Division is conducting extensive investigations to obtain geological information about northern Lake Michigan. Under the direction of Lee H. Somers, research geologist and diving supervisor, over six hundred dives have been made. The data being gathered includes descriptions of bottom features, sediment studies, measurement of strike and dip of bedrock, and one unique activity: the location and sampling of submerged tree trunks and peat deposits related to lower lake levels.

Last year, a team of GLRD divers recovered a number of submerged tree stumps in 32 feet of water off Beaver Island in Lake Michigan. Radiocarbon dating techniques established that the trees had been above water 6,700 years ago, thus indicating the lake level at that time.

Diving conditions in Lake Michigan are usually bad. Inclement weather most of the year, visibility ranging from a few feet to 6 inches, water temperatures below 50°F at diving levels and high surface currents add to the already difficult data gathering process. To cope with poor diving conditions and add a new margin of safety, the GLRD group incorporated a wireless communication channel to serve as an invisible safety line and data transmission link. Each team of geo-divers was equipped with a wireless transmitter/receiver unit. A third geologist acted as communicator and recorder at the surface station. On site, the divers transmit their data, discuss their observations between themselves and the surface communicator. The dry geologist, being warm and comfortable in his normal environment, is free to analyze and record the divers' remarks, ask pertinent questions and provide information if necessary.

During 1969 a great advance was made in the University of Michigan diving program. Through the cooperation of Dr. Joseph MacInnis, a shallow water habitat manufactured by Technautics Corporation of New York was designed to accommodate three to five divers and to serve as a comfortable underwater station in the uncomfortable water. With the aid of this station, diver-scientists are able to extend their diving time to several hours a day.

The most detailed scientific diving expedition ever undertaken has been the $2.5 million Tektite program. Using saturation techniques, four scientist/divers lived in a two-chamber habitat in 46 feet of water off St. Thomas for 60 days. Dr. H. E. Clifton, a geologist with the U.S. Geological Survey's Marine Laboratory, was the group's geo-diver. During the 434 diving excursion hours the group spent outside the habitat, Clifton completed a field map of the vicinity and conducted experiments to determine the nature and degree to which bottom sediments are reworked by marine organisms.

Other geo-divers associated with the project conducted studies of geological conglomerates. Conglomerates are rounded pebbles and boulders bound together by a clay-like substance. Simultaneous geological data was collected from the surface by other investigators. Correlation of the data indicates that the sea level was at one time lower than the present.

Industrial Geo-Diving

Geological diving activity is by no means restricted to institution and government-sponsored programs. Data acquisition by geo-divers plays an important role in many industrial oceanographic studies. Richard Greenbaum, president of Applied Oceanographics of San Diego, feels that geo-diving is one of his company's most reliable methods of obtaining bottom information in moderate-depth water. Typical of the company's work is their study of natural and man-made obstructions that affect sand transport through the littoral zone. A natural seasonal sand migration exists along most coastal areas that is controlled by wave action. In winter, storms and higher surf conditions cause the sand to move out from the beach. In summer, with lower surf conditions, the sand migrates back into the shore. Figure 9-3 shows a geo-diver tracing the path of bottom transport.

In some areas, natural bottom configurations such as submarine canyons allow sand to slip into deeper water where it is lost forever. Prior to man's interference with the natural ecology of the coastline and its contributory sources, this sand was replaced by a continual supply from the rivers and streams. Today most of this sand source has disappeared through pollution, damming and other civilized processes. The net result is a constant erosion of our beaches and shorelines. According to Mr. Greenbaum, there are substantial deposits of sand in the nearshore areas at depths of 50 to 60 feet, and he has suggested "mining" it to our beaches. This idea resulted from geo-diving in the nearshore environment, and considering the economic value of beach front property, it will probably be applied within the next decade.

The Archaeologist Underwater

The archaeologist is not truly a scientific diver; his presence in the sea is devoid of any association with the underwater world except that diving equipment is necessary for access when his "dig" is covered with a transparent medium.

Like the life and physical sciences, underwater history pursuits enjoyed a great boom in recent years, influenced primarily by the availability of SCUBA. Indirectly, SCUBA has also caused professional archaeologists many sleepness nights because underwater archaeological sites are now equally accessible to the sport diving enthusiast.

Skin diving enjoyed its initial boom in the Mediterranean, partially due to modern SCUBA development in France and partially to the good diving conditions in the sea. The ancient trade routes crisscrossed this area, and the shallow nearshore areas are peppered with the remains of countless sunken ships. With them lie answers to the riddles of history.

"Diving archaeologists" are relatively new underwater. However, recovery of artifacts proving to be keystones in reconstruction of the cultural climates of ancient civilizations dates back to the early twentieth century. Discoveries made by diving archaeologists, as well as by commercial divers engaged in other tasks, resulted in the acquisition of some of the most valuable ancient Greek statuary ever recovered. Between 1900 and 1925 Greek sponge divers, under the auspices of the Greek Ministry of Culture, salvaged a museum full of authentic Greek sculpture dating into the Hellenistic Period. This collection contains the only authentic bronzes to survive from that time. The sea is not a hospitable hostess to metal objects left in her care, but the electrolytic action of sea water seems to have been less detrimental to art than have man's utilitarian activities. At least statues that sank to the bottom were not melted

down for the manufacture of cannon or crucifix, according to the vogue of the day.

A considerable gap exists between the idea of "salvage" and the present concept of underwater archaeology. To the underwater historical detective, recovered artifacts have aesthetic appeal, but their central value is that they help pull together all the pieces of the site and facilitate fitting new information into the context of our knowledge of the period—knowledge that in most cases is spotty and to a great extent postulated. The salvor, on the other hand, is interested in recovering antiquities for their market value, or for the satisfaction of possession.

For those whose acquaintance with underwater antiquities is based on adventure cinema or landlocked artists' conceptions of treasure troves,

Figure 9-3. Sand and sediment transport is a major concern of the geo-diver. This photo shows Dr. Robert Dill of the Naval Undersea Research Center, San Diego, using dye to trace the part of bottom transport. (Photo: U.S. Navy.)

the remains of an ancient wreck would be uninteresting and generally disappointing. Except for the hardest stones and inert minerals, the sea extracts a rental that makes the original forms all but unrecognizable.

In his definitive volume, *History Under the Sea*, Mendel Peterson points out not only the condition of ancient wrecks and how to find them, but what to do with artifacts after they are recovered. According to Peterson, gold is one of the few materials able to survive long submergence in sea water virtually unscathed. Next on the survival list is thick brass or copper. Silver will survive if it is in the presence of a sacrificial anode such as iron or steel, otherwise its fate is questionable. The ferrous metals seem to survive in proportion to the amount of nickel they contain—going down the survival list in order: steel, wrought iron, cast iron. Wood lasts according to its density and its environment. If it is buried in a closely packed bottom that will protect it from marine borers, it will last through milleniums; exposed to the open sea, it survives only in direct proportion to its density.

Generally, there are no direct indications of an ancient wreck underwater. Location must be by indirect methods: a large coral outcropping in the shape of a hull, a trail of ballast stones leading from a shallow reef, or an indication on a ferrous or non-ferrous metal detector can be the only clues to the whereabouts of a ship's remains.

The salvor is interested only in recovery for profit or possession, and once the location is found, the course of action is simple—brute force is used to uncover what is there. However, to the underwater archaeologist, the monetary value of what is recovered is relatively unimportant. His interest lies in more subtle considerations. What are the dimensions and exact shape of the vessel? What is her actual age and where does she fit into the history of that period? What does the hull contain that would shed more light on the culture and capabilities of her people? In short, he is seeking exactly the same information as on a land dig.

Direct application of terrestrial archaeological methods to underwater "digs" is a physical impossibility for the underwater archaeologist. Obtaining the information he requires is a difficult task and the tools at his service are extremely crude. The amount of information he manages to obtain is usually proportional to his inventiveness. Airlifts are used to uncover the skeleton of the hull and a laborious procedure of measuring frame remnants and extrapolating them to their probably original form is a major segment of his work. In several recent projects, a metal pipework frame was constructed over the uncovered site to serve as a photographic grid. The site was then photographed much in the same manner as if it were an aerial sector. For deeper work a small submersible, the *Asherah,* was used as a constant altitude photographic platform and to attempt photogrametry. The results were not completely successful, but they did represent a discrete improvement over hand-drawn sketches.

It is doubtful that the technology of underwater archaeology will enjoy any great advances in the next few years. The reason is simple and common—money. All archaeological exploration must be financed from academic or philanthropic funds, and these funds are usually not at a level commensurate with development and implementation of new methods.

Diving Engineers and Diving Technicians

The diving engineer and the diving technician could actually be included in either the Scientific Diving or the Commercial Diving sections of this book. They serve in both categories with equal importance. The technician and the engineer perform the same functions underwater as they do at the surface. The engineer converts topside engineering developments into usable underwater equipment; the technician performs much of the actual work in scientific and engineering diving.

All successful companies involved in the development of man-rated underwater instrumentation or equipment use the services of diving engineers. The probability of an equipment manufacturing company developing a satisfactory underwater instrument or system without involving an engineer who has had the experience of working underwater is extremely remote.

The primary function of the diving engineer is to design, test and evaluate equipment, systems and structures for underwater use. His areas of interest cover the full spectrum of engineering disciplines, and there are few engineering techniques that do not have some application in the underwater world.

One of the most progressive groups in the underwater engineering is the Ocean Engineering group at the Naval Civil Engineering Laboratory, Port Hueneme, California. Their activities range from testing and evaluating equipment designed by industrial manufacturers to in-house development of systems for deep diving and underwater salvage. During 1969 a typical activity of NCEL diving personnel was the evaluation of closed and open circuit hot water systems to maintain thermal balance in divers in cold water operations.

A warm water open cycle system developed for the U.S. Navy Supervisor of Salvage included boiler, hoses and suits. The system was tested in the ocean at Keyport, Washington, and in the cold water tank at San Diego, California. In the same program NCEL designed and supervised the fabrication of a closed cycle self-contained Isotopic Swimsuit Heater System, which utilized a plutonium 238 heating element. This closed cycle system heated water and pumped it through a wet suit undergarment fitted with closely spaced tubing. A standard wet suit was worn over the undergarment to retain heat supplied from the heat exchanger. Although the concept of a self-contained heating arrangement was promising, the problem of radiation dosage proved to be a limiting factor in this system.

Thermal stress is one of the major problems in deep, mixed gas diving, and NCEL has expended much effort developing diver equipment in this area. In addition to the Open Circuit Hot Water System and the Isotopic Swimsuit Heater System, NCEL diving engineers tested prototype chemical heating vests made of open-cell foam sandwiched between two layers of closed-cell neoprene foam. The cellular layer was filled with a mixture of chemicals that had a melting point slightly below normal skin temperature. When the vest was submerged in cool water, the chemical crystalized and released heat. Initial experiments with the prototype vest and the number of chemical compounds showed that this approach to warming

divers in cold water has considerable promise. Further investigation is now underway to improve these chemical heat sources and thus extend cold water dive periods without the incumberance of backpacks or umbilicals. NCEL tests show that mittens filled with a ½-inch layer of chemical provide two hours of warming in ice water.

During the same year NCEL evaluated several electrical resistance diver heating units and a steam-plus-heat exchanger. This prototype heating system supplied heated sea water to as many as four divers wearing open cycle diving suits. Designed for diver comfort at ambient water temperatures as low as 40°F, the system is usable to 600 feet. It has an oil fire closed circuit steam generator and a heat exchanger through which the sea water flows. The hot sea water is blended with cold sea water and delivered through a hose to a manifold which supplies the individual diver's hoses at the diving site. This is one of the few systems to successfully complete the tests.

Besides diver heating problems, NCEL was considering several other major underwater engineering problems at the same time. One of these was the development of an underwater Construction Assistance Vehicle designed to provide divers with a work platform for handling tools, power sources and construction equipment. Translated into the common vernacular, the Construction Assistance Vehicle is an underwater pickup truck for divers. The submersible is powered by an electro-hydraulic power train and is designed for operation to 120 feet. Its basic function is to transport divers, tools, power supplies and equipment to ocean bottom construction and salvage sites. The vehicle utilizes off-the-shelf hardware such as electro-hydraulic power packages, oil immersed lead acid batteries and hydraulic motor driven propulsion units. It will be used to test and refine the work vehicle concept so that operational criteria and specifications can be written for a vehicle capable of operating to the limit of diver depths. Engineering designs for the experimental vehicle are complete and fabrication is now underway.

Activities at NCEL and at all other government laboratories engaged in scientific or engineering diving require the services of a standard diving locker with trained marine diving technicians to support all functions. The diving staff may consist of either all military or all civilian personnel. The NCEL diving locker, established in 1967, is the only active Navy Construction Battalion (CB) diving locker and the first to be devoted to underwater construction problems. All staff personnel are regular U.S. Navy. The new team, called an Underwater Construction Team (UCT), is the forerunner of standard operational units to be set up in conjunction with all Navy Construction Battalions.

Diving at most other U.S. Navy research laboratories is structured quite differently than at NCEL. For example, at the Naval Undersea Research and Development Center (formerly NEL) in San Diego, the diving locker is staffed entirely by civilian divers. From NURDC headquarters at San Diego, scientific diving activity in Southern California, Alaska and the Hawaiian Islands is supervised.

The number of scientific and diving technician personnel varies, but normally thirty-five to forty divers use the facilities. Of the thirty-seven divers currently listed on the NURDC diving roster, four are qualified to depths of 200 feet, and the remainder to 130 feet. About 80% of the diving performed at NURDC is done with standard SCUBA equipment. The remainder is accomplished with tethered systems, usually lightweight fiberglass helmets.

The nature of the diving performed at these government laboratories varies of course with the particular specialization of the laboratory. But as is often the case in a research laboratory, the majority of the underwater testing and work is performed by people at the technician level rather than scientists and engineers.

10.
Sport
Diving

The beginning of sport diving goes back at least 40 years. It is impossible to pinpoint in time, but sport diving as we know it today is not a recent development. The Bottom Scratchers of San Diego, one of the first sport diving clubs, originated around 1938. Using home-made equipment, the Bottom Scratchers brought spearfishing, or underwater hunting, to the United States from its birthplace in the Mediterranean. As early as 1934, Guy Gilpatrick in his book, *The Complete Goggler*, recognized that sport diving would some day be an important segment of the outdoor world. Since then, sport diving has become a family activity, with exceptionally fine equipment and facilities available throughout the world. Figure 10-1 shows a group of sport divers exploring in the Bahamas.

The exact number of sport diving participants is always a question. Records indicate that over 5 million people have been trained as divers, but a conservative estimate of the number of people who practice the sport regularly is somewhere around 500,000.

No spectators exist in diving—to enjoy it you must be a participant. There are, perhaps, as many levels of participation and activities as there are individual divers. The sport is not competitive, except for spearfishing, and the sport diver is free to pursue his interests at his own pace.

If it is difficult to pinpoint the origins of sport diving, it is even more difficult to determine the motivations of people who take up the sport. Certainly one of the most common motivations is adventure. Although this may seem rather trite in our sophisticated society, it is, in fact, this very sophistication which encourages sport diving and exploration in the medium that has been a perennial subject of fishermen's tales. The underwater environment that offers complete isolation from our pressurized civilization and provides unlimited opportunity for recounting personal adventures to awestruck non-participants is very attractive today. These things attract participation, but they are only indirectly associated with sport diving. Only after the student passes the neophyte stage and spends enough hours underwater to be physically and psychologically acclimated to the environment can he begin to enjoy the benefits of sport diving.

Sport diving is sometimes classified under the general title of skin diving. However, skin diving is only one half of the field. There are actually two main categories within sport diving: skin diving and SCUBA diving. Within these two general categories, individual activity is limited only by the imagination and the diver's interests.

Skin Diving

Skin diving is the basic skill in sport diving. Specifically, skin diving means free diving: leaving the surface of the water, diving to a specific depth and performing some function with a minimum amount of equipment. Performance levels in skin diving range from amateurish "snorkeling" to competitive free diving and spearfishing.

Figure 10-1. The exact number of people who regularly enjoy sport diving is unknown—the best estimate is about 500,000. This photo shows a group of sport divers exploring a deep reef in the Bahama Islands under the watchful eye of an instructor. (Photo: IUES, Dave Woodward.)

Skin diving is the only phase of sport diving that lends itself to competition. It is the original form of sport diving and is practiced today by the most sincere advocates of the sport. SCUBA diving, on the other hand, requires less effort on the part of the participant, and is the most popular form of sport diving today.

Each year thousands of people are introduced to the underwater world through snorkeling, which is usually only the first step in becoming addicted to skin and SCUBA diving. The only equipment required is a face mask and snorkel, and the person can paddle along the surface getting a view of the shallow, nearshore underwater environment. Snorkeling is a pleasant activity for those who are not willing to put forth the effort and discipline necessary to master the breath-holding requirements of free diving. The rewards, of course, are in proportion to the effort.

Underwater Hunting

The single largest skin diving activity is underwater hunting, or spearfishing. Equipped with a face plate, a pair of fins, snorkel and an underwater weapon, such as a long, thin spear or a spear gun, the spearfisherman meets a fish on its own terms. He must ventilate his lungs at the surface, take a deep breath, descend to the bottom, stalk the fish, impale it on his spear and bring it to the surface. Since the work involved in actually capturing fish by skin diving precludes overfishing, depletion of fishing areas by spearfishermen represents no threat at all.

Spearfishing can be performed as an individual food gathering process or through clubs and competitive meets. Underwater hunting contests have grown into a worldwide competition and international meets are held each year.

Free diving spearfishermen fit our twentieth century concept of sportsmen considerably better than other hunters and the degree of skill they can acquire through practice is remarkable. An expert spear fisherman can leave the surface, dive to 80 or 90 feet, spear a fish his own size and bring it to the surface without assistance. Records for the size of catch are changing every year, but it is not uncommon for an expert to spear and land fish that weigh more than 400 pounds.

Skin divers have caused ecological perturbations underwater that have resulted in serious consequences or endangered species. Most of these instances have been connected with the taking of "shells." Building a beautiful personal collection of shells is certainly a harmless hobby, but when profit enters the picture, it is no longer a sport. In some areas where no regulations exist, certain species of mollusk have been so heavily collected for their valuable shells that an imbalance has developed.

Typical of this profiteering is the hunting of Pacific triton, whose shell brings $35 on the market. In a 10-year period, divers have removed nearly 100,000 of these from the vicinity of the Great Barrier Reef of Australia. Their absence is considered one of the primary causes of the population explosion of the coral-eating Acanasther planci, the "Crown of Thorns" starfish that is threatening the Pacific Atolls. Regulations prohibiting collection of these animals will probably be placed in effect during 1972.

Even though free divers normally collect expensive shells the activity should actually be considered under commercial diving. Any aspect of diving that involves performance for money is of course immediately out of the sport classification and in the realm of amateur-professional diving.

SCUBA Diving

SCUBA diving is the most popular form of sport diving. The use of air SCUBA equipment in underwater sports requires training and mature judgment, since the greatest danger to the sport diver is himself. Fortunately, few places exist where a person can rent or purchase diving equipment without showing evidence of having had minimum training in its use. In all probability this rule will continue to be stringently enforced. Popular diving programs on television have underrated the potential hazards of diving. In reality, diving is safer than these films indicate, but it does require a specific level of proficiency for survival, and basic safety procedures must be observed.

Perhaps the most popular underwater activity performed with SCUBA equipment is sight-seeing—at least, this is the initial phase undertaken by most new divers, and one that even the most professional divers never tire of. As the diver gains experience, however, his powers of observation and underwater awareness increase, and with many divers there is a desire to let everyone see it. Recording underwater beauty and phenomena has become a popular pastime and, in some cases, a profession among the sport diving community. Courses in underwater photography are taught throughout the world, and within the past five years standardized 35 mm cameras well within the price range of nearly all SCUBA divers have come on the market. Each year, underwater film festivals are held in several locations throughout the world. Both professional and amateur underwater photographers are given the opportunity to show their creativeness.

Underwater photography has been an important tool in communicating the beauty of diving and popularizing the sport. It has also provided a vehicle for many enthusiastic sport divers to obtain semi-professional status. In fact, underwater photography presents one of the few opportunities left for a man to turn his hobby into a satisfying avocation. Once he becomes professional, of course, the diver is no longer a sport diver but is classified as a diving technician. The term "technician" may be construed as an affront to such a creative and aesthetic pursuit as photography, but the underwater photographs of technical equipment and construction details are the ones that sell—not beautifully back-lighted photographs of gorgonians.

Diving for Food

Food gathering is an aspect of sport diving that has assumed wide importance. The skin diver looks upon the SCUBA diver who fishes with breathing apparatus, much as you or I would regard a hunter stalking game in a zoological garden. But this seems to have had little effect on the amount of spearfishing divers perform while wearing apparatus. Hunting shell fish and crustaceans such as lobster, and cephalopods such as octopus and abalone is a popular pastime that serves to augment the gourmet tastes of those with

Figure 10-2. Every sport diver has an inborn desire to be a salvor. A find such as this old anchor shank is enough to excite even the most experienced diver. (Photo: IUES, Dave Woodward.)

beer pocketbooks. Taking this type of food is normally limited by state regulations, and on the whole the sport diving population has exhibited a sportsmanlike attitude in keeping within these regulations. The limits set for a single diving expedition are usually generous enough to permit the taking all the fish protein that can be consumed by a normal family in one week. However, the ability of a diver to find this much and take this much is questionable.

Salvage Diving

Every sport diver seems to have an inborn desire to be a salvor. He will pick something up from the bottom, struggle bringing it into the shoreline, and savor his find as though it were a jewel-encrusted crucifix from a fifteenth century galleon. Without a doubt, he would walk by the same object laying in the street. Figure 10-2 shows a diver who has spotted an old anchor shank.

Sighting the remains of a vessel on the bottom is enough to excite even the most dour personality. It makes no difference whether the wreck has been down for one day, one year, or one century.

Wreck diving has become a major pastime within the sport diving community and there are, in fact, clubs that specialize in nothing else. In many cases, "wreck divers" work in conjunction with local archaeological and anthropological experts to recover portions of submerged history. In other cases, sport divers turn salvaging into apparently profitable operations. These operations are generally given wide publicity, but they are oddities. Even more odd is the amateur salvage operation that turns out to be profitable.

The cost of salvage runs into numbers that are no longer sport, but business, although the probability of breaking even by recovering the treasure is remote. True, many sport divers manage to turn amateur salvaging into profit, but the work and the level of remuneration is most often mundane. The greatest return on amateur salvage ventures is in the recovery of objects lost in recent years, such as anchors, chains, motors, et cetera.

All in all, amateur salvage is an interesting pastime, which in extraordinary cases can lead to a find that produces some monetary gain. In most cases, however, the diver must be satisfied with the pleasure of the chase.

Diving for Enjoyment

Sport diving can be enjoyed at all economic levels. For those who live in areas where water conditions are poor and do not have the means to get to good water, the enjoyment is considerably less—but, nevertheless, there is enjoyment. Affluent members of our society can pack their gear and fly to the Caribbean or Cozumel or out through the Pacific Archipelagos, where the underwater environment becomes more varied and more beautiful.

Compared to other leisure activities, basic sport diving is not particularly expensive. The original investment for all necessary gear is approximately $225—there are no green fees and no entrance fees. If the participant wishes, he has no other expenses except maintaining his equipment and purchasing compressed air for his tanks. However, this is not the way the majority of active divers operate.

Airlines now feature special skin diving transportation schemes, and luxurious resort hotels and facilities catering specifically to divers are spotted around all the exotic diving areas of the world. Ships of some lines feature special cruises into the tropical waters, and diving is generally accepted throughout the world as a reputable sport practiced by a large group of people who can afford the very finest. For the traveling person, sport diving provides a common meeting ground anywhere in the world.

Skin diving clubs date back to the earliest practice of the sport. At that time people with mutual interests simply grouped together to share experiences and technology. In the early days,

Figure 10-3. Spearfisherman Wes Andrew shows one of his "small" 150-pound catches taken near San Diego, California. He has taken several black seabass that weighed in near 500 pounds.

almost all skin diving equipment had to be handmade, and many of the larger equipment manufacturers in business today are headed by people who began making equipment as a hobby in the late 1930s and 1940s.

Diving clubs are widespread today. With a club, the diver of moderate circumstances can pay a small initiation and maintenance fee, which usually entitles him to the use of a diving boat, air at reduced prices, trips to places he could not afford to go by himself and a built-in supply of qualified diving partners. These diving clubs vary from small groups of ten or twelve interested people who meet alternately at different homes and discuss their plans, to clubs having memberships of hundreds of people and facilities that match the most luxurious country clubs.

If, however, there was ever a skin diver's paradise, it is the Underwater Explorers Club at Freeport, Grand Bahama. Opened in 1966, the club has facilities for every area of interest in the underwater world. It features a complete photographic laboratory, marine biology laboratory and pro shop, as well as a recreation area, a library, a bar, a museum, a swimming pool, a 17-foot deep training tank and a fleet of fast boats that operate to the nearby reefs. The club was started by Al Tillman, a founder of the Los Angeles County SCUBA Diver Training Program, and Dave Woodward. It is adjacent to a new modern hotel, the Oceanus, and has introduced more people to sport diving than any other similar facility. One of the most attractive things about the Underwater Explorers Club is that the annual membership dues are only around $10 per year. For this fee and the cost of the particular service, the member can use all the facilities. The club's operators have combed the nearby reefs and are able to provide any type of diving the member wants and is qualified to do.

Qualifications are an important aspect of the Underwater Explorers Club. Unlike other organizations that carry "explorer" in their title, the Underwater Explorers Club is not restricted to those who have accomplished some great underwater feat—it exists to open up the underwater world of the Bahamas to divers from all over the world. The club has an excellent history of safety and pleasure, and all operating personnel are qualified SCUBA instructors. It is possible to fly to Freeport and become a qualified sport diver within a couple of weeks, with 20 to 30 hours of actual sport diving experience in the best waters of the world.

During the 1970s sport diving will continue to expand as one of the major leisure aquatic pastimes. With the standardization of instruction and the self-enforcing of safety restrictions, the sport will be one of the safest and most pleasurable releases from the emotional pressures of our civilization.

11. Diver Training

For each area of diving activity, there is a discrete form of diver training. Until recently, military diver training was the main source of personnel for diving activities. The explosive growth of commercial, scientific and sport diving requirements has created such a vacuum, however, that the military can no longer fill the gap. Like diving technology, diver training went through a metamorphosis in the 1960s. As late as 1958, for example, there was no acceptable national standard syllabus for sport diving. Many training courses to qualify sport divers had no entrance requirements, and in many cases the entire course consisted of as little as 12 hours, with no training in the ocean! These inadequacies have for the most part been rectified, but diver training per se is still changing rapidly.

Sport Diver Training

The most active and widely publicized form of diver training is in sport diving. Several million people have undergone this training, although the active participants number considerably fewer.

Control of sport diver training standards in the United States falls under the influence of two organizations: the National Young Mens Christian Association and the National Association of Underwater Instructors. These organizations have established minimum standards of course content, student performance and instructor qualification. Most qualified instructors carry both certifications.

The average sport diver training course consists of 36 hours of lecture and pool work, including three ocean training dives. At the conclusion of this course, the neophyte diver should be capable of diving without endangering himself or his diving partner. He should be capable of intelligently evaluating the safety of diving situations; he should understand diving physiology well enough to conduct himself safely underwater; he should be familiar enough with the ocean to understand potential hazards; he should be sufficiently adept at handling his equipment so that it presents no problem to him. He must be trained to be aware of his own and his partner's physical limitations. The degree to which the individual student reaches this level varies with his intelligence, his physical and psychological adaptiveness, his desire to dive and, most importantly, how demanding the instructors are.

Under the curriculums established by either the YMCA or NAUI, the certified sport diver is qualified to dive in the ocean to 60 feet. If he follows his training, including keeping himself in reasonably good physical condition, he can expect years of interesting, safe enjoyment.

Scientific Diver Training

Scientific diver training differs from sport diver training in several ways. First, scientific diver training is usually more directly controlled. The trainees are members of an academic staff or employees of a laboratory. Of course, when the training organization exerts more direct control over the individual, it also has more direct responsibility for the diver's performance after training. Secondly, the scientific diver needs to dive deeper than the sport diver and his training must be more rigorous.

Scientific diving has little to do with science. The trainee is normally a scientist or technician within a specific scientific discipline. In some training courses, techniques directly related to a scientific discipline are included, but generally the course is related only to equipping the trained scientist to operate safely and competently underwater.

At the Scripps Institution of Oceanography, the basic course consists of 100 hours of training. Considerable emphasis is placed on physical conditioning, and the performance levels for both entrance and graduation are considerably higher than for sport diving training courses. The course at Scripps is only available on a "need to dive" basis and is not open to the general public. Subsequent to graduation as a qualified diver, a Scripps diver must periodically requalify to demonstrate maintenance of his skill. A new diver is restricted to a specific depth and qualification for deeper diving is based on experience and further testing by the Diving Officer.

Most major academic institutions involved in oceanography or ocean engineering programs have diver training courses. Again, these programs are normally restricted to those in the institution who have a genuine need to learn diving techniques. U.S. Navy laboratories require that many scientists pursue their activities in the underwater environment. The Naval Undersea Research and Development Center at San Diego has the largest group of civilian scientific divers at a Navy laboratory in the United States. In addition to diving activity, the lab also has specific responsibility for training. Their program, like all other Navy laboratory programs, is at approximately the same level as the second class U.S. Navy SCUBA diver. It is, however, aimed at equipping the diver with more self-responsibility, since most scientific diving is not conducted in the presence of the supervisory personnel involved in military operations.

Military Diver Training

The U.S. Navy Diver Training Program is administered on a service-wide basis and is graded into varying levels of performance capability and training. (See Figure 11-1.) The primary levels are SCUBA Diver, Second Class Diver, First Class Diver, Aquanaut and Master Diver. Only enlisted personnel in the U.S. Navy can qualify in the Master Diver capacity.

The SCUBA certification program lasts four weeks and is designed to bring the diver up to a level equivalent to the basic "100 hour" training course used in Scientific Diver training. Second Class Diver schools use a 10-week program and are designed to make the diver competent in shallow water diving gear and basic heavy gear, to a maximum depth of 150 feet. The First Class Diver course is a 14- to 16-week course that includes deep air diving techniques to 225 feet, and mixed gas hard hat diving to 320 feet.

In addition to the actual diving taught in these schools, the student is also trained in explosives, salvage techniques, and repair and maintenance of diving gear. The Aquanaut rating, a relatively recent addition to Navy classifications, is an advanced First Class Diver program specifically for training divers for saturation diving and the use of mixed gas SCUBA equipment.

The U.S. Navy Second Class Diver training program and the SCUBA training program form the basis for nearly all scientific, sport and standard commercial diver programs. The standard training course for both classifications is given in Appendix V.

Figure 11-1. Military diver training includes a rigorous "toughening-up" schedule. This UDT candidate is going through a "diving" exercise during hell week. (Photo: U.S. Navy.)

Training for Commercial Diving

The greatest change in training techniques has taken place in commercial diver training in the past two years. Commercial diving has always used an apprentice/journeyman training program. (See Figure 11-2.) Commercial diving schools teaching basic diving techniques have been operating since the 1920s. Increasing demands for trained personnel with well rounded capabilities have also led to the creation of diver training programs at a vocational level in many junior colleges throughout the United States.

There are six major commercial diver training schools in the United States. The level of training offered and the material covered varies with the individual school.

One of the oldest training centers is the Coastal School of Deep Sea Diving in Oakland, California. Their basic course is a 12-week, 427-hour program. Besides the basic curriculum in air diving, additional courses in mixed gas diving and submersible operation are offered. The basic course at Coastal consists of:

1. Introduction—History and Physics of Diving
2. Psychology and Physiology of Diving
3. Diving Accidents—Prevention and Treatment
4. First Aid—Safety and Precaution.
5. Diving Equipment—Air Supply and Upkeep
6. Hookah Diving—Classroom and Practice
7. Shallow Water Diving—Classroom; Classroom and Practical
8. Practice Dives—Dry Runs
9. Diving Equipment Maintenance
10. Diving and Salvage Tools—Upkeep and Repair
11. Seamanship and Rigging
12. Diving from Floats
13. Diving Problems and Mathematics
14. Advanced Training Lectures or Practical Demonstrations:
    a. Demolition and use of explosives
    b. Use of photography for salvage
    c. General underwater photography
    d. Velocity tools and pneumatic tools
    e. Underwater television lectures
    f. Commercial construction methods
    g. Diving bell and robot familiarization
15. General Salvage—Ships and Aircraft
16. Above Water—Cutting and Welding Instructions

Figure 11.2 (above). A commercial diver is only as good as his journeyman skill. New trends in training equip them to work underwater. Specialized training, such as welding, starts on the surface. Figure 11-3 (opposite page). The new breed of marine technician is being trained in basic oceanography to better understand the new working environment. Training includes both classroom and practical shipboard experience. (Photos: Santa Barbara City College.)

17. Underwater Cutting and Welding Instructions
18. Qualifying Dive—200 Feet—Preparations and Dive
19. Supervision of Diving Operations—Plan and Execute Dive
20. Small Boat—Handling and Moorings
21. Job Inspection Report
22. Supervise Study and Examination
23. Final Examination

Tuition for basic courses varies with the school, and prices as of January, 1970, range from $895 to $1,100. In many cases basic commercial diver training courses are available to veterans and through other federal and state training and rehabilitation programs. The length of courses for mixed gas diving depends on the school. Normally training periods range from 35 to 60 hours, and tuition averages about $1,250 for this specialized segment.

Immediate placement is available for nearly all graduates of commercial diving schools, but the level of their entrance into the diving industry is of course dependent upon their general aptitude and capability for underwater work. Use of diving equipment and the ability to get to a work site are important functions of the commercial diver, but his value to the diving contractor is his ability to perform specific work. Some training courses give specialized welding and burning training, but the new diver must still undergo on-the-job training and begin his career as a tender.

The complexity of emerging underwater technology is placing great demands on the educational and background levels of training. Within the 1970s both the title and job classification of *Diver* will probably disappear in offshore activity. They will be replaced by a new classification: the Marine Diving Technician. These divers will be graduates of a two-year college vocational program and will hold an Associate Degree of Science in Marine Technology.

The marine diving technician (Figure 11-3) will be widely based in all activities associated with underwater construction; he will have a high level of capability in underwater construction and offshore resource recovery technology. Generally he will be the type of new employee a commercial company can fit into any job at the operating level with minimum training. Many city colleges throughout the country have initiated these programs and are receiving encouragement and aid from the diving industry, as well as from state and federal governments. In all probability, the leading commercial diving schools will join with junior colleges to provide the practical aspects of diving equipment training, while the college will provide the peripheral skills necessary for a comprehensive underwater background.

Representative of the Marine Diving Technician programs initiated in the last 2 years is the curriculum at Santa Barbara City College. The program consists of three types of courses. The first is designed to give the student a basic understanding of the physical environment in which he will be working. The second develops basic skills required of the diving technician, and the third consists of general education courses designed to increase the students' knowledge and communicative ability.

For admission the candidate must be a graduate of an accredited high school, have a satisfactory academic record, or possess a state equivalency certificate. He must satisfactorily complete the school and college aptitude tests and pass a comprehensive physical examination. In addition to the physical and academic entry requirements, he must demonstrate a reasonable level of proficiency in aquatic activity.

The satisfactory level of aquatic ability is demonstrated by performance of the following tests:

1. Swim underwater without fins for a distance of 75 feet without surfacing
2. Swim underwater without fins for a distance of 150 feet, surfacing not more than four times
3. Swim 1,000 feet in less than 10 minutes without fins
4. Demonstrate swimming with snorkel and fins and without face mask
5. Skin dive to a depth of 10 feet and recover a 10-pound object
6. Demonstrate the ability to rescue a struggling swimmer and carry him 75 feet on the surface

Candidates for the marine technology program are admitted to the college each September. During the 2-year course of instruction, the students are encouraged and assisted in obtaining part-time employment with operating diving firms. Credit for this work is given towards the Certificate of Completion. During the 2 years of lectures, labs and diving, the Santa Barbara student receives the following instruction.

First Year

Marine Technology 1. Introduction to marine technology. An introductory course designed to acquaint the student with the technical skills required in marine technology, and survey of job opportunities and career potentials in the various areas of marine technology. The course is designed to familiarize the student with basic rigging techniques and procedures as well as navigation, piloting and small boat handling.

Marine Technology 2. Basic Diving (SCUBA and Hookah), Introduction to diving. This course is designed to parallel recognized 100-hour basic diving courses. Satisfactory completion results in a 1-year certificate in diving, supervised by the Santa Barbara City College Diving Control Board. Certification may be extended beyond this 1-year period provided the individual maintains his student status and fulfills the requirements for maintaining his certification.

Marine Technology 3. Advanced Diving. An introduction to the use and care of heavy gear, mixed gas equipment, decompression chambers, bells and submersibles. The physics and physiology of mixed gas and the treatment of diving accidents and diseases are included in this course. (All diving training takes place in water not over 25 feet deep.)

Marine Technology 4. Fundamentals of Marine Engines and Compressors. A course designed to familiarize the student with the principles of operation and maintenance of marine engines and diving compressors. Emphasis is placed on understanding function and theory, proper operation, trouble shooting and replacement of components.

Biology 2. Biological Oceanography. A study of the relationship between marine plants, animals and the physical characteristics of the world's oceans. The course emphasizes the forms of marine organism identification and ecological interrelationships.

Physics 11 and 12. Technical Physics. A 2-semester course in technical physics, with stress on the experimental approach. The methods of keeping laboratory records, analysis and interpretation of data, and the preparation of reports.

Drawing and Blueprint Reading. A practical course in the principles of basic drawing and blueprint reading, with emphasis on the marine environment.

Welding 3. The principles and practical applications of oxyacetylene and electric arc welding. The course includes familiarization with all types of equipment, tools and materials, safety practices and specific welding techniques.

Second Year

Marine Technology 5. Underwater Construction. A practical application of underwater construction techniques. Stress is placed on the student to develop skill in using tools and performing underwater work that is commonly encountered by construction divers.

Marine Technology 6. Underwater Operations. A course designed to familiarize the students with submarine wellheads, pipelines, cables and general marine construction techniques. Emphasis is placed on the diver's role in these projects, and the application of the skills and knowledge acquired in preceding courses.

Biology 5. Marine Biology. An introductory course in the study of marine plants and animals with emphasis on the ecological adaptation of organisms to their habitats in the littoral zone. Lectures and labs consist of field work during low tide periods.

Economics 2. Marine Law and Economics. Fundamentals of marine law, international law important to marine operations, and the economic aspect of sea operations.

English 18. Technical Report Writing. A step-by-step approach to technical writing. The course is designed for engineers or technicians who need to communicate facts and ideas to others.

Electronic Technician 10. Fundamentals of Electronics. An introductory course designed to familiarize the students with electrical principles, components and new technology. The laboratory consists of formal experiments that parallel the lectures.

Electronic Technician 11. A study of the principles and equipment pertaining to many areas of electronics and communication. Basic theory and application of amplifiers, oscillators, microwave power supplies, transmitters and receivers, and test instruments. Laboratory consists of formal experiments that parallel the lectures.

Speech 5. Business Speech. An introduction to speech, stressing participation in group discussion, parliamentary procedure, extemporaneous speaking and conference speaking, including selling techniques, interviewing, et cetera.

Machine Shop 11. An introduction to machine shop practice and operations including setups and operational techniques. The course is designed to familiarize the student with all types of tools and instruments that are basic to the standard machine shop.

History 5. Growth of American Civilization. A survey of the leading social, economic, political and diplomatic traditions that have shaped American civilization from the colonial period to the present.

Speech 7. Fundamentals of Speech. Instruction in the essentials of effective oral communication, stressing speech organization and idea content. Students analyze classical speeches and create original speeches for class presentation.

Biology 98. Special projects in biology. An opportunity for personal study among those students interested in pursuing scientific diving careers.

Photography 1. Introduction to Photography. Fundamentals of the camera, techniques of camera operation, characteristics of film, characteristics of photocopy and dark room chemistry.

Photography 2. Fundamentals of the dark room process, categories of developers, safety in the dark room, chemistry of film processing, drying and storage techniques.

The immediate success of the program and the acceptance of graduates into responsible jobs within the diving community indicate that this type of training will set the pattern for all future professional diving instruction. The major commercial diving companies are cooperating fully by providing both technical guidance and employment opportunities for students during and at the conclusion of training.

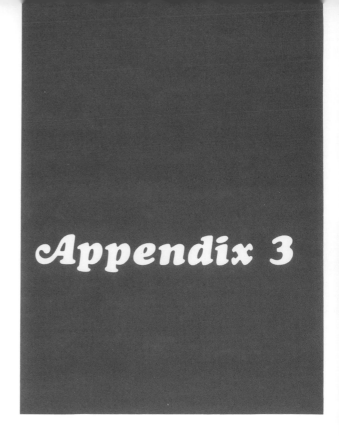

Appendix 3

COMMERCIAL DIVING COMPANIES

United States & Vicinity

Acme Diving Service
3142 Locust Ave.
Norfolk, Va.

Action Divers
3310 Belle Chasse Hwy. S.
Belle Chasse, La.

Adams West Rigging Company
24119 S. Avalon
Wilmington (Los Angeles), Calif.

A-1 Professional Divers
River Road
Kenner, La.

Atlantic Steel & Salvage Corp.
3795 N. W. S. River Dr.
Miami, Florida

Aquatic Divers Inc.
3596 Dalbergia
San Diego, Calif.

Aquatic Services
209 E. 6th
Vancouver, B.C.

Ashenbrenner Diving Service
12538 Ventura Blvd.
Los Angeles, Calif.

Associated Divers Inc.
154 Norman Firestone
Santa Barbara, Calif.

Associated Marine Divers Inc.
P.O. Box 23363
New Orleans, La.

Beachcomber Underwater Salvage & Repair
1815 Clement Ave.
San Francisco, Calif.

B.C. Marine Shipbuilders Ltd.
Foot of Victoria
Vancouver, B.C.

Benson's Diving Service
1265 N.W. 97th
Portland, Oregon

Blue Water Industries Inc.
403 N. Shoreline
Corpus Christi, Texas

Bongiovanni, Sam J.
6 Sherling Ave.
Metaire, La.

Burrard Dry Dock Company
122 W. 42nd
Vancouver, B.C.

Byrd Commercial Diving
3345 N.W.S. River Dr.
Miami, Florida

Burda & Burda & Associates
11106 S. Michigan
Chicago, Illinois

Chapman Derrick
Raymond International, Inc.
Bourse Blvd.
Philadelphia, Pa.

Childers Diving Corporation
201 W. Water
Los Angeles, Calif.

Chisholm Divers
700 Henry Ford Ave.
Long Beach, Calif.

Commercial Diving Company
7814 Arcadia Rd.
Pensacola, Florida

Commercial Divers Inc.
407 S. W. 11th St.
Portland, Oregon

Commercial Divers Inc.
1629 K N.W.
Washington, D.C.

Copeland's Marine Divers
1303 Ayers
Corpus Christi, Texas

Coastal Inspection & Diving Company
3112 Texas
Bridge City (Houston), Texas

Coastal School of Deep Sea Diving
350 29th Ave.
Oakland, Calif.

Coughlin Diving Service
104 Margaret Ave.
Baltimore, Md .

Crandall Dry Dock Eng. Inc.
238 Main
Boston, Mass.

Cribbin Diving Service
433 Catalina Ave.
Norfolk, Va.

Croftin & Morris Divers Inc.
248 Huntsman Road
Norfolk, Va.

Daspit Bros Marine Divers Inc.
2401 Delille
Chalmette (New Orleans), La.

Dean Marine Divers Inc.
309 Cherry Blossom Lane
Gretna, La.

Deep Sea Divers Inc.
1104 Hancock
Gretna, La.

Deeta Marine Divers
104 E. Hillsboro Ave.
Tampa, Florida

Devine Diving & Salvage Company
3405 N.E. 82nd
Portland, Oregon

Diving Unlimited
8716 La Mesa
La Masa, Calif.

Donohugh Boat Service
Bert 117
San Pedro, Calif.

Duane Marine Corporation
Staten Island, N.Y.

Evans Professional Divers
601 Bark Rd.
New Orleans, La.

Explosive Services Inc.
3000 General DeGaulle Dr.
New Orleans, La.

Frolich Bros. Marine Divers Inc.
313 W. Girod
Chalmette (New Orleans), La.

Gage's Diving Center
Key Largo, Florida

G & W Divers, Inc.
Gemini Drive
Houma, La.

Garrison & Divers & Marine Contractors
4310½ E. Madison
Seattle, Washington

Glen Walker Inc.
142 Jement Dr.
Somerdale, N.J.

Global Divers & Contractors
201 Weswego
New Orleans, La.

Great Lakes Dredge & Dock Company
228 N. LaSalle St.
Chicago, Illinois

Great Lakes Sub-Marine Divers
6110 W. 26
Cicero (Chicago), Illinois

Gulf Coast Commercial Divers
6 Carroll Rd.
Mobile, Alabama

Hanson Diving Salvage & Mooring Service
201 Descanso
Los Angeles, Calif.

Harbor Construction Company, Inc.
398 Wickenden
Providence, R.I.

Harbor Diving Service
Englishton Rd.
Old Bridge, N.J.

Haulover Marine Service
699 N.E. 164 St.
Miami, Florida

Hayden Engineering Company
4249 91st St.
Seattle, Washington

Hitner Inc.
2905 Guilford
Philadelphia, Pa.

Hockett Leiter Marine Contractors
3415 N.W. 66th
Seattle, Washington

Hurlen Marine Construction Company, Inc.
410 S. Fontanelle
Seattle, Washington

Hydrotech Service Inc.
2229 Hickory
Harahan, La.

Independent Divers Inc.
River Rd. S.
Kenner, La.

Industrial Diving Company
2050 W. 1st
Vancouver, B.C.

Industrial Marine Divers Company
2050 W. 1st
Vancouver, B.C.

International Divers
136 Aero Camino
Santa Barbara, Calif.

International Underwater Contractors
33 25 127th St.
Flushing, N.Y.
16262 E. Whittier Blvd.
Los Angeles, Calif.

I U C Gulf Inc.
2245 11th Street
Harvey, La.

Jacken Contractors Ltd.
Foot of 20th St.
Vancouver, B.C.

Jacksonville Commercial Divers
455 Lenox Sq.
Jacksonville, Florida

Jacobson Brothers Inc.
2431 N.W. Market
Seattle, Washington

J & J Marine Diving Company Inc.
P.O. Box 4117
Pasadena, Texas

Jarrel Inc.
44 Boardman Rd.
Charleston, S.C.

Logan Diving & Salvage
530 Goodwin
Jacksonville, Florida

Maiatich & Sons
19 & Sylvan
Burlington, N.J.

Mainland Divers
Pitt Bridge
Vancouver, B.C.

Marine Diving & Salvage Company
8348 Lenore
Houston, Texas

Marine Exploration Company
2995 N.W. S. River Dr.
Miami, Florida

Marine Services
147 Jefferson
Chula Vista, Calif.

Maritime Explorations Ltd.
19th 8 Atlantic Ave.
Norfolk, Va.

McCall Diving Service Inc.
5801 Academy
Corpus Christi, Texas

Merrihue, H. J. , Marine Diving
& Salvage
River Road
Kenner, La.

Murphy Pacific Marine Salvage Company
44 Whitehall St.
New York, N.Y.

New England Divers Inc.
44 Water
Beverly, Md.

Newport Diving & Marine Contracting Ltd.
725 Seymour
Vancouver, B.C.

N Y Submarine Contracting Company, Inc.
179 Ellis St.
New York, N.Y.

Norris Inc. Divers
352 Bar Harbor Rd.
Baltimore, Md.

Northwest Diving Ltd.
1365 Coleman
Vancouver, B.C.

Noyes Brothers Divers
1900 S.E. 15th St.
Ft. Lauderdale, Florida

Ocean County Diving Ltd.
1407 E. Washington St.
Toms River, N.J.

Oceaneering International
Sterns Wharf, Santa Barbara, Calif.
866 Cardova, Vancover, B.C.
624 Front, Morgan City, La.

Ocean Systems Inc.
801 Ave E, Rivera Beach, Florida
IMC Bldg., Houston, Texas
Isaac Newton Sq., Reston, Va.
17 Marine Center Bldg., Santa Barbara, Calif.
Hwy. 90, E. Morgan City, La.

Pacific Marine Engineering
1412 Chester Ave.
Arcata, Calif.

Packer Diving & Ocean Engineering
Hwy 90
W. Morgan City, La.

Parker, T.R. Diving Service
2229 Warmouth
San Pedro, Calif.

Pelican Marine Divers
2228 N. Von Braun
Harvey, La.

Peterson Diving Service
7456 Main
Chicago, Illinois

Podesta Divers & Construction Inc.
Pier 3½
San Francisco, Calif.

Raymound International Inc.
2 Pennsylvania Plaza
New York, N.Y.

Rev-Lyn Contracting Company
350 Broadway
Lynn, Mass.

Rike Diving Company
771 47th
Brooklyn, N.Y.

Roy Yorks Diving Service
Walden Rd.
Jacksonville, Florida

Salmon's Dredging Corporation
Cherry Hill Lane
Charleston, S.C.

Sanford Marine Services Inc.
River Rd.
Berwick, La.

Sausalito Underwater Search
Bridgeway & Johnson
Sausaslito, Calif.

Seaboard Marine Divers & Consultants Ltd.
7942 Winston
Vancouver, B.C.

Skipper's Diving Company
408 E. Wright
Pensacola, Florida

Southcoast Divers
206 Terminal Island
Los Angeles, Calif.

Southwest Diving Company
12994 Trail Hollow
Houston, Texas

Stroud Diving & Hydrography Inc.
3542 Morin
Jacksonville, Florida

Sub Marine Construction
8932 Ozark
Chicago, Illinois

Sub Sea International
1600 Canal
New Orleans, La.

Suncoast Divers Inc.
3433 Gandy Blvd.
Tampa, Florida

Taylor Diving & Salvage Company Inc.
Engineers Rd.
Belle Chasse, La.

Tiedemann & Company Inc.
74 Trinity Place
New York, N.Y.

Tooker & Sons Inc.
3333 Richmond
Staten Island, N.Y.

Triton Engineering & Constuction Ltd.
866 Cordova
Vancouver, B.C.

Undersea Systems Inc.
110 W. Main Bay Shore
New York, N.Y.

Underwater Advisors Inc.
1133 Broadway
New York, N.Y.

Underwater Techniques Inc.
721 St. Davids Ave.
Warminster, Pa.

Underwater Limited Inc.
2945 N.E. 2nd Ave.
Miami, Florida

Universal Divers Inc.
Berth 234
Terminal Island (Los Angeles), Calif.

Universal Diving Service
524 W. Montecito
Santa Barbara, Calif.

Vancouver Divers & Contractors Ltd.
122 W. 42nd
Vancouver, B.C.

Ventures International Inc.
1641 Redwood Dr.
Harvey, La.

Videospection Engineering Ltd.
1661 W. 8th
Vancouver, B.C.

Vogt & Son Marine Divers
1930 Karl
Arabi (New Orleans), La.

Walton Company, Inc.
201 Marginal St.
Boston, Mass.

West Coast Diver's Service
2533 Government
Victoria, Vancouver, B.C.

West End Diving & Salvage Company
4714 Bridgeton
Chicago, Illinois

Yule Marine Contractors Ltd.
Foot of Carroll
Vancouver, B.C.

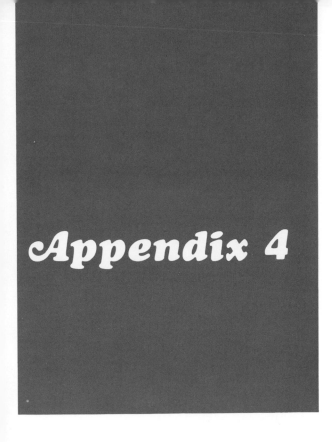

Appendix 4

SCIENTIFIC DIVING PROGRAMS

Universities & Colleges

Cape Fear Technical Institute
Wilmington, N.C.

The Catholic University of America
Washington, D.C.

The City University of New York
New York, N.Y.

Columbia University
New York, N.Y.

Cornell University
Ithaca, N.Y.

Duke University
Durham, N.C.

Florida Atlantic University
Boca Raton, Florida

Florida Institute of Technology
Melbourne, Florida

Florida State University
Tallahassee, Florida

Fullerton Junior College
Fullerton, Calif.

Harvard University
Cambridge, Mass.

Highline College
Midway, Wash.

The Johns Hopkins University
Baltimore, Md.

Humboldt State College
Arcata, Calif.

Lehigh University
Bethlehem, Pa.

Long Island University
C.W. Post College
Greenvale, N.Y.

Long Island University
Southampton College
Southampton, N.Y.

Massachusetts Institute of Technology
Cambridge, Mass.

Naval Postgraduate School
Monterey, Calif.

New York University
Bronx, N.Y.

North Carolina State University
Raleigh, N.C.

Northeastern University
Boston, Mass.

Nova University
Fort Lauderdale, Florida

Oceanographic Institute of Technology
Long Beach, Calif.

Oregon Institute of Marine Biology
Eugene, Oregon

Oregon State University
Corvallis, Oregon

Pomona College
Claremont, Calif.

San Diego State College
San Diego, Calif.

San Jose State College
San Jose, Calif.

Santa Barbara City College
Santa Barbara, Calif.

Southeastern Massachusetts
Technological Institute
North Dartmouth, Mass.

Stanford University
Pacific Grove, Calif.

Syracuse University
Syracuse, N.Y.

Texas A&M University
College Station, Texas

University of Alaska
College, Alaska

University of Arizona
Tucson, Ariz.

University of California
Berkeley, Calif.

University of California
Davis, Calif.

University of California
Santa Barbara, Calif.

University of California
Scripps Institution of Oceanography
La Jolla, Calif.

The University of Chicago
Chicago, Illinois

University of Connecticut
Storrs, Conn.

University of Delaware
Newark, Del.

University of Florida
Gainesville, Florida

University of Georgia
Athens, Ga.

University of Hawaii
Honolulu, Hawaii

University of Illinois
Urbana, Illinois

University of Maine
Orono, Maine

University of Massachusetts
Amherst, Mass.

University of Miami
Miami, Florida

University of Michigan
Ann Arbor, Mich.

University of North Carolina
Chapel Hill, N. C.

University of New Hampshire
Burham, N.H.

Appendix IV—Scientific Diving Programs   233

University of Pacific
Stockton, Calif.

University of Rhode Island
Kingston, R.I.

University of Southern California
Los Angeles, Calif.

University of Southern Mississippi
Hattiesburg, Miss.

University of South Florida
Tampa, Florida

University of Texas
Austin, Texas

University of Washington
Seattle, Wash.

University of West Florida
Pensacola, Florida

University of Wisconsin
Madison, Wis.

Virginia Institute of Marine Science
Gloucester Point, Va.

Walla Walla College
College Place, Wash.

Yale University
New Haven, Conn.

Naval Laboratories

Experimental Diving Unit
Washington Navy Yard, Washington, D.C.

Naval Civil Engineering Laboratory
Port Hueneme, Calif.

Naval Medical Research Institute
Bethesda, Md.

Naval Ordinance Laboratory
White Oaks, Md.

Naval Research Laboratory
Washington, D.C.

Naval Ship Research & Development Center
(Formerly U.S.N. Mine Defense Laboratory)
Panama City, Florida

Naval Undersea Research & Development Center
San Diego, Calif.

Submarine Medical Research Center
Submarine Base, New London, Conn.

Government Agencies

NOAA
Office of Environmental Systems
Rockville, Md. 20852

Smithsonian Institution
Office of Oceanography & Limnology
Museum of Natural History
Washington, D.C.

Sandy Hook Marine Laboratory
P.O. Box 428
Highlands, N.J.

Tiburon Marine Laboratory
P.O. Box 98
Teburon, Calif.

Bureau of Commercial Fisheries

Biological Laboratory
P.O. Box 155
Auke Bay, Alaska

Biological Laboratory
Milford, Conn.

Biological Laboratory
75 33rd
St. Petersburg Beach, Florida

Biological Laboratory
Federal Building, Gloucester St.
Brunswick, Ga.

Biological Laboratory
Sabine Island, Gulf Breeze, Florida

Biological Laboratory
P.O. Box 3830
Honolulu, Hawaii

Biological Laboratory
West Boothbay Harbor, Maine

Biological Laboratory
Oxford, Md.

Biological Laboratory
P.O. Box 6
Woods Hole, Mass.

Biological Laboratory
1451 Green Rd.
Ann Arbor, Mich.

Biological Laboratory
Building 302
Fort Crockett, Galveston, Texas

Biological Laboratory (Marine Mammals)
Sand Point Naval Air Station
Seattle, Wash.

Biological Laboratory
2725 Montlake Blvd.
Seattle, Wash.

Exploratory Fishing & Gear Research Base
P.O. Box 2481
Juneau, Alaska

Exploratory Fishing & Gear Research Base
State Fish Pier
Gloucester, Mass.

Exploratory Fishing & Gear Research Base
239 Frederick St.
Pascagoula, Miss.

Exploratory Fishing & Gear Research Base
2725 Montlake Blvd.
Seattle, Wash.

Fishing Oceanography Center
P.O. Box 271
La Jolla, Calif.

Ocean Research Laboratory
South Rotunda Museum Building
Stanford, Calif.

Tropical Atlantic Biological Laboratory
74 Virginia Beach Dr.
Miami, Florida

Industrial Organizations

A.C. Defense Research Laboratory
Goleta, Calif.

Applied Oceanographics Inc.
San Diego, Calif.

Battelle Memorial Laboratory
Columbus, Ohio

Biotechnology Inc.
Falls Church, Va.

Dunlap and Assoc.
Darien, Conn.

Dillingham Corporation
San Diego, Calif.

General Oceanographics
San Diego, Calif.

Interstate Electronics Corp.
Anaheim, Calif.

Lockheed Ocean Laboratory
San Diego, Calif.

Marine Advisors, Inc.
La Jolla, Calif.

North American Rockwell Inc.
El Segundo, Calif.

Oceanautics Inc.
San Diego, Calif.

Westinghouse Ocean Research Laboratory
San Diego, Calif.

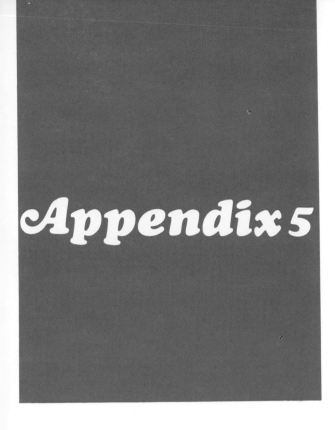

Appendix 5

Unit I—Indoctrination

U.S. Navy Diver Training—2nd Class and Scuba

Topic	2nd Class Periods			SCUBA Periods		
	T	P	T	T	P	T
1. Registration and Introduction	2	0	2	2	0	2
2. Physical Examination and Oxygen Tolerance Tests	0	4	4	0	4	4
	2	4	6	2	4	6

Unit II—Physics and Medical Aspects of Diving

Topic	2nd Class Periods			SCUBA Periods		
1. Physics of Diving	3	0	3	3	0	3
2. Anatomy and Physiology Related to Diving	2	0	2	2	0	2
3. Primary and Secondary Effects of Pressure on the Human Body	2	0	2	2	0	2
4. Divers' Diseases and Injuries	2	0	2	2	0	2
5. Use of the Treatment Tables	1	3	4	1	3	4
6. Air Decompression Tables and Decompression Procedures	4	4	8	4	4	8
7. Environmental Hazards of Diving	1	0	1	1	0	1
8. The Recompression Chamber and Student Chamber Runs	1	5	6	1	5	6
9. Artificial Respiration and Resuscitation	1	1	2	1	1	2
	17	13	30	17	13	30

Unit III—Swimming

		T	P	T	T	P	T
1.	Swimmer Screening	1	1	2	1	1	2
2.	Pool Training	0	6	6	0	6	6
3.	Swimming in Open Water	1	8	9	1	8	9
4.	Skin Diving (without SCUBA)	0	2	2	0	2	2
		2	17	19	2	17	19

Unit IV—Buoyant Ascent Training

		T	P	T	T	P	T
1.	Buoyant Ascent Procedures	1	2	3	1	2	3
2.	Buoyant Ascent - 18 and 36 feet	0	5	5	0	5	5
		1	7	8	1	7	8

Unit V—Underwater Swimming with Open Circuit SCUBA

		T	P	T	T	P	T
1.	Types of SCUBA	1	0	1	1	0	1
2.	SCUBA Maintenance	2	12	14	2	12	14
3.	Charging SCUBA	1	7	8	1	7	8
4.	Open Circuit SCUBA Pool Training	2	7	9	2	7	9
5.	Procedures for Open Water Swimming with Open Circuit SCUBA	1	0	1	1	0	1
6.	Open Water Swimming with Open Circuit SCUBA	1	18	19	1	18	19
7.	Open Circuit SCUBA Intermediate Dive	0	6	6	0	6	6
8.	Open Circuit SCUBA Qualification Dive	0	6	6	0	6	6
		8	56	64	8	56	64

Unit VI—Underwater Searches and Work with Open Circuit SCUBA

		T	P	T	T	P	T
1.	Underwater Search Procedures	1	11	12	1	11	12
2.	Underwater Work with SCUBA	1	6	7	1	6	7
		2	17	19	2	17	19

Unit VII—Related Subjects for Diving

		T	P	T	T	P	T
1.	Tides and Currents	1	0	1	1	0	1
2.	Compass	1	1	2	1	1	2
3.	Rubber Suit Familiarization and Maintenance	1	0	1	1	0	1
4.	Diving Requalifications, Logs, Records, and Reports	1	1	2	1	1	2
		4	2	6	4	2	6

Unit VIII—Use of Tools Underwater

		T	P	T	T	P	T
1.	Tools Used Underwater	1	0	1	1	0	1
2.	Practical Application in the Use of Tools Underwater	0	7	7	0	7	7
		1	7	8	1	7	8

Unit IX—Deep Sea Diving Orientation

	T	P	T
1. Deep Sea Diving Outfit; Function and Nomenclature	6	0	6
2. Deep Sea Diving Techniques and Procedures	2	6	8
3. Familiarization Diving with Standard Deep Sea Diving Equipment	0	26	26
	8	32	40

Unit X—Diving with Deep Sea Diving Equipment

1. Application of Diving Techniques and Procedures in Open Water	0	8	8
2. Underwater Work in Standard Deep Sea Diving Equipment	0	40	40
	0	48	48

Unit XI—Diving with Lightweight Diving Equipment

1. Lightweight Diving Outfits; Functions and Nomenclature	2	0	2
2. Lightweight Diving Techniques and Procedures	2	10	12
3. Underwater Work Using Lightweight Diving Equipment	0	24	24
	4	34	38

Unit XII—Helmet and Dress Repair

1. Diving Dress Repair	0	3	3
2. Belt and Shoe Repair	0	1	1
3. Maintenance of Lifeline and Airhose	0	1	1
4. Air Helmet Repairs	0	2	2
5. Lightweight Diving Mask Repairs	0	1	1
	0	8	8

Unit XIII—Marlinspike Seamanship

1. Marlinspike Seamanship Fundamentals	2	0	2
2. Practical Application of Marlinspike Seamanship	0	6	6
	2	6	8

Unit XIV—Underwater Cutting

1. Introduction to Underwater Cutting	1	0	1
2. Oxygen-electric Cutting Techniques	2	0	2
3. Practical Application of the Oxygen-electric Method of Cutting Underwater	0	15	15
	3	15	18

Unit XV—Review and Examinations

1. Review and Examinations   1   7   8

Total theory periods	54	37
Total practical periods	266	123
Total periods for the course	320	160
Total hours for the course	266	133
Total weeks for the course	8	4

Unit I—Indoctrination

Topic 1: Registration and Introduction

A. Objectives
1. To register the student and provide him with the required materials for the course.
2. To inform the student of the mission and regulations of the school.
3. To develop proper student appreciation and attitudes towards the training offered by the school and to stimulate his pride in becoming a Navy diver.
B. Outline of Instruction
1. Complete registration forms.
2. Issue manuals and materials.
    a. Discuss availability of reference books.
    b. Reading assignments accomplished before class presentation.
3. Mission, purpose, and regulations of school.
    a. Mission.
    b. School regulations.
    c. Other pertinent regulations.
        (1) Uniform.
        (2) Liberty.
        (3) Restricted areas.
    d. Divers special pay.
    e. Scholastic requirements.
4. Student-instructor relationship.
5. Need for divers in the Navy.
    a. Salvage work.
    b. Ship repair.
    c. Submarine rescue.
    d. SCUBA divers.
    e. Research.

Topic 2: Physical Examination and Oxygen Tolerance Test

A. Objectives
1. To determine the ability of the student to meet the physical requirements for the course of instruction.
2. To determine the student's susceptibility to oxygen poisoning.
B. Outline of Instruction
1. Preliminary physical examination by hospitalmen.
2. Completion of medical examination by diving medical officer.
3. Recompression chamber run.
    a. Take chamber down to equivalent of 112 feet.
    b. Return to 60 feet.
    c. Administer pure oxygen for a period of 30 minutes at 60 feet.
4. Evaluate student's ability to withstand pressure.

Unit II—Physics and Medical Aspects of Diving

Topic 1: Physics of Diving

A. Objectives
1. To provide the student with a thorough knowledge of the physics of air and water pressure applicable to diving.
B. Outline of Instruction
1. Normal air.
    a. Definition.
    b. Composition.
    c. Properties.
    d. Characteristics.
        (1) Weight.
        (2) Occupies space.
        (3) Compressibility
        (4) Other characteristics.
    e. Laws affecting the gases which compose air.
        (1) Boyle's law.
            (a) Definition.
            (b) Example and application.
        (2) Charles' Law.
            (a) Definition.

      (b) Example and application.
- (3) Dalton's Law.
  - (a) Definition.
  - (b) Example and application.
- (4) Henry's Law.
  - (a) Definition.
  - (b) Example and application.

2. Water.
   a. Composition.
   b. Characteristics.
   - (1) Weight.
   - (2) Occupies space.
   - (3) Compressibility.
   - (4) Taste.
   - (5) Color.

3. Terminology and values used in pressure (partial, atmospheric, gauge, and absolute).
   a. Absolute pressure.
   - (1) Definition.
   - (2) Example and application.
   b. Other pressures related to absolute.
   - (1) Definition.
   - (2) Example and application.

4. Buoyancy in water.
   a. Archimedes' Principle.
   b. Example and application.

5. Definitions.
   a. Buoyancy.
   b. Density.
   c. Area.
   d. Volume.

6. Summary.
   a. Composition of air.
   b. Characteristics of air and water.
   c. Laws governing gases.
   d. Pressure, absolute and relative; how to derive.
   e. Computation of pressure at various atmospheres.
   f. Buoyancy in water.
   g. Effect of pressure on gas absorption.

Topic 2: Anatomy and Physiology Related to Diving

A. Objectives
   1. To inform the student of the effects of pressure and changes of pressure on the human body.
   2. To provide the student with a knowledge of the anatomy and physiology of the circulatory and respiratory system of the human body for a better understanding of what happens when pressure is applied or removed.

B. Outline of Instruction
   1. Anatomy and physiology—the study of various organs and parts of the living body, their functions and activities.

a. Anatomy of the circulatory system.
   - (1) Heart.
   - (2) Arteries.
   - (3) Veins.
   - (4) Capillaries.
b. Physiology of the circulatory system.
c. Anatomy of the respiratory system.
   - (1) Lungs.
   - (2) Trachea.
   - (3) Bronchi.
   - (4) Nose and throat.
d. Physiology of the respiratory system.
   - (1) Law of diffusion of gases and liquids (Henry).
   - (2) Law of partial pressure (Dalton).
e. Body cavities containing air.
   - (1) Lungs.
   - (2) Ears.
   - (3) Sinuses.
   - (4) Intestines.

Topic 3: Primary and Secondary Effects of Pressure on the Human Body

A. Objectives
   1. To provide the student with an understanding of the primary and secondary pressure phenomena.

B. Outline of Instruction
   1. Primary pressure phenomena (the mechanical effects of pressure).
      a. Effects of pressure applied equally to the body.
      - (1) Blood pressure.
      - (2) Skull.
      b. Effects of pressure applied unequally to the body.
      - (1) Lungs (squeeze and traumatic air embolism).
      - (2) Ears.
      - (3) Sinuses.
      - (4) Face.
      - (5) Teeth.
      - (6) Intestines.
      - (7) Spontaneous pneumothorax.
   2. Secondary pressure phenomena (the disturbances in gas equilibrium, ie, the solution and dissolution of gases in the body tissues).
      a. Narcotic effect of oxygen.
      - (1) Theories
      b. Toxic effect of oxygen.
      c. Toxic effect of carbon dioxide.
      d. Nitrogen absorption and elimination.
      e. Effect of pressure in excess of one (1) atmosphere on body tissues.
      f. Principles involving prevention of caisson disease.

Topic 4: Divers' Diseases and Injuries

A. Objectives
1. To provide the student with a knowledge of the symptoms and causes of the various diseases and injuries that occur in diving and underwater swimming.
2. To provide the student with a knowledge of the prevention and treatment of these diseases and injuries.
B. Outline of Instruction
1. Anoxia.
2. Carbon dioxide excess.
3. Squeeze.
4. Compressed air illness (bends, caisson disease).
5. Air embolism.
6. Nitrogen narcosis.
7. Oxygen toxicity.
8. Spontaneous pneumothorax (pneumo-air thorax-chest).

Topic 5: Use of the Treatment Tables

A. Objectives
1. To teach the student the use of treatment tables.
2. To impress upon the student the importance of selection of proper treatment tables.
3. To provide practical experience in the selection and use of treatment tables.
B. Outline of Instruction
1. Treatment Table 1.
2. Treatment Table 1A.
3. Treatment Table II.
4. Treatment Table IIA.
5. Treatment Table III.
6. Treatment Table IV.
7. Recurrences.
   a. During treatment.
   b. After treatment.
8. Review case histories. Select cases which illustrate both proper and improper selection and use of treatment tables.
9. Practical application of treatment tables through use of hypothetical cases with varying symptoms.

Topic 6: Air Decompression Tables and Decompression Procedures

A. Objectives
1. To teach the student the various methods of decompression.
2. To provide practice in the practical application of decompression tables.
B. Outline of Instruction
1. History of decompression.

2. Decompression.
   a. Definition.
   b. Methods.
      (1) Stage.
      (2) Regular.
      (3) Surface.
3. Navy Standard Decompression Tables (Using compressed air).
   a. Repetitive groups.
   b. Surface interval credit table.
   c. Repetitive dive time table.
   d. No decompression dives.
   e. Exceptional exposures.
4. Practical application of decompression tables by use of theoretical dives.

Topic 7: Environmental Hazards of Diving

A. Objectives
1. To provide the student with a knowledge of the various environmental hazards encountered in diving.
2. To acquaint the student with the best methods of combating these hazards.
B. Outline of Instruction
1. Drowning.
2. Exposure to climate (surface).
3. Exposure to cold water.
4. Surf, tide, currents.
5. Contaminated water diving.
6. Underwater blast.
7. Nuclear radiation.
8. Marine life.
   a. Vegetable.
   b. Animal.

Topic 8: The Recompression Chamber and Student Chamber Run

A. Objectives
1. To provide the student with a knowledge of the characteristics of the recompression chamber.
2. To provide for practical operation of the recompression chamber by the students.
B. Outline of Instruction.
1. Types of recompression chambers.
   a. Two (2) lock chamber.
   b. One (1) lock chamber.
2. Air supply for chambers.
   a. Capacity.
   b. Pressures and depths.
   c. Ventilation.
   d. Supply and exhaust valves.
   e. Gauges.
   f. Diving amplifiers.

3. Precautions in use of the chamber.
   a. Lighting.
   b. Door.
      (1) Sealing and releasing.
   c. Explosive fires.
      (1) Causes.
      (2) Prevention.
   d. Maintenance of chamber.
      (1) Fireproof materials.
      (2) Heaters and fans.
      (3) Oxygen manifolds.
   e. Ventilation.
4. Practical use of recompression chambers.
   a. Practice maintaining steady rate of ascent (60 ft. per minute).
   b. Use of man lock, treatment lock, and medical lock.
   c. Simulated treatment using oxygen.
   d. Observe safety precautions.

Topic 9: Artificial Respiration and Resuscitation

A. Objectives
   1. To provide the student with a basic understanding of artificial respiration, resuscitation and other first aid measures.
   2. To review the student's understanding of artificial respiration and resuscitation.
B. Outline of Instruction.
   1. Shock.
   2. Artificial respiration.
      a. Drowning.
      b. CO_2 poisoning.
      c. CO poisoning.
      d. Mouth to mouth method.
   3. Use of mechanical resuscitation (student participation).
   4. Hemorrhage (bleeding).
   5. Fracture.
   6. Burns.
   7. Wounds.

Unit III—Swimming

Topic 1: Swimmer Screening

A. Objectives
   1. To demonstrate proper underwater recovery strokes.
   2. To screen the students on their ability to perform underwater recovery strokes.
   3. To acquaint the students with the proper method of wearing a lifejacket for SCUBA swimming.

B. Outline of Instruction
   1. Swimming without fins.
      a. Side stroke.
      b. Breast stroke.
      c. Pool side drill.
      d. Swimming test.
   2. Lifejackets.
      a. Types.
      b. Wearing.
      c. Inflating.

Topic 2: Pool Training

A. Objectives
   1. To improve the student's swimming techniques.
   2. To teach the student use of swimming equipment.
   3. To develop the student's confidence in his ability and his equipment.
   4. To increase the student's endurance.
B. Outline of Instruction
   1. Pool training without fins.
      a. Side stroke.
      b. Breast stroke.
   2. Pool training with fins.
      a. Flutter kicks.
      b. Time swim.
      c. Proper water entry.
      d. Clearing mask.
      e. Surface diving.
      f. Equalizing pressure on ears.
      g. Develop distance swimming.

Topic 3: Swimming in Open Water

A. Objectives
   1. To acclimate the student to open water swimming.
   2. To increase the student's swimming ability.
   3. To increase the student's confidence.
   4. To qualify the student in a 1,000 yard (without fins) swim for record purposes.
B. Outline of Instruction
   1. Swimming techniques in open water.
      a. Strokes.
      b. Increase distance capability.
         (1) 500 yards (without fins)
         (2) 500 yards (with fins)
         (3) 750 yards (without fins)
         (4) 750 yards (with fins)
         (5) 1,000 yards (without fins, qualification swim)
         (6) 1,000 yards (with fins)
   2. Safety precautions, open water swimming.
      a. Swim in buddy pairs.
      b. Ensure lifejacket is in operating condition.

    c.   Establish trouble pick up signals between swimmers and boat.

    d.   Watch for sea pests.

  3.   Night swimming in open water.

    a.   Need for night swimming

    b.   Review safety precautions and procedures.

    c.   Issue and instruct students on the use of flares.

    d.   Marine life at night.

    e.   Visibility restriction.

Topic 4: Skin Diving (without SCUBA)

A. Objectives.

  1.   To develop the student's ability to skin dive without SCUBA.

  2.   To prepare the student for open water buoyant ascent.

  3.   To develop the student's confidence in the water.

  4.   To develop the student's ability to equalize in the water.

B. Outline of Instruction

  1.   Caution students about forcing their equalization.

  2.   Students enter water.

  3.   Dive under boat.

  4.   Have timed breath holding contest (do not exceed 60 seconds).

  5.   Students "bring up bottom" at 12-15 feet.

  6.   Boat moves to deeper water.

  7.   Students "bring up bottom" at 18-25 feet.

  8.   Utilize buddy system.

Unit IV—Buoyant Ascent Training

Topic 1: Buoyant Ascent Procedures

A. Objectives

  1.   To provide the student with a knowledge of buoyant ascent procedures.

  2.   To demonstrate proper free ascent (old method) procedures without lifejacket.

  3.   To demonstrate the proper method of buoyant ascent.

  4.   To provide practical experience in buoyant ascent procedures at a shallow depth.

B. Outline of Instruction

  1.   Review job analysis sheet.

  2.   Demonstrate successful and unsuccessful free ascent.

  3.   Demonstrate buoyant ascent.

  4.   Students make buoyant ascent.

Topic 2: Buoyant Ascent—18 and 36 foot

A. Objectives

  1.   To provide the student with practical experience in buoyant ascent methods.

  2.   To increase the student's confidence in his ability to safely ascent from a depth in the event of equipment failure.

B. Outline of Instruction

  1.   Thoroughly rebrief students on procedures.

  2.   Follow job analysis sheets for 18 and 36 foot buoyant ascents.

Unit V—Underwater Swimming with Open Circuit SCUBA

Topic 1: Types of SCUBA

A. Objectives

  1.   To inform the student of the characteristics of the types of SCUBA used by the Navy.

  2.   To provide the students with an understanding of the basic principles of operation of the different types of SCUBA.

B. Outline of Instruction

  1.   Open-circuit SCUBA.

    a.   Procedures of operating and operational characteristics.

      (1) Breathing compressed air.

      (2) Exhalation given off to the sea.

      (3) Unattached by air hose or lifeline.

      (4) Consists of:

        (a) One or more compressed air cylinders (2000 or 3000 psi).

        (b) Demand regulator.

        (c) Suitable mouthpiece or mask to breathe through.

    b.   Principles of operation.

      (1) H-p cylinder (one or more).

      (2) Suitable manifold if more than one cylinder.

        (a) On and off valve which applies air to the demand regulator and also allows recharging.

      (3) A reserve valve on the manifold or regulator valve.

      (4) Demand regulator reduces air to a usable divers pressure.

      (5) Demand regulator to operate with minimum effort; regulator operates on partial vacuum.

      (6) Employment.

        (a) Quick inspection.

        (b) Minor underwater damage control.

        (c) Search.

    (7) Limitations.
        (a) Not to exceed 130 feet except in emergencies.

2.   Closed-circuit SCUBA.
    a.   Definition and characteristics.
        (1) Closed-circuit (rebreather) uses breathing oxygen; never industrial type oxygen.
        (2) No exhalation is given off to the sea.
        (3) CO_2 absorbent baralyme or soda-lime to absorb CO_2 exhaled by operator.
        (4) Breathing bag acts as a low pressure volume tank furnishing gas to the diver at ambient pressure.
        (5) Mounted on vest or strap arrangement so as to keep swimmer on even keel.
        (6) Valves of various types are used for supply.
        (7) Other valves.
            (a) Over pressure valve (pop-off) may be located in one of four places (bag, canister, valve assembly, or mask).
            (b) Surface breather used on several different units.
            (c) On and off valve (respiratory valve)—to close breathing bag off the atmospheric air while not in use.
            (d) Water drain valve (spit valve)—to drain any water that might be in mask.
    b.   Principles of operation.
        (1) Oxygen supply: 1 or 2 small cylinders (small, as compared to open circuit cylinders).
        (2) More efficient than other types, uses 100% of gas carried by swimmer.
        (3) Breathing bag.
        (4) Canister.
        (5) Face mask breathing types.
        (6) Operating procedure.
        (7) Breathing cycle.
        (8) Types of closed circuit.
            (a) Draeger.
            (b) U.S. Navy
            (c) CDBA.
            (d) Pirelli.
        (9) Two classes.
            (a) Circulatory.
                1. Draeger
                2. U.S. Navy
            (b) Pendulum.
                1. CDBA
                2. Pirelli

      (10) Limitations.
        (a) 25 ft. for 75 minutes.
        (b) Oxygen poisoning is caused by breathing oxygen under pressure. It results in convulsions.
      (11) Employment.
        (a) Sneak attack.
        (b) Allows for communications underwater.

3.   Semiclosed circuit SCUBA.
    a.   Definition and characteristics.
        (1) Semiclosed circuit SCUBA—uses mixed gas supply, helium/oxygen or nitrogen/oxygen.
        (2) Partial rebreathing occurs—a certain amount of gas is intentionally discharged.
        (3) More efficient than open, less than closed circuit.
        (4) CO_2 absorbent (baralyme or soda-lime) to absorb CO_2 exhaled by operator.
        (5) Mounting on vest or strap arrangement so as to keep swimmer balanced.
        (6) Basic components.
            (a) Breathing bag.
            (b) Carbon dioxide absorbent canister.
            (c) Mouthpiece, mask, or both.
            (d) Breathing tubes.
            (e) Check valves for the breathing circuit.
            (f) Gas supply.
            (g) An emergency reserve gas supply.
        (7) Two special components.
            (a) A reliable automatic inject system for mixed gas.
            (b) An adjustable exhaust valve for the excess breathing medium.
        (8) The injector is the heart of the semiclosed system.
            (a) Consists of a pressure regulator and a metering orifice with special filter.
    b.   Principles of operation.
        (1) Mixed gas supply—2 small cylinders (small, as compared to open-circuit cylinders). MK V has two sets of bottle sizes, CDBA has one set.
        (2) Breathing bag.
        (3) Canister.
        (4) Face mask breathing tubes—MK VI uses two breathing tubes and mouth-

piece. CDBA has mask and single breathing tube.
(5) Breathing cycle.
(6) Types of semiclosed circuits.
    (a) CDBA.
    (b) U.S. Navy, MK VI.
(7) Two classes.
    (a) Circulatory.
        1. MK VI
    (b) Pendulum.
        1. CDBA.
(8) Limitations.
    (a) Duration of gas supply governed by depth, which in turn determines the proper gas mixture and flow setting.
    (b) Do not exceed the maximum depth specified for the mixture in use.
    (c) Be constantly alert for failure of the exhaust valve to bubble.
(9) Uses.
    (a) UDT and EOD operations.

Topic 2: SCUBA Maintenance

A. Objectives
1. To provide the student with a knowledge of the maintenance and repair procedures for open-circuit SCUBA.
2. To provide practical experience in maintaining and repairing open-circuit SCUBA.
B. Outline of Instruction
1. Mouthpiece open-circuit SCUBA.
    a. Disassemble and inspect regulator.
    b. Repair and reassemble regulator.
    c. Disassemble cylinder valve assembly.
        (1) Overhaul and replace damaged parts.
    d. Reassemble cylinder valve assembly.
2. Face mask open-circuit SCUBA.
    a. Disassemble and inspect assembly.
    b. Repair and reassemble mask assembly.
    c. Disassemble and inspect first stage regulator assembly.
3. Student application.

Topic 3: Charging SCUBA

A. Objectives
1. To teach the student the methods of charging SCUBA.
2. To provide practice in charging SCUBA.
B. Outline of Instruction
1. Types of high pressure systems.
    a. High pressure motor driven 3000 psi air compressor.

    b. High pressure gasoline driven 3000 psi air compressor.
    c. High pressure air flasks.
2. Equipment used with high pressure systems.
    a. Oil separator
    b. Air filter.
    c. Stop valves.
    d. Relief valve.
    e. Pressure gage.
    f. Manifold.
    g. Charging lines.
3. Procedures for charging (student application).
    a. Valves.
    b. Check temperatures.
    c. Desired pressures.
    d. ICC pressure stamped on bottle.
    e. Hydrostatic tests required.
    f. Securing from charging.

Topic 4: Open-circuit SCUBA Pool Training

A. Objectives
1. To teach the student the proper swimming techniques when using open-circuit SCUBA.
2. To teach the student the procedures for ditching, donning, and clearing SCUBA underwater.
3. To provide practical application of techniques and procedures used in swimming with open-circuit SCUBA.
B. Outline of Instruction
1. Donning gear.
    a. Life jacket.
        (1) No quick release.
    b. Belt and knife.
    c. Bottles.
        (1) Quick release on strap.
        (2) Carry by manifold on bottles.
        (3) Lay flat (never lean against anything).
2. Proper water entry with SCUBA.
3. Swimming techniques.
    a. Flutter kick.
    b. Do not use arms.
4. Clearing air hose and masks.
    a. Clearing masks.
        (1) Face mask open-circuit.
        (2) Mouthpiece open-circuit.
    b. Clearing hose.
        (1) Free flow.
        (2) Rolling on side.
5. Ditching and donning.
    a. Ditching.
    b. Donning.

Topic 5: Procedures for Open Water Swimming with Open-Circuit SCUBA

A. Objectives
1. To acquaint the student with procedures of loading LCPRs with SCUBA gear.
2. To acquaint the student with procedures of disembarking from LCPR in swim area.

B. Outline of Instruction
1. Loading LCPRs
   a. By pair number
   b. 2 men bottle
   c. Manifold toward outboard side
   d. Check swim float and buddy line.
2. Saddling up.
   a. When ordered.
   b. Ready to enter water.
   c. Enter in order by pairs.
      (1) Float man hook up to compass man.
   d. Swim.
   e. Report to beach timer.
   f. Leave water.
   g. Load truck.
   h. Last two men each boat.
   i. Return aboard boat.
   j. Return to pier.
   k. Disembark boat.
   l. Return to building.
   m. Wash gear.

Topic 6: Open Water Swimming in Open-circuit SCUBA

A. Objectives
1. To develop student's confidence in the use of SCUBA.
2. To increase the student's time submerged and distance covered underwater.
3. To develop student's proficiency in using the underwater compass.
4. To test the student's ability to swim submerged using SCUBA, and to maintain a compass course.

B. Outline of Instruction
1. Open water swimming with SCUBA.
   a. Assemble gear (use check sheet).
   b. Check and record bottle pressures.
   c. Don equipment prior to reaching point of commencing swim.
   d. Check own and buddy's gear.
   e. Hook buddy line to buddy.
   f. Hook up swim buoys.
   g. Enter water together.
   h. Check compass point.
   i. Submerge and commence swimming.
   j. When reaching designated area report to beach timer.

   k. Be alert to assist other swimmers.
2. Increase time submerged and distance covered.
3. Underwater compass.

Topic 7: Open-circuit Intermediate Dives

A. Objectives
1. To provide the student with a knowledge of the proper procedure to make a 60 ft. dive in open-circuit SCUBA.
2. To provide practical experience in planning and making a 60 ft. dive in open-circuit SCUBA.

B. Outline of Instruction
1. Prepare for descent.
2. 60 ft. dive.
   a. Buddy system.
   b. Descent.
   c. Circle line.
      (1) Job.
      (2) Compass line search.

Topic 8: Open-circuit SCUBA Qualification Dives

A. Objectives
1. To provide the student with a knowledge of the proper procedures to make a 130 ft. dive in open-circuit SCUBA.
2. To provide practical experience in planning and making a 130 ft. dive in open-circuit SCUBA.
3. To qualify the student in a 130 ft. dive for record purposes.

B. Outline of Instruction
1. Prepare for descent.
2. 130 ft. dive.
   a. Buddy system.
   b. Descent.
   c. Circle line search.
   d. Ascent.
3. Second dive following same procedure.
4. Decompression stop.

Unit VI—Underwater Search and Inspection

Topic 1: Underwater Search Procedures

A. Objectives
1. To provide a knowledge of underwater search and inspection procedures.
2. To provide practical experience in underwater searching and inspecting.

B. Outline of Instruction
1. Trunk line search method.
   a. Used by UDT and others.
   b. Method of laying out trunk line.
   c. Procedure used in trunk line search.

2. Running jackstay method.
    a. Uses of this method.
    b. Means of rigging for jackstay method of search.
    c. Search procedure using jackstay method.
3. Circling line method.
    a. Use of this method.
    b. Means of rigging circling line.
    c. Procedure of search.
    d. Advantages and limitations of this method.
4. Underwater ship bottom inspection.
    a. Hogging line method.
    b. Means of rigging and methods of search.
    c. Evaluation of this method of search.
5. Methods used in inspecting of:
    a. Propellers
    b. Shafts.
    c. Glands.
    d. Rudder.
    e. Struts.
    f. Bottom area.
6. Fittings found on bottoms of most ships.
    a. Sea-chests.
    b. Pump intakes.
    c. Underwater sound dome.
    d. Shaft strut bearing.
    e. Rudder.
    f. Screws
    g. Fathometer housing.
7. Compass line search.
    a. Uses of this method.
    b. Procedures for search.
    c. Advantages and limitations of this method.
8. Planning Board.
    a. Means of rigging.
    b. Procedure of search.
    c. Advantages and limitations of this method.
9. Clump line.
    a. Means of rigging.
    b. Procedure of search.
    c. Advantages and limitations of this method.

Topic 2: Underwater Work, Using SCUBA

A. Objectives
    1. To provide the student with practical experience in making minor repairs underwater, using SCUBA.
B. Outline of Instruction
    1. Student perform jobs on surface.
    2. Student dives with buddy and accomplishes the following:
        a. Pipe job.

b. Two-man flange job.
c. Pontoon job.

Unit VII—Related Subjects for Diving

Topic 1: Tides and Currents

A. Objectives
    1. To emphasize the necessity for knowing the state of tides and existing currents and their effect on free swimmers.
    2. To provide the student with a knowledge of tides and currents.
    3. To acquaint the student with the means and methods of obtaining information concerning tides and currents.
B. Outline of Instruction
    1. Theory of tides.
        a. Rise and fall.
        b. Effect of moon.
        c. Effect of sun.
        d. Daily tides.
    2. Types of tides.
        a. Spring.
        b. Neap.
        c. Mean low.
    3. Tidal currents.
        a. Definition.
        b. Types.
        c. Buoys, channels, and tidal effect on current.
        d. Locating swift currents.
        e. Calculate current on location.
            (1) Examples.
    4. Ocean currents.
        a. Causes.
        b. Characteristics.
        c. Directions.
        d. Local area.

Topic 2: Compass

A. Objectives
    1. To familiarize students in the use of the underwater wrist compass.
    2. To acquaint the students with the safety precautions to be followed while using the compass.
    3. To give students practice in the use of the underwater wrist compass.
B. Outline of Instruction
    1. Compass.
        a. FSNG 6605-290 4111.
    2. Safety precautions.
    3. Use of compass.
        a. Two methods.

4.   Dry land drill (practical).
5.   Procedure in water.

Topic 3: Rubber Suit Familiarization and Maintenance

A.   Objectives
1.   To acquaint the trainee with the types of exposure suits; their care, use and maintenance.
B.   Outline of Instruction
1.   Dry suit.
2.   Wet suit.

Topic 4: Diving Requalification, Logs, Records, and Reports

A.   Objectives
1.   To acquaint the student with the various logs, forms, records, and reports connected with diving.
2.   To acquaint the student with the procedures for diving requalification.
B.   Outline of Instruction
1.   Qualification.
2.   Requalification.
3.   Diving records system.
4.   Diving log book.
5.   Record of dive forms.
6.   Diving duty summary.
7.   Report of decompression sickness and all diving accidents.
8.   Service record entries.

Unit VIII—Use of Tools Underwater

Topic 1: Tools Used Underwater

A.   Objectives
1.   To provide the student with a knowledge of the care and use of tools used underwater.
2.   To teach the student safety precautions to be observed when using tools underwater.
B.   Outline of Instruction
1.   Nomenclature and use of tools.
   a.   Hand tools.
   b.   Pneumatic tools.
   c.   Special tools.
      (1)   Cement gun.
      (2)   High-velocity stud-driver.
2.   Underwater use of tools.
   a.   Sledge hammer and chisels.
      (1)   Removal of rivets, bolts, small piping.
   b.   Pneumatic chipping gun and chisel.
      (1)   Removal of rivets, bolts, piping.
      (2)   Cutting of steel skin.
      (3)   Removal of manila and wire impacted in screws.

   c.   Hacksaw.
      (1)   Cutting of piping, metal stock, manila, wire.
   d.   Pneumatic drill.
      (1)   Placing of holes for tapping.
      (2)   Venting of gas filled compartments.
      (3)   Holes for hook bolts.
   e.   Hammer and nails.
      (1)   Building and repairing of wood patches.
      (2)   Manufacture of templates underwater.
   f.   Divers deadman.
      (1)   To increase body pressure in using pneumatic drill.
   g.   Saws, pneumatic, hand, two-man.
      (1)   Remove piling, timbers.
   h.   Cement gun.
      (1)   Lay concrete patches; close off various openings.
   i.   High-velocity stud-driver.
      (1)   Drive threaded studs in plate, concrete.
3.   Maintenance of tools.
4.   Safety precautions.
   a.   Handling tools.
   b.   Using tools.
   c.   Lanyards.
   d.   Tool bag.

Topic 2: Underwater Work in Standard Deep Sea Diving Equipment

A.   Objectives
1.   To provide the student with practical experience in using tools underwater.
B.   Outline of Instruction
1.   Drill and tap (deep sea equipment).
   a.   Angle of pneumatic drill.
   b.   Proper leverage.
   c.   Drilling holes.
   d.   Tapping holes.
   e.   Use tapped hole by bolting flange to plate.
2.   Sledge rivets (deep sea equipment).
   a.   Each student remove one rivet, using hand sledge.
   b.   Procedure for swinging sledge underwater.
   c.   Method of holding chisel on rivet.
3.   Cutting rivets with pneumatic hammer (deep sea equipment).
   a.   Method of holding pneumatic hammer.
   b.   Selection of proper chisels.
   c.   Cutting rivet from plate.
   d.   Smooth plate surface after removing rivet.
4.   Building a box (deep sea equipment) (SCUBA for course 2).

a. Secure all pieces of wood together with light line.
b. Place wood under work bench or other structure to prevent it from floating away.
c. Nail sides of box together.
d. Nail on ends of box.
5. Hacksaw job (SCUBA for course 2).
a. Place work bench in open tank.
b. Diver enters water with hacksaw, angle iron, and C-clamp.
c. Diver secures angle iron to bench with C-clamp.
d. Diver cuts one-inch piece off end of angle iron, removes C-clamp, and returns to the surface with tools and materials.

Unit IX—Diving Orientation

Topic 1: Deep Sea Diving Outfit; Functions and Nomenclature

A. Objectives
1. To provide the student with a knowledge of the functions and nomenclature of each part of the deep sea diving outfit.
2. To teach the student the procedures for testing and maintaining all parts of the deep sea outfit.
B. Outline of Instruction
1. History and development of diving equipment.
2. Types of diving outfits and their uses.
3. Nomenclature, function, and construction of the standard deep sea diving outfit.
a. Helmet and breastplate.
b. The dress.
c. Shoes.
d. Helmet cushion.
e. Belt and jock strap.
f. Diver's knife.
g. Diver's air hose.
h. Amplifier and lifeline cable.
i. The nonreturn valve.
j. The air control valve.
k. The air regulation exhaust valve.
l. The supplementary relief valve (spit cock).
m. The diver's air manifold.
n. Oil separator.
o. Couplings, air hose double male; double female.
p. Reducers, type "S".
q. Reducers, type "T".
4. Tests and maintenance of diving equipment.

Topic 2: Deep Sea Diving Techniques and Procedures

A. Objectives
1. To teach the student correct basic diving techniques, using deep sea diving equipment.

2. To teach the student correct diving procedures when using deep sea diving equipment.
3. To emphasize the importance of using standard diving techniques and procedures.
4. To provide practice in dressing and undressing a diver.
5. To provide practice in using standard diving signals.
B. Outline of Instruction
1. Diving signals and communications.
2. Tending the diver.
3. Dressing and undressing the diver.
4. The procedure for descent.
5. Working on the bottom.
6. The procedure for ascent.
7. Log keeping.
8. Safety precautions.
9. Students practice dressing and undressing.
10. Students practice using hand signals.

Topic 3: Familiarization Diving with Standard Deep Sea Diving Equipment

A. Objectives
1. To provide practice in diving techniques, using deep sea diving equipment.
2. To develop student confidence in deep sea diving equipment and in his ability to dive.
B. Outline of Instruction
1. Familiarization dive in tank or shallow water.
a. Signals (hand).
b. Manipulation of all valves.
c. Practice basic techniques and procedures.
d. Telephone communication.
e. Keeping diving logs.
2. Working dive in tank or shallow water.
a. Pipe job.
b. Knot tying.
c. Single flange assembly.

Unit X—Diving with Deep Sea Diving Equipment

Topic 1: Application of Diving Techniques and Procedures in Open Water

A. Objectives
1. To provide practice in the application of techniques and procedures, using deep sea equipment in open water.
2. To develop the student's confidence in the use of underwater equipment.
B Outline of Instruction
1. Open sea diving.
a. Diving station.
(1) Check equipment.
(2) Provide for standby diver.

(3) Review signals and tending procedures.

(4) Air supply and standby air supply.

(5) Take soundings.

b. Familiarization dives.

(1) Enter and leave water by stage.

(2) Enter and leave water by ladder.

(3) Procedure for decompression stops.

(4) Diver tending another diver.

(5) Swimming on surface.

c. Free searching project.

(1) Diver backs out across bottom, using circle line for guide.

(2) Diver eliminates desired area of bottom by moving in a series of increasing arcs, using hand signals and telephone for directional control from surface.

(3) Upon locating object, diver sends for square mark on lifeline and air hose, and sends object to surface.

Topic 2: Underwater Work in Standard Deep Sea Diving Equipment

A. Objectives

1. To provide practical experience in working underwater in standard deep sea diving equipment.

2. To gain proficiency in the use of basic tools underwater.

3. To qualify the student in a 50-foot dive for record purposes.

B. Outline of Instruction

1. Tooker patch job.

a. Diver searches for and locates tooker patch assembly (consists of a mock porthole frame in a cylindrical steel tank).

b. Diver sends for tooker patch by square mark (patch is a folding type porthole patch with steel strong back).

c. Diver installs tooker patch and sends for air blow hose and connects it to tank.

d. Diver surfaces and compressed air is blown to tank. If patch is secured properly the assembly will surface.

2. Single flange job.

a. Diver searches for and locates single flange.

b. Using open-end wrenches, diver disassembles single flange and sends gasket to surface for verification.

c. Upon verification of gasket, diver retrieves it, reassembles single flange, and returns to surface with flange in hand.

3. Pile sawing.

a. When descending, make sure saw teeth are pointed away from you.

b. Both divers climb on stage. Each assumes a position on either side of the timber to be cut.

c. Place saw in a position approximately ¾ inch from top of the timber.

d. Each diver holds free hand under saw blade, taking short strokes to start cut.

e. Cut timber.

f. When nearing completion of cut, shorten strokes.

g. Retrieve timber and saw.

h. Return to surface.

4. Overhead patch.

a. Diver places himself on stage.

b. Secure tool bag to stage.

c. Loosen nuts and stow in tool bag.

d. Remove patch and gasket.

e. Send gasket to surface for inspection by the instructor.

f. Return gasket to diver.

g. Diver replaces gasket and patch.

h. Tighten nuts.

5. Angle descending line.

a. Place eye splice of 21 thread on diver's arm.

b. Place diver on descending line.

c. Take heavy strain on descending line and place on bitts.

d. Diver proceeds to descend by straddling the descending line.

e. After diver reaches bottom, the tenders will bend the shackled plate to the 21 thread.

f. Diver and tenders will then proceed to yard, and stay the shackled plate to the descending weight.

g. When diver receives the plate, shackle it into descending line.

h. Diver returns to surface.

6. Hogging line.

a. Place hogging line around ship's bottom.

b. Place diver on hogging line.

c. Diver makes inspection of ship's bottom, using hogging line to keep himself placed against the bottom of the ship.

d. After inspection, diver returns to the surface.

7. Blank removal.

a. Stage with blanks attached is lowered to the bottom.

b. Divers are placed on stage.

c. Divers remove blanks and gaskets.

d. Send gaskets to surface for inspection by the instructor.

e. Return gaskets to divers.

f.   Replace gaskets and blanks.
g.   Tighten all nuts.
h.   Return divers to surface.
8.   Excavating.
   a.   Check nozzle holes to make sure they are clear.
   b.   Diver takes hose and nozzle to bottom.
   c.   Diver plants himself firmly on the bottom with feet wide apart.
   d.   Call for pressure in hose.
   e.   Use sweeping motion.
   f.   When hole is required size, call topside to cut the pressure.
   g.   Pull hose nozzle clear of hole.
9.   Qualifying dive.

Unit XI—Diving with Lightweight Diving Equipment

Topic 1: Lightweight Diving Outfits; Functions and Nomenclature

A.   Objectives
   1.   To teach the student the nomenclature, function, and operation of all parts of the standard lightweight diving outfit.
   2.   To teach the student the proper procedures for checking, testing, and maintaining lightweight diving equipment.
   3.   To instill in the student a feeling of confidence and trust in the equipment.
B.   Outline of Instruction
   1.   History and development.
      a.   Lightweight diving outfit.
      b.   Use and comparison with deep sea equipment.
      c.   Limitations (60 feet normal working limit).
   2.   Use of lightweight diving equipment.
      a.   To perform work, inaccessible to divers using deep sea equipment.
      b.   Emergency diving.
      c.   Routine diving not requiring prolonged exposure.
      d.   Damage control.
   3.   Nomenclature and function.
      a.   Mask.
      b.   Dress
      c.   Belt.
      d.   Air hose.
      e.   Lifeline.

Topic 2: Lightweight Diving Procedures and Techniques

A.   Objectives
   1.   To teach the student the use of lightweight diving equipment and lightweight diving procedures.

2.   To develop student's confidence in lightweight diving equipment and in lightweight diving.
3.   To develop skill in the proper method of entering the water, using hand signals and valves, and accomplishing different tasks, using lightweight diving equipment.
B.   Outline of Instruction
   1.   Safety precautions.
      a.   Exhale when ascending.
      b.   Never remove lifeline.
      c.   Ditch mask only as a last resort.
   2.   Lightweight dive using mask, bathing suit, and weighted belt.
      a. Instruction before entering water.
         (1)   Proper method of securing lifeline to diver.
         (2)   Location of air control valve.
         (3)   Location of exhaust valve.
         (4)   Proper use of weighted belt.
         (5)   Proper method of securing lifeline to diver.
      b.   Dress diver and commence dive.
         (1)   Proper water entry.
         (2)   Observe hand signals.
         (3)   Proper ditching of weights.
         (4)   Ditching mask on bottom, exhaling on way up to surface.
         (5)   Student free dives to bottom and dons mask and belt.
         (6)   Student returns to surface when signal from instructor is given.
   3.   Lightweight dive wearing suit, mask, and belt.
      a.   Students assemble by open tank and receive instruction on proper method of dressing.
      b.   Dress student and commence dive.
         (1)   Observe water entry.
         (2)   Observe dumping air before descent.
         (3)   Diver completes pipe job.
         (4)   On signal from instructor, diver returns to surface.
         (5)   Repeat procedure until each student has made a dive.
   4.   Single flange.
      a.   Dress diver in suit, mask, and belt.
         (1)   Suspend flange in vertical position in center of tank.
         (2)   Rig hogging line under flange.
         (3)   Diver enters water, sits on hogging line.
         (4)   Diver disassembles and assembles flange.
         (5)   Repeat dives until each diver has completed job.
   5.   Hacksaw job.
      a.   Dress one diver in suit, mask, and belt.
         (1)   Place work bench in open tank.

(2) Diver enters water with hacksaw, angle iron, and C-clamp.

(3) Diver secures angle iron to bench with C-clamp.

(4) Diver cuts small piece off, removes C-clamp, and returns to the surface with tools and material.

(5) Repeat until each student has completed job.

Topic 3: Underwater Work Using Lightweight Diving Equipment

A. Objectives
1. To provide practical experience in open sea diving, using lightweight diving equipment.
2. To provide the student with an opportunity to utilize the underwater skills developed during tank diving.
3. To provide experience in some of the more difficult types of underwater tasks, which may be encountered.

B. Outline of Instruction
1. Safety precautions.
    a. Keep all lines clear.
    b. Enter holes, feet first.
    c. In the event the mask must be ditched, exhale on the ascent.
    d. Avoid sudden movements that might pull mask off.
    e. Use proper tending techniques.
2. Free searching project.
    a. Diver backs out across bottom, using circle line for guide.
    b. Diver eliminates desired area of bottom by moving in a series of increasing arcs, using hand signals for directional control from surface.
3. Single flange job.
    a. Diver searches for and locates single flange.
    b. Using open end wrenches, diver disassembles single flange and sends gasket to surface for verification.
    c. Upon verification of gasket, diver retrieves it, reassembles single flange, and returns to surface with flange in hand.
4. Tooker patch job.
    a. Diver searches for and locates tooker patch assembly (consists of a mock porthole frame in a cylindrical steel tank).
    b. Diver sends for tooker patch by square mark (patch is a folding type porthole patch with steel strong back).
    c. Diver installs tooker patch, sends for air blow hose, and connects it to tank.
    d. Diver surfaces and compressed air is blown

in tank. If patch is secured properly the assembly will surface.
5. Blank removal.
    a. Stage with blanks attached is lowered to the bottom.
    b. Divers are placed on stage.
    c. Divers remove blanks and gaskets.
    d. Send gaskets to surface for inspection by the instructor.
    e. Return gaskets to divers.
    f. Replace gaskets and blanks.
    g. Tighten all nuts.
    h. Return divers to surface.
6. Pile sawing.
    a. When descending, make sure saw teeth are pointed away from you.
    b. Both divers climb on stage. Each assumes a position on either side of the timber to be cut.
    c. Place saw in a position approximately ¾ inch from top of the timber.
    d. Each diver holds free hand under saw blade, taking short strokes to start cut.
    e. Cut timber.
    f. When nearing completion of cut, shorten strokes.
    g. Retrieve timber and saw.
    h. Return to surface.
7. 17-stud pontoon.
    a. Lower pontoon to bottom.
    b. Loosen and remove nuts. Place them in tool bag.
    c. Signal for square mark.
    d. Remove gasket and send to surface for inspection by the instructor.
    e. Return gasket to diver.
    f. Replace gasket and patch.
    g. Tighten nuts.
    h. Call for and secure blow hose to pontoon.
    i. Clear diver and blow pontoon.
8. Hook and tee bolt patch.
    a. Pontoon is lowered to bottom.
    b. Tee bolts and patch are carried to the bottom in tool bag by diver.
    c. Diver connects bolts and patch to pontoon.
    d. Diver sends for blow hose, using a square mark signal and connects to pontoon.
    e. Diver surfaces and compressed air is blown to pontoon. If patch and tee bolts are secured properly, pontoon will surface.

Unit XII—Helmet and Dress Repairs

A. Objectives
1. To teach the student the proper methods of testing the deep sea diving dress.

2. To teach the student the proper methods and techniques for repairing the deep sea diving dress.
3. To acquaint the student with the tools and materials used in the repair and testing of the deep sea diving dress.
4. To provide practice in making tests and repairs to the deep sea diving dress.

B. Outline of Instruction
1. Causes of leaks and failures.
2. Testing the diving dress.
   a. Use of breastplate.
3. Patching the diving dress.
   a. Glove and cuff patching.
   b. Crotch patch.
   c. Knee and elbow patch.
   d. Heel and toe patch.
   e. Repairing the collar.
4. Relationship of safety and good equipment.
5. Means of preservation of diving equipment.
6. Student application.
   a. Test and patch diving dresses.

Topic 2: Belt and Shoe Repairs

A. Objectives
1. To teach the student the proper method and procedures for repairing diving belts and shoes.
2. To provide practice in making repairs on belts and shoes.

B. Outline of Instruction
1. Repair of diving shoes.
   a. Straps.
   b. Soles.
   c. Toes.
   d. Preservation and maintenance.
2. Repair of diving belts.
   a. Buckles.
   b. Weights.
   c. Preservation and maintenance.
3. Student application.
   a. Repair shoes and belts

Topic 3: Maintenance of Lifeline and Air Hose

A. Objectives
1. To teach the student the proper method for making, maintaining, and testing air hose and lifelines.
2. To provide practice in making and testing air hose and lifelines.

B. Outline of Instruction
1. Lifelines.
   a. Makeup.
   b. Maintenance.
   c. Testing.

2. Air hose.
   a. Makeup.
   b. Maintenance.
   c. Testing.
3. Air hose connection.
4. Checking for safety.
5. Telephone line; care and maintenance.
6. Student application.

Topic 4: Air Helmet Repairs

A. Objectives
1. To teach the student the equipment and material used to repair diving helmets.
2. To teach the proper method and procedure for repairing the air diving helmet.
3. To provide practice in repairing the air helmet.

B. Outline of Instruction
1. Repair of helmet and breastplate.
   a. Renew port of faceplate glass.
   b. Replace supplementary exhaust valve (spit cock).
   c. Replace air exhaust valve
   d. Replace studs in breastplate.
   e. Install telephone transceivers in helmet.
   f. Renew helmet gasket in breastplate.
   g. Install jack plug and anchor plug on telephone and lifeline cable (demonstrated by instructor only).
2. Tools and equipment used to repair air helmets.
3. Student application
   a. Repair air helmets.

Topic 5: Lightweight Diving Mask Repair

A. Objectives
1. To teach the student the proper procedures and methods for repairing lightweight diving masks.
2. To acquaint the student with the tools and equipment used to repair lightweight diving masks.
3. To provide practice in repairing lightweight diving masks.

B. Outline of Instruction
1. Repair of valves.
   a. Nonreturn valve.
   b. Air supply valve.
   c. Exhaust valve.
2. Straps.
3. Face plate.
4. Face mask.
5. Relationship of safety and good equipment.
6. Preservation and upkeep of equipment.
7. Tools and equipment used to repair lightweight diving masks.
8. Student application.

Unit XIII—Marlinespike Seamanship

Topic 1: Marlinespike Seamanship Fundamentals

A. Objectives
1. To provide the student with a knowledge of the construction, use, and care of fiber and wire rope.
2. To teach the student the purpose and use of splices in fiber and wire rope.
3. To teach the student the purpose and use of terminal fittings on wire rope.

B. Outline of Instruction
1. Fiber rope.
   a. Types.
      (1) Manila.
      (2) Sisal.
      (3) Hemp.
      (4) Cotton.
   b. Sizes - how measured.
   c. Care and maintenance.
2. Wire rope.
   a. Types.
      (1) Steel.
      (2) Bronze.
      (3) Copper.
   b. Sizes - how measured.
   c. Care and maintenance.
3. Synthetic rope.
   a. Nylon.
      (1) Sizes.
      (2) Care and maintenance.
4. Splices.
   a. Types.
      (1) Eye splice.
      (2) Short splice.
      (3) Long splice.
   b. Application of various splices.
   c. Strength of splices.
   d. Safety factors.
5. Wire rope clips.
   a. Use.
   b. Method of application.
   c. Strength.
6. Terminal fittings.
   a. Types.
   b. Strength
   c. Method of application

Topic 2: Practical Application of Marlinespike Seamanship

A. Objectives
1. To provide practice in the application of marlinespike seamanship.

B. Outline of Instruction
1. Splices, fiber rope.
   a. Eye.
   b. Short.
   c. Long.
2. Splices, wire rope.
   a. Liverpool.
   b. Salvage.
   c. Gun factory.
3. Knots and hitches.
   a. Becket bend.
   b. Bowline.
   c. Bowline on a bight.
   d. Clove hitch.
   e. Overhand knot.
   f. Passing a stopper.
   g. Rolling hitch.
   h. Round turn and two half hitches.
   i. Slip knot.
   j. Square knot.
   k. Mousing.
   l. Seizing.
   m. Catspaw.
4. Terminal fittings.
5. Wire rope clips.
6. Shackles.
7. Practical application in knot typing and splicing.

Unit XIV—Underwater Cutting

Topic 1: Introduction to Underwater Cutting Equipment

A. Objectives
1. To acquaint the student with the various methods of underwater cutting.
2. To provide the student with a knowledge of the types and uses of equipment necessary for underwater cutting.

B. Outline of Instruction
1. History of underwater cutting.
   a. Oxy-hydrogen first used in 1926 during salvage of the S-51.
   b. Used extensively during war to aid in repair and salvage of ships.
   c. In 1942, Navy started extensive development of oxygen-arc method of underwater cutting.
2. Nomenclature, general underwater cutting.
3. Methods of underwater cutting.
   a. Oxy-hydrogen method.
      (1) May be used with shallow water diving outfit.
      (2) Makes a neater cut.
      (3) Requires more skill.
      (4) Problems of gas supply.

b. Oxy-arc method.
   (1) Less skill required.
   (2) Can be used to cut laminated plating.
   (3) Positive action, no flame adjustment.
   (4) Oxygen is the only gas necessary.
   (5) Diver must be fully insulated, because of danger of electrical shock.
c. Metallic arc cutting.
   (1) No source of fuel gas required.
   (2) Can be used to cut nonferrous metals.
   (3) Standard surface electrode holders and electrodes may be adapted for emergency use underwater.

Topic 2: Oxy-arc Cutting Techniques

A. Objectives
  1. To provide the student with a knowledge of the nomenclature, care, and maintenance of oxy-arc underwater cutting equipment.
  2. To teach the student oxy-arc underwater cutting techniques.
B. Outline of Instruction
  1. History of oxy-arc underwater cutting.
    a. Improved oxy-arc torches and electrodes and standardization of operating techniques developed as early as 1942.
    b. World War II brought about the need for simplification of underwater cutting methods.
    c. Oxy-arc cutting has a positive action; no flame adjustment is necessary.
    d. Oxygen is the only gas necessary.
    e. Less skill and practice is required.
  2. The torch (holder for electrodes).
    a. Navy standard oxy-arc torch.
    b. Light in weight.
    c. Simple in construction.
    d. Emergency torch may be fabricated aboard ship from pipe fittings.
    e. Present day torch is designed to accommodate both the steel tubular electrode and the ceramic electrode.
  3. Electrodes.
    a. Steel tubular.
     (1) May be drawn from standard stock.
     (2) With A.C. source of power, steel tubular electrodes are more satisfactory than ceramic electrodes.
     (3) Better for cutting thicker plates.
     (4) Rate of burn off is very fast.
     (5) Made of tubular steel, 5/16 inch outside diameter, 1/8 inch bore; flux or paper coating for insulation.

    b. Ceramic electrodes.
     (1) Standard stock issue.
     (2) Made of highly refractory, silicon carbide.
      (a) Core, 1/2 inch diameter.
      (b) Bore, 1/8 inch diameter.
      (c) Length, 8 inches.
      (d) Sprayed with 1/32 inch mild steel, covered with fiberglass cloth, and wrapped with masking tape.
     (3) Short length makes it easy to manipulate in confined spaces.
     (4) Light in weight makes it ideal for air freight shipment.
     (5) The skilled diver can cut more than 10 times the amount of metal with one ceramic electrode than he can with one steel tubular electrode.
  4. Welding generators.
    a. D.C. power source is preferred.
    b. 300 ampere capacity should be minimum.
    c. Starting and proper setting.
  5. Welding cable.
    a. Size 0 and size 2/0 recommended.
    b. Ground cables may be made up of the same size cable.
     (1) Securely attach ground to clean plate near work.
  6. Safety switch.
    a. Standard stock No. 17-S-4580.
    b. Knife type.
    c. Unfuseable.
    d. Single pole.
    e. Single throw.
  7. Oxygen in cylinders.
    a. Care in handling.
  8. Gauges for oxygen cylinders.
    a. Care in handling.
  9. Safety precautions in oxy-arc underwater cutting.
    a. Use of the complete deep sea diving outfit, including gloves, when employing electric power for underwater welding and cutting.
    b. Ensure against explosions of accumulated gases in the compartment.
    c. Safety switch on the ON position only when the diver is ready to start cutting.
  10. Technique for oxy-arc underwater cutting.
    a. Steel tubular electrode.
     (1) Seat electrode in holder firmly against rubber gasket.
     (2) Hold electrode perpendicular to surface to be cut.
     (3) Call for "Switch On".
     (4) Move electrode along desired path of cut, using a "dragging" motion.

(5) Pressure is exerted in two directions, inward to compensate for burn off, and forward to advance the cut.

(6) When cut is finished or when changing electrodes, call for "Switch Off."

(7) Technique for cutting 1/4 inch or thinner plate differs.

    (a) Hold the electrode at a 45 degree angle when cutting; effective thickness of the plate is thus increased.

(8) Compressed air is recommended in place of oxygen when cutting cast iron or nonferrous metals.

b. Techniques for using the ceramic electrode.

(1) Hold electrode perpendicular to surface to be cut.

(2) Call for "Switch On" when ready to cut.

(3) Move electrode in desired path as soon as cut is started, maintaining a light contact with the work.

(4) Do not exert any inward pressure.

(5) When not cutting or when changing electrodes, call for "Switch Off".

(6) Compressed air is recommended in place of oxygen when cutting cast iron or non-ferrous metals.

(7) Cutting these metals is a melting process.

Topic 3: Practical Application of the Oxygen-arc Method of Underwater Cutting

A. Objectives

1. To provide practical experience in oxy-arc cutting procedures.

2. To develop student skill in oxy-arc cutting.

    a. Using tubular steel electrodes.

    b. Using ceramic electrodes.

B. Outline of instruction

1. Setting up the equipment.

2. Accomplish the following projects, using both tubular and ceramic electrodes.

    a. Cut "I" beam, heavy angle, railroad iron.

    b. Cut three (3) feet of flat plate, free hand.

    c. Cut circular hole in vertical plate. Instructor designate size of holes.

    d. Make precision cut, using specified template.

    e. Pierce four (4) holes, spacing designated by instructor

    f. Cut heavy wire rope.

Unit XV—Review and Examination

Topic 1: Review and Examination

A. Objectives

1. To provide for continuous evaluation throughout the course.

2. To evaluate student achievement in knowledge and skill as related to the objectives of the course.

3. To summarize and review the course of instruction.

B. Outline of Instruction

1. Summarize and review all units of instruction.

2. Class discussion.

3. Administer final examination.

4. Review examination.

5. Reteach weak areas.

Bibliography to Part III

Andersen, B.G. 1968. *Diver performance measurement: underwater navigation, depth maintenance, weight carrying capabilities.* ONR Res. Task NR 196-068. Groton, Conn: Gen Dynamics EB Div.

Bass, G.F. and Throckmorton, P. 1961. Excavating a bronze age shipwreck, *Archaeology* vol. 14., no. 2.

Carlisle, J.G. Jr., Turner, C.H., Ebert, E. E. 1964. *Artificial habitat in the marine enviornment.* Sacramento: Dept. of Fish and Game.

Chabert, G. 1969. Constructing a new sea-loading system off Egypt. *Ocean Industry* vol. 4, no. 4.

Clark, E. 1951. *Lady with a spear.* New York: Harpers, Heinemann.

Clarke, W.D. 1967. *The nearshore waters, California's resource of the future.* San Diego: Westinghouse Res. Lab. Sci. Paper 67-1C7-OCEAN-P1.

COMEX. 1968. *Deep diving techniques at the service of industry.* Paris: Compagnie Maritime D'Expertises.

Conference for National Cooperation in Aquatics. 1962. *New science of skin and scuba diving.* New York: C.N.C.A.

Cousteau, J. Y. 1953. *The silent world.* London: Hamish Hamilton.

———. 1965. *World without sun.* New York: Harper & Row.

CSL, Univ. of Fla. 1967. *Quarterly report of the Dept. of Speech, University of Florida* vol. 7, no. 1. Gainesville: University of Florida.

Davis, R.H. 1962. *Deep diving and submarine operations.* 7th ed. London: Saint Catherine Press Ltd.

Dinnen, J. 1969. *Capt. Mitchell reviews Navy salvage and diving operations.* Washington: DATA.

Diolé, P. 1953. *The undersea adventure.* Sedgwick & Jackson, L'aventure sous-marine. Paris: Albin Michel.

———. 1954. *4000 years under the sea.* New York: Julian Messner.

Dugan, J. 1965. *Men under water.* Philadelphia: Chilton.

Ellsberg, E. 1956. *On the bottom.* New York: Dodd, Mead & Co.

Haas, H. 1947. *Drei Jäger auf dem meeresgrund.* Zurich, Orell Fussle, London: Jarrolds.

———. 1952. *Manta.* Berlin: Ullstein. (English: *Under the red sea.* London: Jarrolds.)

———. 1958. *We come from the sea.* London: Jarrolds.

Hollien, H., Brandt, J. and Thompson, C. 1967. *Preliminary measurements of pressure response to low frequency signals in shallow water.* CSL/ONR Rpt. #9.

———, Brandt, J. and Malone, J. 1967. *Underwater speech reception thresholds and discrimination.* CSL/ONR Rpt. #7.

Kenny, J. E. 1969. Scientific diver. Part 1: the geo-diver. *Ocean Industry* vol. 3, no. 12.

Laffont, M. and Delauze, H.G. 1969. The Janus experiment—working at 500 feet. Special report: working in the sea. *Ocean Industry* vol. 4, no. 2.

Limbaugh, C. 1961. Cleaning symbiosis. *Scientific American* pp. 42-49.

Link, A. 1965. Outpost under the ocean. *National Geographic* vol. 127, no. 4.

Link, M.C. 1960. Exploring the drowned city of Port Royal. *National Geographic* vol. 117, no. 2.

Lonsdale, A.L. and Laplan, H.R. 1964. *A guide to sunken ships in American waters.* Arlington, Va: Compass Publications.

MacInnis, J.B. 1966. Living under the sea. *Scientific American* vol. 214, no.3.

Mitchell, E. 1970. U.S. Navy diving activity. Personal communication.

National Geographic Society Research Reports, 1963. Washington: National Geographic Society.

Neushel, M., Clarke, W.D. and Brown, D.W. 1967. Subtidal plant and animal communities of the Southern California Islands. *Proceedings, symposium on the biology of the California Islands.* Santa Barbara: Santa Barbara Botanic Garden.

Office of Naval Research. 1967. *Project Sealab Report, an experimental 45-day undersea saturation dive at 205 feet.* ONR Report ACR-124. Washington: U.S. Government Printing Office.

Peterson, M. 1965. *History under the sea.* Washington: Smithsonian Institution Press.

Rackl, H. 1968. *Diving into the past.* New York: Charles Scribner's Sons.

Schaible, E.L. 1970. *Naval special warfare.* Personel communication.

Searle, W.F. Jr. 1969. How deep sea work capability is growing. *Ocean Industry* vol. 4, no. 4. and Kunz, H.S. *Test procedures for supervisor of salvage sponsored work projects for Sealab III.* Unnumbered report. Office of Supervisor of Salvage, United States Navy Ships Systems Command. Washington: U.S. Government Printing Office.

Stenuit, R. 1966. *The deepest days.* New York: Coward-McCann.

Taylor, J. 1966. *Marine archaeology*. New York: Thomas Y. Crowell.

Tibby, R.B. 1968. USC range at Santa Catalina Island. Undersea Technology Conference. ASME pub. 69-UNT-7.

Thorne, J. 1969. *The underwater world*. New York: Thomas Y. Crowell.

Turner, C.H., Ebert, E.E. and Given R.R. 1964. *An ecological survey of a marine enviornment prior to installation of a submarine outfall*. Sacramento: California Dept. of Fish and Game.

_____, _____, and _____. 1968. *The marine environment offshore from Point Loma, San Diego County*. Sacramento: California Dept. of Fish and Game.

U.S. Marine Corps. *Amphibious reconnaissance*. pub. FMFM 2-2. Washington: U.S. Government Printing Office.

U.S. Navy. 1968. *United States Navy ship salvage manual*, vol. 1. Office of the Supervisor of Salvage, U.S.N. Ship Systems Command. Washington: U.S. Government Printing Office.

U.S. Navy. 1969a. *NCEL ocean engineering program, March 1968 to March 1969*. Port Hueneme, California: NCEL.

_____. 1969b. *Technical research programs, fiscal 1969, Naval civil engineering labratory*. Port Heuneme, California: NCEL.

_____. 1970. *United States Navy diving manual*, part 1, NAVSHIPS 250-538. Washington: U.S. Government Printing Office.

Weltman, G., Egstrom, G.H., Elliott, R.E. and Stevenson, H.S. 1968. *Underwater work measurement techniques: initial studies*. ONR Report 68-11. Los Angeles: University of California, Biotechnology Laboratory.

Part 4
The Future

12. The Diving Business

Viewing the diving business as a complete entity is no less complex than surveying the diverse activity within the field. Diving is growing so rapidly that any attempt to correlate the business today with that of 10 years ago would be meaningless. Although we can trace diving back to antiquity, it really did not become a big business until the mid-1960s. In the last 5 years it has grown more than in the previous two thousand— and it will double in the next 5 years.

Each segment of diving activity—commercial, military, scientific and sport—is in some manner interrelated from the business standpoint. In some cases similarity of basic equipment and techniques makes common market areas. The nature of the mission in each area and the specialized equipment needed for this work does, however, provide sufficient delineation for separate consideration. In this chapter we will investigate the characteristics of each major market segment, its size as an equipment market and is probability for expansion.

Commercial Diving

The least understood aspect of diving, commercial diving is also the most difficult market to service. It is eminently unresponsive to standard advertising and marketing techniques, the least forgiving of mistakes and the most clannish. The commercial diving market can be summed up in two words—expanding and wary.

Its continued expansion is assured by rising requirements for ocean resources. Regardless of any ecological regulations formulated within the next quarter century, the industrial and post-industrial societies are irrevocably tied to energy from hydrocarbons. The energy industries in turn are tied to the oceans, and commercial diving is an important contributor to the overall business.

Several years ago the main emphasis in diving was placed on extending man's depth capability underwater. Although a rash of deep water living experiments produced valuable data, they were not pertinent to commercial work in the sea. The principal depth of diving activity is now less than 350 feet. True, there are exploratory jobs in deeper water, but considered as part of the total commercial diving revenue, they amount to less than 1% of total diving income. This will not change drastically within this decade.

In 1971, there is no question about the commercial diving industry's capability to work at depths of 600 feet; saturation diving jobs at 300 feet are now commonplace in the Gulf of Mexico oil fields. There is no accurate way to project the amount of diving work that will be generated by the offshore energy industry, but with the close relationship between offshore production and the diving business, it is proper to take the level of offshore oil recovery as a barometer for the diving business.

In his *Ocean Industry* article, "Potential of the Sea," Donald Taylor said,

> A number of well-known geologists—writing in *Ocean Industry*—have pointed out that most favorable prospects in the oil industry lie beneath the water. Certainly, recent articles on the Gulf of Mexico have indicated there are many prospects in deep water awaiting exploration. Reliable spokesmen have estimated that the offshore region, which now supplies about 16% of the world's crude and natural oil liquids, should supply one third of the total demand 20 years from now. This means that in the late 1980's, offshore production will exceed the present total production.

The wariness of the commercial diving marketplace is firmly based on experience. To understand this, it is only necessary to look back to the early 1960s. At that time, offshore oil field diving was severely threatened by a growing attitude in oil technology to "engineer-out" the diver from undersea systems. Determined to change this attitude, the diving industry cast about to bring more technology into the field. Although there was no paucity of technology ready for application, few companies outside the field really understood the problems of working in the sea. (There are still relatively few!)

After several years of unsatisfactory products and aborted R&D attempts, the diving business developed a "show-me-first" attitude. This attitude still exists today and, generally, not without good cause. In most industrial efforts, product failure results in loss of profit. In diving, the cash loss is often accompanied by loss of life. A diving

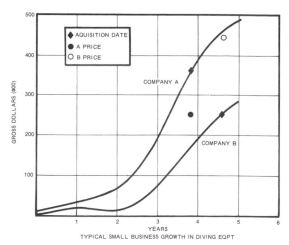

Figure 12-1. Typical small business growth in diving equipment.

company is therefore not as apt to take a chance on a new product or technique as an ordinary commercial company.

Gaining a foothold within the marketplace is a long process—one seldom accomplished by standard consumer advertising or marketing methods. These methods help bring new products or services to the potential customer's attention, but they must be accompanied by the willingness and ability to demonstrate competence over a reasonable length of time. Because of this, few large companies have successfully entered the field directly. The normal procedure is to acquire a small enterprise that has already undergone the painful and expensive process of proving capability. These larger companies then bring their internal technological and manufacturing capability to bear (also often a painful process) with the hope of improving both product and profit. Sometimes it works.

"Garage operations" that stick out the rough years usually profit handsomely. The pattern of business development for these companies, even though their product or service mix may be quite different, shows a remarkable similarity. Figure 12-1 shows the growth patterns of two actual companies, called *A* and *B*, that went through the process of growth and then acquisition. Both started with light capitalization; both were product oriented. Company *A* was engaged in development and manufacture of diver life—

support equipment; Company *B* produced underwater electronic equipment for diver use.

Company *A's* products were more closely associated with standard technology of the diving industry and it is obvious in looking at the graph (Figure 12-1) that acceptance and gross business were built at a rapid and constant rate. Company *B*, on the other hand, went through a longer period of proving their capability because their products were based on a new concept. After two years of trial, their products were accepted and business growth took off on a healthy climb.

Company *B* represents a typical experience. The apparently satisfactory start of business in the first year represents introduction of a highly technical product into a market segment having a specific need for the product. The unnerving downslope of gross business during the second year represents a period of field evaluation and product changes in response to the users' needs. An analysis of small companies in this industry shows that those who actually fail do so at this point for one of two reasons. Either they are not sensitive to the actual needs of the market and are therefore unable to respond to the requirements or they are so undercapitalized that they are financially unable to respond. In the case of Company *B*, additional capital and new management were brought in at the two-year point. The disparity of acquisition price for the two companies indicates that the management of Company *B* was considerably more adept at negotiation.

For people eying the diving business as a potential market or business, some primary questions must be answered. How large is commercial diving? What is a conservative projection of its growth in the near future? How much of a market is it for services and hard goods?

Answering the last question first, the principal market for commercial equipment is in the Gulf region around New Orleans. Only about 28 diving companies are located there, but they account for 80% of the commercial diving business in the United States. Of this 80%, nearly 95% is concentrated in the 18 largest companies.

For 1971 the offshore commercial diving business will have a gross income near $28 million, $20 million of which will be made by two companies.

Including offshore and harbor diving, the total gross income of the commercial diving business will be approximately $35 million for 1971.

As in any service business, salaries are the principal expense. Equipment expenditures vary from year to year, but over several years the average percent of gross business spent on equipment falls between 16-20%. As the major companies invest more in saturation systems, each with a price tag of $250,000 to $350,000, the percentage approaches the higher margin on a current year basis.

The total equipment purchase estimate of commercial firms engaged in offshore work is $5.1 million for 1971. The remainder of the equipment market, which is widely dispersed over the United States, will produce about $1.1 million in sales.

If the diving industry is to keep pace with even the conservative projections of the offshore resource recovery industry, it must maintain a 25% a year expansion over the next decade. If the optimistic industry predictions were accepted, the expansion rate will have to be closer to 30% a year. Using conservative forecasts, Table 12-1 shows both the gross and anticipated equipment expenditures.

As offshore activity rises, the prospects for the diving business become increasingly healthly

Table 12-1
Commerical Diving Growth
1971-1980 (in millions)

Year	Offshore Gross	Harbor & Other Gross	Joint Equipment Expenditures
1971	$ 28.0	$ 7.0	$ 6.2
1972	35.0	7.7	7.6
1973	44.0	8.5	9.3
1974	55.0	9.4	11.5
1975	68.0	10.4	14.0
1976	86.0	11.4	17.4
1977	108.0	12.6	21.8
1978	135.0	13.8	26.6
1979	169.0	15.2	33.2
1980	210.0	16.7	40.8

and based on sound business considerations. Nearly all major producers agree that the proportion of offshore work actually performed underwater will increase in the next decade. Storage and delivery systems produced completely underwater will be common. As this comes to pass there will be an ever expanding need for diving services. Both the services and the equipment requirements will more than double in the next 5 years. By 1980, diving services will grow to 7.5 times the current level, and the requirements for equipment will be multiplied by a factor of at least 6.5.

Sport Diving

In the field of commercial diving, the major deterrents to development of a broad bank of market data are the small number of companies and the closely held fiscal information in a competitive business. In sport diving, the situation is more in line with standard statistical methods of market analysis. The market as a whole is similar to any consumer market in the leisure field. It is feasible to develop a model of the average consumer, to extrapolate his spending characteristics and even predict how long he will continue to be a viable entity within the market.

Each year *Skindiver Magazine,* using standard statistical methods, makes a complete market survey in the sport diving field. For this chapter, their statistical model of the sport diver was checked against a cross section of 100 divers trained around San Diego over a 10-year period. The figures of the *skindiver* survey agreed in all basic areas with the "100 survey" within 5%.

Only one area was not covered by the *Skindiver* survey—how many people participate in the sport of skin diving? Using their audience research, *Skindiver* estimated a total readership of 465,000. Depending on the quoted source, however, the size of the world skin diving market is placed between 750,000 and several million. The disparity in numbers is partially a result of definition. Who should be qualified as a skin diver? The person who regularly practices, but what is "regularly practicing"? Should the vacation diver, although he is usually a one-time customer, be in-

Table 12-2
Skindiver Training 1970-1971

	1970	Projected 1971
YMCA	10,000	12,500
NAUI	40,000	55,000
LA	12,000	10,000
NASA	19,000	26,000
All Others	18,400	16,000
Total	99,400	119,000

cluded? For our purposes, different criteria for estimating the actual number of skin divers will be used.

According to the *Skindiver* survey, 75.6% of all current participants have taken training in an official program. Although there have been recent shifts in these percentages the figures in Table 12-2 present a reasonably correct estimate of people trained.

Applying the factor for those skin divers who do or do not train formally would bring these totals to 130,000 for 1970 and 158,000 for 1971. The 75.6% figure of people who take training is applicable when analyzing the skin divers who joined the sport during the last 10 years but does not present a true picture of the future or current ratio of trained and nontrained diver. Through self-imposed regulations, improved communication and wide availability of training courses and instructors, the percentage of new divers taking training before actually starting the sport is now 85%-90%.

Using these figures, there will be between 133,000-142,000 new skin divers in 1971. Figure 12-2 shows the number of skin divers who actually participate in skin diving in the United States.

According to the 1970 *Skindiver* survey, the average participant in the United States has been in the sport for 3.1 years. Computed one year later, using both the *Skindiver* survey and the "100 survey," indications are that the average experience is now closer to 3.6 years. Excluded from the computations is the once a year vacation diver who may later become a full-time participant. The number of U.S. skin divers in 1971 will fall somewhere in the range of 500,000-615,000, with the highest probability of about 560,000.

Using training as an index and excluding such possibilities as economic depression, direct extrapolation at the current rate of increase puts the number of U.S. skin divers at slightly over 1,000,000 in 1975. In other words, in a 5-year period the number of participants will double.

In terms of dollars and cents, this growing underwater population converts into big business, both in training and in new equipment purchases. The charge for a standard skin diving course vary according to the training organization and its location. Generally, prices range from $35-$50 with a mean of $39. In 1971, if the conservative projections are met, the newest crop of skin divers will pay $4.5 million for training.

Compared to the amount beginning divers will spend for new equipment, training costs are very small. According to the *Skindivers* survey, each new student spends $190 on his initial equipment purchase, or a total near $23 million in 1971. For those staying with the sport for 3 years, the equipment investment rises to $450. The "100 survey" verified these figures and showed that a sport diver maintaining interest for 7 years or more invests about $800. Excluding major items such as boats, which are only indirectly associated with the sport, 40% of long-term divers invested $1,200 or more during a 7-year period.

The overall market for skin diving training and equipment reaches a surprisingly high sum. With 1971 spending of $4.5 million for training, $23 million for initial equipment purchase and continuing purchases of $15 million, direct spending in the marketplace for the sport's basic commodities provides many good business opportunities. Related spending for services directly associated with the sport can now be considered good business as well. These peripheral services are also helping to cure one of skin diving's basic problems—the dropout.

There is no question about the statistics—the longer the sport diver continues in the field, the broader his underwater interests become and the

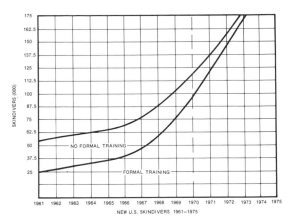

Figure 12-2. New U.S. skin divers, 1961-1975.

more money he spends. For the sport diver, interest appears to progress in a definite cycle.

The sport diver's first activity is underwater sightseeing. Depending on his location, this involves $100-$600 in equipment for year-round capability. From sightseeing, the diver normally branches into spearfishing, in most areas a disappointing and short-lived pastime. The next step is usually underwater photography. This phase may include the amateur photographer syndrome, i.e., the continual purchase of progressively more professional and expensive equipment. The problem is that somewhere between the sightseeing and the semi-professional stage, many skin divers lose interest in the sport.

Loss of interest is a problem that has been fought very actively by the industry. They have tried to change the public image of the sport from a daredevil/adventurer activity to that of healthy family recreation. Much headway has been made by diving clubs which are actively supported by the reatil skin diving business. In addition to increasing the services available such as group trips and boat charters, retailers have helped get the skin diver through that phase when he has seen almost everything in his local waters and is tiring of just "going diving."

Nearly every skin diving store in southern California runs at least one charter a week, carrying up to 10 divers at an average of $12 a person. Multiplication shows that 200 x 10 x 12.00 x 52 weeks is $1.2 million a year. Extrapolated over the entire skin diving business, even this peripheral

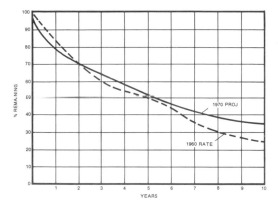

Figure 12-3. Activity life span of skin divers.

service takes on aspects of a going business. But, its real value to the industry is not the income it produces, but the fact that it maintains the divers interest and keeps him active.

Figure 12-3 shows the relative dropout rate of divers trained in 1960 and that projected for those trained in 1970. It is a little early to evaluate the validity of the 1970 projection, but it is based on sound thinking and some success in earlier efforts. From the figure, it is obvious that there is no hope of correcting the large dropout rate immediately following course completion and purchase of the basic equipment, but this sharp loss has been clearly identified. These are people who decide that diving is not their cup of tea.

In the earlier chapter on skin diving, the reasons people started in the sport were discussed. During training, almost all pyschologically unsuited people drop from the course, but there remains a certain percentage who simply delude themselves about their motivation to pursue the sport. Feeling it incumbent to complete the training and to make a show of actually diving in the ocean, they drop out at the earliest opportunity. These are not the people the industry wants to keep active. They are looking for people who, if given a continually interesting program, would never think of dropping out.

The diver who stáys in the sport is not only a purchaser of equipment but also of services. Of the 560,000 divers participating during 1971, 86.2% of them will scuba dive an average of 26.2 days

according to *Skindiver*. In the southern California "100 survey" the average was closer to 35 dives a year, but this is subject to distortion because of the good year-around diving conditions. As a national average, the *Skindiver* figures are undoubtably more meaningful.

Based on these figures, even a small service such as providing compressed air results in impressive totals. At an average charge of $1.50 a tank, the air required to keep skin divers breathing underwater comes to sales of $19 million annually. If the same percentage is maintained until 1975, skin divers will blow $34 million worth of bubbles into the water.

It is difficult to pinpoint other expenditures in the field. The best estimates are based on reports of skin diving shops: $2.4 million for equipment maintenance, $5.7 million for boat charter and special services and $1.85 million for small item purchases.

The current and anticipated sport diving market is shown in Table 12-3.

Sport diving's prospects of meeting or exceeding the predictions are positive for several reasons. First, the sport is in keeping with the life style developing in this country, particularly with the younger generation. No other sport is as directly involved with nature. Secondly, the cost of the sport is well within the reach of a large portion of our society. In fact, it is even cheaper to start diving in 1971 than it was in 1960. The initial investment has not really changed. Through mass and better production methods the individual item costs have not risen at the same rate as other hard goods in the United States. And last, but certainly

Table 12-3
Skindiving Expenditures
1971 and 1975 (in millions)

	1971	1975
Major equipment purchases	$38	$ 70
Services purchases	29	66
Training	4.5	9.2
Total	$71.5	$145.2

a primary consideration, is the availability of more diversions within the sport and the generation of a new set of services designed to satisfy the well-heeled diver.

Scientific Diving

The scientific diving market is small compared to commercial or sport diving businesses. Its importance to our environmental survival and its growth potential, however, justifies a close examination. One of the difficult problems in describing the market is again the parameters that qualify a diver as a scientific diver. Few people in this field are primarily engaged as diving technicians. As with the military diver, diving capability is normally in addition to some other basic qualification. For example, a laboratory technician working in marine biology might spend as little as 20% of his working time engaged in diving—a practicing scientist even less. Generally, a diving technician is employed by the research institution as an electronic or mechanical technician and he dives when his particular technical expertise is required underwater.

There are exceptions to this general rule. At some U.S. Naval laboratories engaged principally in underwater research, full-time scientific divers are employed. Those involved throughout the whole system, however, number no more than 100. The part-time divers, scientists and technicians numbered about 2,000 in 1970. The rate at which new scientific divers come into the field is rapidly increasing, even under the austere budgets allotted to undersea research in the past several years. The current estimate is that there will be approximately 1,200 new scientific divers in 1971, and that the rate will increase 20% each year for the foreseeable future. If this growth rate is realized the field will expand to 12,000 divers by 1975. In the context of existing facilities this is an awesome number.

In 1971, the scientific diver's basic equipment is similar to that of the sport diver; it takes $500-$600 to outfit him. These divers usually operate out of a common diving locker and all but their personal equipment is maintained and used on a community basis. The average capital invest-

Table 12-4
Scientific Diving Equipment Expenditures
1971-1975 (in millions)

Year	Equipment	M&R	Total
1971	$1.40	$ .34	$1.74
1972	1.70	.48	2.18
1973	2.20	.65	2.85
1974	2.50	.87	3.37
1975	3.00	1.16	4.16

ment in a locker to support 60 divers ranges from $60,000-$100,000 or an average of $1,200 per man.

Each year approximately 10% of the total capital investment in the diving locker is spent for maintenance and repair. This figure does not include salaries of personnel associated with the operations. At a university diving locker with a capital investment of $100,000 approximately $10,000 is allotted to maintenance and repair; at governmental laboratories, the percentage is slightly higher; at commercial research firms the figure is lower. During the next 5 years the growth rate of the scientific diving field will remain relatively constant. Table 12-4 shows the total expenditure projected for products and services excluding personnel salaries.

It should be pointed out that these statistics represent the baseline spending for some 110 scientific diving programs in academic, governmental and industrial activities. Included are such items as small craft specifically acquired for diving purposes, but excluded are larger craft that serve as both classical research platforms and as diving craft. Also excluded are research projects such as the Sealabs, which although scientific in nature are more correctly classified under government projects.

Differentiating between government programs and purely scientific activity is academic. Until several years ago, most basic underwater research was sponsored through the Department of Defense. One of the principal centers controlling these funds was the U.S. Office of Naval Research.

Since fiscal 1968, however, there has been a gradual shift in this research to other government departments. Although the shift took place mainly because of budget tightening, it should be considered as a pattern which will continue.

Several other government departments are emerging as major sponsors of underwater research related to man's activity in the ocean—the Department of the Interior, the Department of Health, Education and Welfare and the Department of Commerce. They are concerned with maintaining an environment suitable for human habitation, the conservation of resources and basic oceanographic research. While the Navy was at the forefront in general underwater oceanographic research in the mid-1960s, it has given up its preeminent position because available funds are barely enough to maintain the forces necessary to prosecute its basic mission. Their research funds are now concentrated in the development of techniques, equipment and systems specifically tailored to meet current requirements. They have and will continue to maintain a leadership position in basic submarine physiological research through their laboratory system. Through the Office of Naval Research, funds do reach the scientific community, both academic and commercial, for research related to diving. These funds come from the same pot as the general Navy funds, but in recent years they too have been dwindling under the pressure of austerity.

The total funding for diving-related research which is not directly consumed within the Naval research community is now less than $3 million annually. The principal part of these funds are expended for salaries and expenses not directly related to diving. The actual funds that do result in procurement and services related to the diving market are included in Table 12-4.

In dealing with the complex funding structure from the federal government to scientific diving from departments other than the Navy, we enter the maze of the U.S. Oceanographic Budget. Tracking funds through this system is not unlike attempting to trace a line through a piece of marble—it is a three dimensional game. The funds seem to disappear from the surface and then re-

appear somewhere further on bearing a completely different tag.

At the upper level of the Oceanographic Research hierarchy we find great machinations taking place in competition for funds and authority. Nearly every major department has cast part of its lot into the sea. The list of government agencies with marine budgets reads like the U.S. Government Organizational Manual and includes the Department of Defense, Department of Commerce, Department of the Interior, Department of Transportation, Department of Health, Education and Welfare, Department of State, Atomic Energy Commission, National Science Foundation, National Aeronautics and Space Administration, Agency for International Development and Smithsonian Institution.

The total marine sciences budget including the Navy's Anti-Submarine Warfare Research and Development was $938 million is fiscal 1970 and estimated at $1,070 million for fiscal 1971; $1,200 million has been requested for fiscal 1972. Reorganization of the nonmilitary aspects of oceanography under the National Oceanic and Atmospheric Administration (NOAA) in 1971 shows some possibility of an organized national effort moving in one general direction, but any attempt at this time to predict the outcome would be the purest speculation.

In 1971, the total funding for projects that are purely scientific and directly associated with basic man-in-the-sea research or uses scientific diving as the major means of investigation amounts to $21.6 million. Of this, $18.8 million or 87% stays within the government research community; the remaining 13% is widely distributed between academic and commercial research institutions.

Although the federal government has been the primary source of scientific diving funds, other revenue sources are assuming some importance. The coastal states are acting to reverse the ravages of oceanic pollution. Now in the early stages, state activity to control the nearshore environment will provide part of the funding necessary to maintain a steady growth of the scientific diving community in this decade. Undoubtedly, the larger part of this money will go directly to the academic institutions, but commercial oceanographic research groups will receive a higher proportion than they have in past years.

In addition to state funding, waterside industry will make larger contributions to preserve the environment and will employ the services of information acquisition firms with underwater capability. On a small scale, this is already underway. Energy companies, particularly publicly owned utilities, have become extremely sensitive to thermal pollution problems and are gathering sufficient information to assure compliance with future regulations.

There will be a general shift in the center of gravity of scientific diving activity away from Department of Defense orientation toward environmental problem solving, but the basic ocean data now being amassed and the availability of funds to prosecute the problems assure a bright future for the field.

Military Diving

The U. S. military forces spend more money on diving than any other group in the world. The effect of their diving activity on the commercial aspects of diving, however, is not of the same magnitude as the level of expenditure. In this book all other parts of the diving business are treated as free enterprises into which capital flows and is distributed to other parts of the economy. In the military, diving is an integral part of defense operations, is not separated from other activity and with the exception of equipment and services procured docs not appear as a line item.

In this section all military diving will be treated as a closed loop within the federal government. In the context of the diving business, the only aspects of military diving that will be considered are those items that break this loop, i.e., the procurement of equipment and services from the industrial community. Those services procured from the diving business are part of commercial diving. This section, therefore, addresses itself primarily to the market for diving equipment in the military.

If the growth of diving and diving expenditure had followed the rate planned during 1965-1968,

military diving would be much larger than it is today. Most of the plans made at that time, including the Sealab experiments, the Large Object Salvage System, the 3800 series swimmer systems and many others of a classified nature, have since been modified, pared down or cancelled. In the light of recent progress in all aspects of Navy diving capability, the modification and cancellation of many of these projects may produce a positive effect even during this period of fiscal doldrums.

The reorganization of diving within the Navy has turned out to be considerably more than an administrative juggling; it has produced a complete change. The "not invented here" attitude has disappeared and has been replaced by an atmosphere of mutual and cooperative assitance between military and industrial diving entities. The change is apparent at nearly all but a few installations that seem to be out of the mainstream of current Naval policy.

The appointment of Admiral Elmo Zumwalt as Chief of Naval Operations has helped to reinforce the policies of the Office of Supervisor of Salvage and Diving to close the gap between state-of-the-art technology and Navy diving. Within a few years, the U.S. Navy will be using standard diving equipment that represents a 25-year leap in Navy diving technology.

Changes have permeated every aspect of Navy diving. All the requirements of cancelled or modified programs have been reassessed in regard to the new Navy missions and roles. Although research and development is at a considerably lower level than in the 1960s its effectiveness, measured in product relevance, reflects a sound base from which to move forward.

Even under a tightened budget, procurement of diving equipment and diving-related systems represents a sizeable market element. To extract total figures from the annual budget it would be necessary to travel three separate routes because basic research funds, research and development funds and procurement funds all take separate paths through the system. Those basic research funds relating to diving technology have been included under scientific diving. All other funds

Table 12-5
Military Expenditures
for Diving Equipment (in millions)

Item	1971	1975
Operational support equipment for tactical diving units	$ 2.7	$ 3.8
Special developments in support of tactical diving	4.7	7.0
Operational support equipment for fleet and EOD units	5.5	7.5
Deep diving research in support of fleet units	4.2	3.5
General research and development in support of diving	1.5	2.0
U.S. Army and other programs	6.3	3.0
Total	$24.9	$26.8

result directly from requirements established by the U.S. Navy's Supervisor of Salvage and Diving (OSSD), which is not only responsible for establishing requirements but also directly controls certification of equipment for general Navy use. The use or purchase of nonstandard or unaccepted equipment requires a waiver from OSSD.

Under the Naval commands system, where money may pass through seven or eight levels before actually leaving the system in the form of a supply contract, the multiple paths of the three types of money branch out into rivulets that finally reach industry. On basic development contracts of a major size, the OSSD maintains technical control and responsibilities for system acceptance, but administration of the contract is normally assigned to the Naval Material Command. Small procurements are usually handled by the Naval district in which the requiring unit resides. The bulk of equipment supply contracts emanate from these local Naval districts.

Procurement for U.S. Marine operating units, with the exception of specialized equipment for unique operations, is handled through routine Navy procurement channels. U.S. Army procurement of diving equipment, once made through Navy channels, is now handled by the Army's own internal material system. Unlike the Navy, where there is a preponderance of small-unit purchases, Army purchases are made in bulk and are much less diverse than those made by the Navy.

In recent years sales to friendly foreign navies have been increasing. Made with U.S. funds through the Military Assistance Program, they are normally handled by the embassy or technical mission of the purchasing nation. Delivery is made in this country directly to the technical mission. Export of certain types of diving equipment is rather complicated because of the "munitions" classification placed on military-applicable units.

The total expenditures for diving related equipment, excluding vessels and shore installations (shown in Table 12-5) are on the order of $25 million for 1971. This time period cuts across two fiscal years and represents the funds remaining in fiscal 1971 and the proposed expenditures for the initial half of fiscal 1972 (July 1, 1971- December 31, 1971). The figures for 1975 are based on speculation of continuing efficiency and a budget of the same magnitude of 1971/1972.

Next to the figures we usually see in national defense budgets, these amounts are certainly modest. A graph depicting the actual rate of spending in the six areas listed in Table 12-6 would be only a confusing criss-crossing of lines. Spending for operational support equipment would be nearly a straight line with a slight upward slope as the tactical warfare lessons learned in Southeast Asia are converted into standard hardware. Special developments to support tactical diving, on the other hand, are on a sharp downward slope that will bottom out in the next 18 months. Programs that have been long underway are being phased out and new, more relevant requirements are being introduced. By 1975 new work more firmly based on actual tactical experience will again bring this area of funding to the initial phases of major program status.

Regarding the modest funding for fleet unit support, it must be remembered that this mainly represents hardware contracts bringing developed equipment into fleet units. The deep diving research in support of fleet units will maintain a nearly constant level with changes in emphasis. A much higher percentage of the funds will go toward meeting EOD requirements.

General research and development to support Navy diving will be let to industrial organizations at increasing rates. It is now apparent that, dollar for dollar, commercial firms can provide excellent results at cost effective prices.

Unlike some areas of the diving business, military diving does not show the exciting growth potential exhibited in most market segments. It does, however, maintain a reasonably consistent level of business that is apparently surviving the relentless sword of government economy.

13. The Future: 1971-2000

Just barely into the 70's the whole field of diving is undergoing explosive growth. Technology and capability are expanding at a rate that is outstripping our ability to apply them. In commercial diving, economics is the controlling parameter of work depth; diving as a military weapons system is important in controlled tactical warfare; scientific diving has proven an important function in controlling oceanic ecology; sportdiving, after 10 years of steady growth, is still one of the few remaining areas in which man can come into contact with relatively unspoiled nature. From all standpoints—as a business, a potential market, a career opportunity or recreation—the outlook for diving during the next 30 years is extremely favorable.

Predicting things to come in a particularly volatile field is risky at best. In diving, where basic knowledge is being acquired at a rate that often proves yesterday's facts to be bad guesses, the degree of risk is even higher. Certain patterns have developed in the last decade, however, that will effect diving in the near future, and it is therefore feasible to extrapolate diving activity to 1976 with some degree of certainty and to 1981 with mild confidence. Projections into the period 1981-2000 are pure speculation based on our current ideas about the best use of the ocean as part of our environment.

1970-1976

Few if any revolutionary innovations will appear between 1970-1976 that have not been treated in previous chapters of this book. The major changes will center around optimizing existing systems and applying available physiological and engineering data. Although this may seem rather negative, the vast research data banks that have been relatively untouched and the generally haphazard methods of applying engineering technology to man-in-the-sea will show that many current problems result from management troubles and lack of communication. Reasons for the lack of communication are treated in other sections of this book. Generally, however, they relate to the competitive manner in which diving grew up and are complicated by the fact that no real communication channel exists where information can be exchanged between the commercial, military, scientific and sport diving communities. This situation started changing in the beginning of this decade and will continue to improve. Organizations such as the Marine Technology Society, the Offshore Technology Conference National Association of Underwater Instructors and commercial diving organizations, now in embryonic stages, are opening new channels for the transfer of information and techniques.

Another factor that will aid the organization of diving technology is the entrance of large industrial firms into the business of providing diving equipment. After several false starts, the industrial community has realized that man-in-the-sea provides a viable opportunity for profit oriented exploitation of the ocean. Quite different from earlier efforts to penetrate the market, the latest influx of industrial talent is based on sound business considerations. In the earlier attempts (nearly all failures) industry felt that if they made a sizeable private investment in the ocean, the government would eventually pick up the tab, as it had in the space industry. After a series of "next years," the message came through—opportunities were there, but only for profit-oriented competence on a competitive basis. Industrial organizations with the resources to invest in applying 20th century technology to the diver's engineering problems will

be responsible for rapid advances in equipment and standardization of techniques.

Before looking at specific changes, it would be well to review diving capability as of January, 1971. The obvious method is simply to consider depth capability (Table 13-1), even though this is not the most important criterion for meaningful analysis.

Compared to the depth capability a decade ago, the figures are impressive, but they do not represent a true picture of diving capability. They are, in fact, the results of experiments. For example, in the commercial field most work is and will continue to be shallower than 400 feet. In military diving the U. S. Navy has a single system capable of 850 feet; therefore, salvage operations at that depth are not an established fleet-wide fact. The listed scientific depth capability results from an experiment demonstrating the practicality of a lock-out type submersible for scientific purposes, and little has been done beyond depths of 200 feet since the 700-foot experiment was conducted.

The true basis for determining diving ability is how well the diver can perform his assigned mission at a particular depth. "Well" in this case means in comparison to his surface work capacity and takes into consideration such factors as economics and safety. In this context we see that in January 1971, there is still much to do. Even if we

Table 13-1
Open Ocean Workings

Activity	Depth Capability	Demonstrated by	Date
Commercial	850 ft.	French	11/70 (COMEX)
Military	850 ft.	USA (U.S. Navy)	10/70
Scientific	700 ft.	USA (Ocean Systems)	3/68

Simulated (Laboratory)

Activity	Depth Capability	Demonstrated by	Date
Great Britan	1,500 ft.	Royal Navy	3/70
France	1,706 ft.	COMEX	11/70
USA	1,050 ft.	IUC	4/68

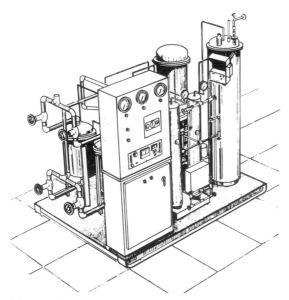

Figure 13-1. CVI Corporation's gaseous helium recovery system will reclaim helium at 99% purity for ½ ¢ per cubic foot. Helium reclamation innovations will be introduced in the near future.

assume that the experimental depth capability is real, we have just scratched the surface.

Commercial Diving

The most obvious changes occurring in commercial diving in the 1971-1976 period will be the organization of diving as a profession, and a marked increase in world-wide standardization of equipment and techniques. There will be no let up in the competitiveness of the business, but most competition will shift from "secret" equipment and methods to operating efficiency. Little change will occur in the type of work performed; offshore oil fields will still be the largest diving services market. Customers will continue to press for more economical diving services, and in response the diving companies will press for better decompression techniques and better equipment. The result will be better decompression tables giving an improved ratio of working/decompression time.

Diver tools will also receive great emphasis. Research conducted during the last several years will be the basis for a whole new generation of tools. These tools will not be merely underwater

adaptations of land tools but will be based on the actual ergometric capability of a neutrally buoyant diver. Most of the new power tools will be electro-hydraulic, with surface-supplied power units for shallow work and self-contained, underwater, electrohydraulic power packages for depths beyond 150 feet. It is doubtful that any revolutionary underwater power packs will be developed. Progress will be based on engineering and physiological data already in existence.

Economy will be the controlling parameter in improved life-support equipment. For instance, a major expense of deep saturation diving is helium, which will be treated as a capital investment rather than as an expendable item. Cost reductions in this area will be made by two methods. The first and most widely used will be effective reclamation of the diver's exhaust gases (see Figure 13-1). Already under development in the late 1960s, this equipment will be available to reprocess breathed mixtures, remove all impurities and compress the helium to high pressure. Several versions are now in the final stages. The most promising uses selective cryogenic condensation for removing impurities, and liquid nitrogen as a refrigerant source. Reported cost of reclaiming the helium at 99% purity is one-half cent per cubic foot.

Another cost reduction method will be completely closed circuit diver life-support systems. Already used by some European diving companies, the systems should be adopted by U. S. firms around 1975. There is still considerable resistance to this equipment, principally because of the relatively complicated electronic circuitry used for controlling oxygen partial pressure, and because the system is based on a different operating scheme than the equipment successfully used for the past 20 years. But the advantage of the closed circuit system from both economic and logistic standpoints will hasten its acceptance. In all probability, the ultimate combination will be cryogenic reclamation for gases used in decompression or saturation chambers and closed circuit apparatus for divers.

Little change is expected in commercial diving mobility requirements, so for simplicity and safety divers will continue to use umbilical connection, even though their life-support equipment could be used in SCUBA configuration. The umbilical cable will carry an emergency gas connection, electrical power for suit and gas heater, and communications connections. There will be a constant move toward reducing the size of the umbilical so as to increase the diver's movement capability, but the umbilical will not be discontinued for the first half of the decade and possibly not until the 1980s.

It is highly unlikely that electronic communication equipment designed to eliminate helium gas voice distortion will gain wide acceptance during the first half of the decade. Some progress was made along this line in the late 60s, but the complexity of the equipment combined with its lack of versatility and high cost have made acceptance extremely slow. Application of microcircuitry electronic techniques will improve both performance and dependablility but will not remove basic objections to the method or the stigma of past failures. One solution under consideration is a combination of less sophisticated electronics and a specially learned lexicon to minimize communication complexity.Work is underway to provide a digital communication system with conversion at the transmitter receiver through a voice reconstruction system. Although this offers promise, wide commercial usage in the next five to seven years is very remote.

Military Diving

The major changes to be realized in the 1971-76 period within the military diving area are already taking shape. Perhaps the most important change, the reorganization of diving into a managable entity, actually took place in the closing years of the 60s.

The first indications that the new organization will be able to cope with the job came late in 1970 when the Navy announced completion of tests on the Mark I Deep Diving System and the authorization to use the system to 850 feet. The U. S. Navy Salvage Office's goal of extending capability to 1,000 feet should be reached early in the decade. The exact value of this capability to reach and salvage vessels is questionable. However, the research and development that permitted the Navy to reach 850 feet provide the basis for rapid improvement in fleet diving equipment and techniques. The funds available will decide the degree of improvement.

Reorganization of the operating diving force within the Navy is not the only change that will effect military diving. The austere financial condition in all defense funding has pressured the Navy to look closely at requirements and to discontinue any research not leading to direct improvements. As a secondary effect, the Naval Laboratory System, which is primarily tasked with conduct or management of Navy diving equipment development, has had to carry out a drastic reorganization in order to be responsive to the Navy's operating needs in terms of available funds. But the result of this pressure has been positive. Irrelevant programs have been discontinued, duplication of effort has been stopped and the atmosphere is quite healthy for sensible, rapid development of equipment to meet the Navy's needs.

One promising development is alleviation of thermal stress on long, deep, mixed gas dives. Solving this problem requires heating the insprled gas to prevent excessive heat loss through the respiratory system and an effective, light method of replacing body heat lost to the helium and the water. The only available system that has proven effective is the open circuit hot water suit. But this system is not satisfactory for very deep diving missions on tethered equipment and is not usable on a free-swimming mission. Development is underway for new, improved closed loop water systems and for workable electrically heated undergarments. Research is also being continued on self-contained heater units, which may be available in prototype form for the free-swimming diver by 1973.

It is doubtful that any major breakthrough will be attained in military life-support systems. The equipment used with the Mark I Deep Dive System will gradually be integrated into fleet operations, so that by 1974 most of the antiquated standard or heavy gear will disappear. In the early part of the decade, new prototype free-swimmer life-support equipment will be ready for evaluation. The equipment, developed under contract to the U. S. Navy by Scott Aviation Company, will be a general-purpose mixed gas unit capable of meeting most tactical diving mission requirements.

Scheduled for 1972, the units could be in general service before 1975.

The U.S. Navy will probably not follow the commercial companies' lead in going to closed circuit breathing apparatus for salvage diving. The Navy is heavily committed to the semi-closed circuit apparatus for operation to 1,000 feet, and sparse funding will prevent any changeover. Very deep diving operations conducted by the Navy are limited, so a switch in deep life-support apparatus could not be justified.

The effectiveness of tactically trained Navy divers in low-profile, revolutionary and guerrilla warfare missions will be the basis for a marked increase in the Navy's support for tactical diving. This increase will lead to improved training facilities, the probable initiation of a center devoted to developing equipment and systems for low-profile warfare and an upgrading of tactical and combat diving status within the Navy.

Scientific Diving

Even with the amount of work to be performed in the shallow nearshore waters it is improbable that any revolutionary changes in scientific diving methods will occur before 1976. The academic generation that will appear on the scene by that time will consist of a much higher percentage of trained scientific divers. Interest and activity for *in-situ* scientific investigation will continue on the upswing.

Concomitant with the growing concern for the environment, ocean ecology will assume considerably more importance in both privately and government-sponsored aquatic research. The problems of nearshore ocean pollution have only been scratched, and answers must be forthcoming before any intelligent plan for reversing the damage can be implemented. Pressure for answers will cause the greatest increase in funding that underwater research has ever experienced. To accomodate the requirements, the scientific diving community will devise new methods to increase the research diver's capability. Unlike earlier work performed in the name of science, these improvements will not be geared toward making headlines but rather toward making the scientist's adap-

tation to the environment more complete and increasing his underwater work time. With the improvements in shallow water decompression techniques that will be available in the early 1970s, scientific divers will be provided with the physical wherewithal to remain underwater for long periods. This should not be construed as meaning through such projects as Conshelf, Sealab or Tektite. The new projects will be routine and will be conducted on a daily basis. Logistic support for the diver will be centered around portable, lightweight habitats designed to give the scientist a regular work day on the ocean floor.

Initial working experiments using portable habitats have already taken place in the inhospitable waters of the Great Lakes. Designed for daylong stays, these habitats will provide a dry environment in which the scientist can analyze his work, eat, and have a place to retreat when the stress becomes oppressive. Since most ecological research will not be conducted in the warm, gin-clear water of the tropics, the underwater "recoupe" stations will go a long way toward making work days in the littoral areas a little more bearable. The habitats will be designed for relatively shallow research; equipped and ready to go they will cost less than $20,000.

It is highly improbable that complex deep habitat systems will be a meaningful part of scientific exploration by man-in-the-sea. These systems are of questionable scientific value and data is collected at tremendous cost, so they are no longer considered feasible. In a speech presented to the Navy League Symposium at Washington D. C. in 1970, Jon M. Lindberg, one of the moving forces in undersea exploration, summed up the feelings of many of his peers when he said, "We have tended, I think, to devote too much of our advanced thinking to glamorous 'hows' of man-in-the-sea and too little to the more practical 'why'." For instance a large amount of money and effort in recent years has been allocated to fixed sea-floor habitat systems such as the French Conshelf, Link's activities, Sealab and Tektite. Some of them have been very successful technically and received generous publicity and acclaim. As a result, the sea-floor dwelling concept has developed its own momentum, and recommendations are widespread for even grander projects in the future. The Stratton Commission recommended that a sea-floor station to house as many as 150 men be installed on the continental shelves. The mission of all these men has not, however, ever been established.

Mr. Lindberg also commented on the nature of undersea living experiments:

All of the early undersea living projects, it is true, were intended to be experimental rather than to pay their way. Much valuable data has been obtained. But when analyzed for operational use, the seafloor concept has serious disadvantages. Logistical problems and hence cost are very great, perhaps an order of magnitude greater than alternate methods. The fixed habitat, being a sessile object, can serve only an area within range if its occupants. For swimming and walking divers, range is limited by umbilical hoses to 100 to 200 feet. Free swimming divers might be considered, but even with perfected equipment the hazards to saturated scuba divers away from their base are great. Small submersibles could be used for transporation, but their maintenance at the habitat would add greatly to its complexity. Who, then, is going to utilize the seafloor dwelling concept after experimental phase is completed? Industry is definitely not interested—costs are too high. Even for scientists whose missions and constraints are quite different, the value of fixed seafloor dwellings is questionable. Competition for research funds is acute and the lack of a wide ranging capability imposes great disadvantages.

If sea-floor habitats are to play any role whatsoever in the future of scientific diving, they will be of either the Tektite of Helgoland type, equipped for long stays in shallow water and greatly simplified for more economical operation. In the context of Mr. Lindberg's speech, the diving community has learned the "how" of sea-floor habitats, but unless someone comes up with a defensible "why" and a reasonable ratio for cost/data, the prognosis for their use in the next decade is very shaky.

The second half of the 70s will see some radical changes in scientific diving. The ideas, in prototype form, will begin making limited appearances around 1973. One aspect of research that will begin changing before 1976 is the sea-floor area a scientific diver can cover. Science fiction movies and TV notwithstanding, the scientific diver in 1971 is limited in horizontal mobility by the strength and condition of his leg muscles. Practical underwater transportation has been a subject of considerable rhetoric and experimentation; and solutions in most cases were impractical from either a cost or a use standpoint. The situation has changed and as a result of several well-funded research projects designed to provide lateral mobility for military tactical swimmers, scientific divers may now be able to drive to work.

Marine scientists at well-endowed research laboratories will soon have reliable underwater transportation available in several forms. Most promising is a civilian version of the second generation military swimmer delivery vehicles. Easy to maintain and operate, the SCUBASUB, Figure 13-2, is the first commercially available wet submersible to mate modern materials technology with the requirements of scientific diving. With it, the diver's horizontal range is extended from 500 to 1,000 yards to several miles, and the effort normally devoted to swimming can be turned to constructive work.

Sport Diving

Little or no change is expected in the activity or equipment used in sport diving. The principal difference will be in the services available. Even in the closing years of the 60s, it was evident that sport diving had a growth rate based on population expansion. As our cities and coastal areas have become more and more populated, sport diving as a pleasurable escape has sustained its following. As a recreational pastime, it contributes little to the technological expansion of man-in-the-sea. Its primary objective will be to achieve more effective manufacturing of more dependable equipment.

General 1971-1976

In summary, the 1971-1976 period will be devoted mainly to improving existing systems and

Figure 13-2. Small wet submersibles like this SCUBASUB 300 will extend the scientific diver's range from hundreds of yards to several miles.

Saving time and energy, these boats will greatly increase diver effectiveness. (Photo: Submergence Technology Corporation, San Diego.)

applying new technology to unsolved or partially solved problems. A primary target for the materials expertise will be the thermal stress experienced on long, deep dives and free-swimming dives in cold water. For diving suits, new basic materials employing a noncompressible structure and a high insulation characteristic will appear on the market during this period. With this portion of the problem solved, the additional task of providing self-contained heater units will fall with the realm of practicality. By 1975, thermal stress problems on mixed gas or long-exposure missions should be a thing of the past.

By 1970 the basic data and prototype equipment needed to economically and safely extend divers' bottom time and work capacity had appeared. The ergometric data relating to diver work capacity make it feasible for ocean engineers to design equipment the diver can successfully manipulate; closed circuit mixed gas breathing apparatus and gas reclaimers capable of reducing diver support costs on deep, long dives have been

successfully tested and will be on the market by 1973; improved physiological information for both compression and decompression is already providing a basis for major revision and improvement of decompression techniques.

Small wet submersibles for transporting military, scientific and marine technician divers will gain widespread use by 1974 and will lead to considerably more sophisticated vehicles by the end of the 70s. New concept craft, such as the Shelf Diver (1968) and the French Argyronete (see Figure 13-3), scheduled for completion in 1972, indicate the role to be played by man-in-the-sea—a role completely different from that envisioned 20 years ago.

Diving 1976-1980

Spurred by a world-wide realization of the ocean's importance as a resource and as a political battleground, diving will undergo a metamorphosis in the second half of the decade. One of the most

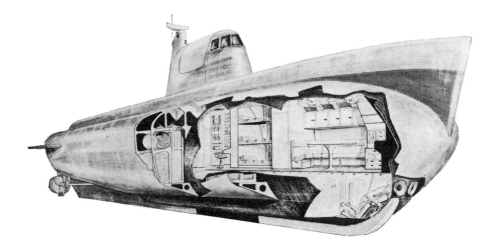

Figure 13-3. Long-range, deep surveys will be conducted by sophisticated submersibles similar to the Argyronete. Using lockout techniques, scientific divers will have access to all the continental shelf. Specifications: surface displacement, 255 tons; diving displacement, 300 tons; over-all length, 82 ft.; over-all width, 22 ft.; pressure hull (wet), 7 ft.; OD x 16 ft. long; pressure hull (dry), 12 ft. OD x 40 ft. long; design depth, 1,970 ft.; crush depth, (3) 3,800 ft.; surface cruising speed, 6 kts.; surface maximum speed, 7 kts.; diving speed, 4 kts.; battery capacity, 1,200 kwh; power available for underwater work (6 hrs/day), 20 kw; with auxilliary power, 100 kw.

controversial questions will be what state or country holds jurisdiction of what part of the sea and to what degree. Fishing rights have always been a matter of negotiation between maritime countries, and they will continue to be a point of international misunderstanding. In fishing, the territorial perrogative of the country claiming ownership is usually violated on a transitory basis—in the recovery of other oceanic resources, principally from the continental shelves; intrusions will be extensive and long-term. With access to all parts of the continental shelves through man-in-the-sea, the floor of the "territorial sea" will be a major bone of contention. At the beginning of the decade, claims for the territorial sea ranged from 3 miles by major powers up to 200 miles by some of the smaller countries. International proposals to extend jurisdiction to the continental shelf limit, normally a water depth of 200 meters, are under consideration. Even if this arbitrary limit is accepted, the probability of misunderstanding and confrontation is high.

Although territorial boundaries of the free seas will be in question, there will be no question about man's ability to reach and work on continental shelves anywhere in the world. Laboratory experiments in the middle of the decade will show that man can work from 2,000-2,500 feet in the ocean. The controlling parameter will again be the economic necessity of putting him there and the cost involved. There may be legitimate reasons for men to work at 1,000-1,500 feet, but motivating "whys" will be few and far between. For ecological work, the nearshore, relatively shallow shelf will remain the area of concern; for purposes of resource recovery, the continental shelf will be the zone of exploration and exploitation.

It is possible to predict with a fair degree of accuracy the technological developments that will be available to diving. Which ones will be adapted by the various segments of the field is of course governed by the economic constraints placed upon them. These constraints will be secondary ramifications of the overall economic atmosphere and directly related to the ability of ocean-oriented business to continue its present growth rate and the manner in which national resources are apportioned to meet national goals.

Some government support will be required if we are to solve physical and ecological oceanic problems. Government participation in ocean-oriented activity will, however, never reach the proportions it did in the aerospace industry. Basic research in the field, particularly in areas not related to direct exploitation, will be sponsored by federal funds. The applied research required to extend man's ability to work underwater will be taken over by private interests. This general trend will also effect the ecological work performed in the sea and most work in the latter part of the decade will be performed by people interested in making a profit. Ocean ecological control will, by this time, not be looked upon as a restraint to living but will be accepted as a process for resource reclamation.

Diving Equipment

Changes in diver equipment will reflect a simplification of diver dress and a better relationship of equipment to both the anthropometric characteristics of the underwater technician and the mission to be performed. In all phases of diving, the trend will be toward completely untethered life-support systems. Mixed gas units will be of the closed circuit type, with control functions performed by microelectronics circuitry. The mean time between failures on these controls will be decidedly better than at the beginning of the decade. Maintenance requirements will be minimal and the basic cost of the units will be proportionally lower than those now on the market. General purpose mixed gas SCUBA design will take advantage of the new material developments of 6,000 psi will be carried in small pressure flasks. Partial pressure of oxygen will be capable of variations to maintain the proper level from the surface to 1,500 feet. Diving time for a single charge will vary little as a function of depth, and a capability of 10 hours at 1,500 feet will be standard. Although this time will exceed the work capacity of man, it will provide a considerable margin of safety and a psychological cushion for the technician.

Deep mixed gas diving will be performed under saturated conditions using deck-mounted saturation systems employing the same functional units used in 1971. Creature comforts will greatly

improve and decompression will be completely preprogrammed and computerized. Computer technology and advanced sensing instrumentation will play an important role near the end of the decade, and decompression accidents resulting from human error or miscalculation will be a thing of the past. Dives will be programmed in software with a sensor override. The sensors will be sophisticated by 1971 standards but normal technology by the end of the decade. Using both sonic and electronic techniques, the miniature sensor package will be attached like a thin belt to the diver and will sense and transmit information about the diver's vital functions and detect formation of inert gas bubbles in his system long before the onset of symptomatic pain. Output from the sensors will be buffered into the computerized decompression control system and will automatically cause a change in program if necessary. In all probability, the computer control will stop the program in process, announce the need for a program change based on the physiological data, ready the program and go into the new schedule as soon as it is released by the diving supervisor.

Adequate sensors, combined with vastly improved navigation and communication equipment, will at last free the diver from the umbilical—by 1980, in fact, there will be no defensible reason for its use. Looking at the field today, it does require some stretch of the imagination to visualize such technology gaining a foothold in a business that generally operates on a "show me" basis. However, taking into consideration that the requirements of diving are bringing in a new generation of technicians familiar with the latest technology and, further, that most larger diving companies are becoming divisions of large, technologically oriented corporations, it is not unreasonable to expect a drastic change in attitude toward acceptance of new methods.

On the basis of the amount of sophisticated electronics that will be an accepted part of diving technology at the end of this decade, the changes will indeed be drastic. Electronic innovations will not be limited to the control and decompression systems but will also form part of the diver-worn equipment. For example, the communication channel used by divers while underwater will be complex by 1971 standards. Several possible avenues of development could lead to a satisfactory standard system within the next 10 years. The most promising route appears to be automatic digital communication—a route already under active development in other facets of advanced communication technology. A successful system will depend on a combination of education within the field, application of phonetic experimental research now underway and intelligent use of state-of-the-art microelectronics technology.

The function of education in this program will be to introduce a complete lexicon for divers.

Few communication situations in underwater work cannot be expressed in a language limited to 500 words. This restricted vocabulary will provide a defined area in which experimental phoneticians can devise a system to electronically recognize specific phoneme patterns, even when they are distorted by a mixed gas environment. Once recognized, the voice signal will be converted to a digital signal and transmitted over an acoustic link to a receiver; at the receiver, the digital signal will be reconverted to a synthetic voice signal for application to a suitable audio transducer. Prototypes of the transmission system and the synthetic voice generator are already available. The remaining task of developing a phonetic recognition device will soon be within the realm of practicability.

Voice communication equipment will also be used for monitoring the diver's physical condition while he performs his underwater mission. The same sensors used in the decompression and compression stages will operate during the active part of the dive to keep control station apprised of any change in the diver's base functions. The data will be transmitted over the channel on a time-sharing basis and displayed on a master console operated by a single technician.

A solution to the critical problem of navigation for free-swimming divers will also form part of this communication channel. The signal transmitted from the diver will pinpoint his location for the dive controller. On free-swimming operations, positions of all divers will be shown on a display system not unlike that used in radar consoles. Navigation information will not be a one-sided affair. Azimuth and range to a specific base point

will be constantly available to the swimmer. Several methods of display are possible; based on current human behavioral studies, a combination audio-visual input will be used. Azimuth will appear at the top of the diver's mask—all he need do is look upward to see his orientation. An audio signal announcing range will reinforce the visual display. The range signal will be presented in such a manner as to recreate a sound source emanating from the base; that is, the signal applied to each ear will vary according to the diver's attitude, with maximum signal applied to each ear when he is swimming toward the base point.

The scientific diving community will continue to be principally engaged in defining the problems and gathering data of the nearshore continental shelves. The amount of data gathering required to intelligently plan reconstruction of the damaged areas of our oceans will provide ample work for the remainder of the decade and a good portion of the 1980s. There will be increased participation of scientific divers in moderately deep research through the use of submersibles built on the style of the Argyonete. These specialized scientific divers will be performing lockout operations at 1,500 feet for collecting data and installing and maintaining instrumentation.

Moderate depth habitats and reliable underwater vehicles will bring scientists to deeper water on the continental shelves. In the early 1970s, mixed gas diving was an economic impossibility for most academic research organizations because of the high cost of acquiring, operating and logistically supporting these operations. Completely closed cycle breathing apparatus will bring the cost of mixed gas SCUBA operation to approximately $2 an hour—well within even the most austere budgetary restrictions. With the standard available instrumentation, these moderately deep operations will be as safe as scientific air SCUBA diving was in 1970.

Diving as a sport will maintain its steady growth in the latter half of the decade. The principal life-support equipment will continue to be open circuit air SCUBA, modified to include some technology developed for other portions of the field. The most noticeable change will be in air storage tanks and diver dress. Air tanks will be considerably smaller and lighter. Working pressure of these vessels will be about 4,000 psi, with a weight only 50 percent of those now in use. There will be no attempt to increase the sport diver's bottom time; the main thrust of improvement will be directed toward simplifying the diving process and adding convenience. Except for skin diving, the present face mask will be out of use, replaced by a light, molded polycarbon full face mask; the bite-type mouthpiece will be relegated to a place of honor in the diving history museum.

The "toys" available to the sport diver will be wide and varied, including a comprehensive assortment of underwater vehicles to completely remove the pain from underwater excursions. Some vehicles have already made their debut, but these are the precursors of a breed of underwater propulsion devices that will enable the sport diver to move around the underwater environment at a comfortable speed of 2 to 3 knots.

1981-2000

By 1980 there will be no question about man being a reentrant mammal; the sea will be part of his regular working environment. True, he will not have access to the deep continental slope and the basins without the aid of submersibles, but his ability to work in all the relatively shallow regions will enable him to implement the programs we can now envision.

Between 1971 and 1980, research on the hyperbaric limit of human tissue will determine man's ultimate depth capability. Present estimates vary from 3,500 to 5,000 feet. These estimates are, for the most part, guesses—educated guesses based on the response of animal tissue to high pressure. Even if the pressure question is answered academically within the next 10 years, the complexity of supporting life at these pressures will be monumental, but not impossible. Hydrogen as the inert carrier in a breathing mixture combined with life-support equipment that provides assisted breathing could answer the problems of respiratory ventilation at those depths. The decision as to whether this working depth will become a reality will again be determined by economics.

Despite the ecological pressure it will generate, recovery of hydrocarbon resources will continue to grow at a rate directly related to the world's population, the GNP of the industrial nations and the ability to the preindustrial nations to consume power and commodities at affluent society levels. The last quarter of this century will see an increasing proportion of resources, both mineral and hydrocarbon, coming from the seas. In the field of offshore oil and gas recovery emphasis will be on using the undersea environment for more direct recovery. Undersea installations will include not only wellhead, separators, pipeline and storage, but all power and equipment necessary to move the oil and gas to transportation terminals as well.

Even if we discount 50 percent of the ideas offered by the futurists, we can expect that as we approach the year 2000 we will have developed the "whys" for what we now call undersea colonization. The first of these colonies will be no more than an undersea oil platform with attendants and marine (diver) technicians spending up to a week in the offshore site. Oil field maintenance, repair and monitoring will be performed by marine technicians using underwater vehicles, probably fourth generation descendants of the swimmer delivery vehicles now entering the field. These vehicles will probably be equivalent to underwater pickup trucks designed to carry the technicians and their equipment to wellheads, control or pumping stations within a range of 5 miles. Accurate navigation will be a problem long solved, and the standard life-support equipment will provide a comfortable environment for the diver during the 6 to 8-hour work day.

Projecting technology into the last 20 years of this century cannot be done by straight-line extrapolation, which would severely underestimate the level of technical competence. According to the visionaries of our time, we must consider the factors of synergism and serendipity in any projection of this length. *Synergism* is the combined action of two or more substances, situations or conditions to achieve an effect of which neither is individually capable. *Serendipity* is the faculty of making fortunate and unexpected discoveries by accident. In substance, the development of man-in-the-sea has proceeded to a point where progress hinges not only on developments within the specific field, but also on those in other scientific disciplines. Work now underway in other fields may produce conditions that will generate scientific breakthroughs. Looking at the future of a field that is undergoing massive and rapid technological advancement, the probability of serendipity, the unplanned discovery, cannot be discounted.

In their book *The Year 2000*, Herman Kahn and Anthony Wiener speculate on the development of technology up to and just beyond the year 2000. They develop several lists of technological breakthroughs, some synergistic, some serendipitous, that might reasonably be expected in the last third of the 20th century. Although a few of these deal directly with the underwater world, it is not difficult to establish interrelationships and to project their effects upon our future aquanauts.

One area receiving primary attention by Kahn, Weiner and other futurists is the availability of new "supermaterials." Projections include nearly every type of substance now on the market and under developement: high-strength and high-temperature structural materials; super-performance papers, fibers, plastics, glasses, alloys, ceramics, intermetallics and cermets. Developments in some of these areas are already beginning to effect the capability of divers. True, in 1971 the interrelationships between materials technology and diving equipment had not proceeded as far as might be expected. But by 1980 this situation will have changed. Little if any brass, steel, or what we now call noncorrosive metals will be used in diver equipment. Closed circuit SCUBA equipment will no longer use chemical filter substances for removal of respiratory waste; ceramic molecular screens will do the job, with the benefit of weight and bulk reductions and an increase in safety and economy. Fabrication of noncorrosive, maintenance-free structures for the underwater environment will be standard engineering practice. Nuclear underwater power packages and economically feasible cryogenic gas supplys will make the autonomous underwater habitat a commercial practicality by the year 2000.

The degree to which underwater colonization will materialize is a purely speculative issue. Not only does it depend on technological advances, it is equally dependent on economic and cultural changes that will take place during the next 20 years. We can certainly expect resource industries that will be completely underwater. Resources will be taken from the sea, processed in underwater plants, transferred to submarine transports and then delivered to shoreline terminals. These plants will be established for mineral recovery from both the sea floor and the water itself. If oceanic pollution is reversed in the next decade, these underwater industries may also include food production. Sea farming to meet the world's growing protein needs has been one of the overworked features used in all attempts to promote ocean exploitation. It now appears to be controlled by a negative synergistic factor; pollution of the ocean is rapidly making the environment unsuitable for foodstuff production. If we cannot reverse the pollution process within this decade, we may find that the ocean not only is unsuitable for artificial farming, but that nothing from the sea is edible.

In the category of tools, underwater shaped charge explosvies will reach a state of sophisticated development. This technology will be widely applied in underwater construction, particularly in cutting and joining large-diameter pipelines. Other joining methods such as adhesives and welding will also undergo radical change. Wet welding of a quality obtainable today only under dry welding conditions will be standard. The welding process will undergo a gradual change, first through the technique of plasma welding, which will produce high-quality joints slowly and expensively, then through lasers, which will produce nearly perfect joining rapidly and economically.

The increasing placement of valuable scientific and industrial complexes in remote locations on the sea floor will cause a whole new set of security problems to emerge. Underwater sensors capable of monitoring large areas will be commonplace. These sensors will not be perimeter-type devices common on land but will be three-dimensional space monitors able to detect infiltration from any direction. Underwater law enforcement will be required on both a national and an international basis and there will undoubtedly be a new breed of

Figure 13-4. By the year 2000, dolphins will be widely used for both survey and patrol missions.   *This 1969 photo shows a dolphin equipped with a radio telemetry instrumentation package.*

security personnel—the underwater Pinkerton, or equivalent.

From a military standpoint, the nearshore environments of all the major powers will gain in importance. Of all the scenarios that can be formulated for games of the future, the most consoling is that nuclear war, at least in its grandios aspects, will be averted. Warfare will be limited to insurgency and counterinsurgency operations—a type of warfare well-suited to the underwater. Infiltration of military and para-military personnel will be accomplished by water, and the special warfare forces will be the principal tactical weapon of both

offense and defense. It may well be that by the year 2000 the underwater environment will be the scene of major confrontation, a place where warfare of the 21st century will be fought.

An Optimistic Scenario for the Year 2000

Looking to the year 2000 in the underwater world we are forced into complete speculation. This is interesting from several standpoints; first, 2000 is so far off that the speculator, within the bounds of sanity, may say things with great impunity because there are no facts to controvert; secondly, it is a fascinating game to take today's technology, synthesize synergisims, allow a margin for unplanned discovery and then translate this imaginary state of technology into the specifics of man-in-the-sea.

The following paragraphs are the result of such a game. Not all the results will suit everyone's taste, particularly those dealing with biomedical and genetic aspects of the world 30 years hence. Nevertheless, even these areas are projected in an optimistic but conservative manner from a base of current knowledge.

Before going underwater, we might look at a map of the world. There is little or no physical change except for the imaginary boundary lines which form the borders of countries touching the oceans. The maps show the boundaries extending out to the 200-meter contour of the ocean bottom. In this new water area the respective coastal country exercises jurisdiction, controlling all resource rights on the surface, in the water and on the sea floor. Beyond the 200-meter contour, the slopes and the basins are under international control. Any country or consortium may use these areas on a lease arrangment through the International Oceans Commission, which grew out of the United Nations in 1985. The boundaries were established by a five-year international survey conducted between 1987-1991, with a final agreement in 1995.

The problems of pollution are not yet solved on an international basis. Effluent and waste from the industrial and postindustrial societies are no longer emptied into the ocean and has not been since the late 70s, but the effects of centuries of aggravated pollution have not been completely reversed. Some 20 years after it was technically feasible, sea farming has begun; the residual pollutants have finally been reduced to a point where they are biologically inactive and the sea is once again a viable source of protein for the world's 6.5 billion people. The prospects of high-yield protein are excellent through the use of several laboratory-created food chains and geneological control of the food species. The only surviving air-breathing water mammalian creatures are those of the cetacean family. All are used extensively in research and survey programs. Interspecies communication is an established fact but not through the expected audio interpretation route; during the mid-80s experiements were successfully conducted implanting electrodes into the mammals' brains, and direct electronic communication was developed using a computer as a two-way translater.

All the major naval powers use dolphins to patrol their water space (see Figure 13-4). Few details of the animal training or conditioning have been released except that certain of their instincts, such as herding, have been suppressed through cryosurgical brain modification, and a large part of their learning and psychological motivation is performed by direct electrical input to the nervous system.

The United States, Japan, Russia, France and nearly all the Scandanavian countries have established inhabited underwater industrial complexes in their water space on the continental shelves. The habitats, maintained at the ambient pressure, are self-contained and self-sustaining. The internal atmosphere is maintained at a comfortable 75°F with a closely controlled composition based on habitat depth. Clean nuclear power provides the basic energy source and cryogenic systems are used to clean and reconstitute the breathing medium. Except for the start-up oxygen and carrier gas, no outside supply is needed. Makeup oxygen is provided from a redundant system; the principal units produce the bulk gas through a forced biological process using a combination of reclaimed carbon dioxide, synthetic plants and artificial light; the other supplementary system, capable of producing all the required oxygen in an emergency, uses the surrounding ocean as a gas source.

The underwater habitants are still gas-breathing. Their visible physical characteristics are no different from those of their ancestors 30 years ago, but extensive internal modification has occurred. Their average IQ is nearly 30 points higher if measured on a 1970 scale; their physiology has been chemically modified to reduce discomfort at lower temperatures in a light gas environment; and, through the use of psychobiological chemical treatment and conditioning, long periods of isolation from the surface environment are taken as a matter of course.

The standard underwater work assignment is three to six months, but many of the personnel voluntarily extend for periods up to one year. As with space platform assignments, no aquanaut is permitted to remain underwater for more than one year without a three-month reassignment to the surface. Tests conducted over the past several years show that the synthetic underwater environment is actually healthier than that on the surface in terms of both physical and mental well-being. The supercontrolled environment represents a maximum of man-created aesthetics, opportunity for educational and vocational advancement. More importantly, it is one of the few places left where a human can derive a sense of accomplishment from his work.

The long-term effects of living in elevated pressure are completely under control; in fact, since 1995 there has not been one fatality that could be traced directly to the secondary medical effects of pressure. The only cumulative effect ever attributed to pressurized living resulted from the processes of rapid compression and decompression, which is no longer a problem. In the transition from surface to ambient pressure or the reverse, aquanauts are placed in a state of medical "hybernation" ranging in length from a day to a few weeks. Compression and decompression are computer-controlled and programmed at an incremental rate of pressure change to prevent even the slowest tissue from developing enough differntial pressure to cause stress. The complete process is an extremely pleasant experience for the aquanaut and includes such mood and attitude controlling devices as preselectable and programmed dreams.

On excursions from the habitat complex, the aquanaut uses an extremely simple-looking life-support system that is actually quite sophisticated by 1971 standards. Based on cryogenic principals, the unit is "fail safe" and capable of supporting even heavy labor for a period of eight to twelve hours. Underwater vision under all turbidity conditions, day or night, is excellent but synthetically produced. Using a broad beam, high-resolution sonar (not mutually exclusive in the year 2000), a three-dimensional, full-color image is presented to the aquanaut by holographic techniques. Except for near-point vision tasks, no artificial illumination is required.

Generally, underwater working conditions are highly desirable; tools are effective and easily managed; heavy work is performed by machines "slaved" to the aquanaut; underwater lateral transportation at speeds up to 50 knots is a reality; habitat living conditions leave nothing to be desired. The only concern to the aquanauts, the only one that has not been successfully treated chemically or soothed by educational propaganda, is the appearance of *Homo aquarius* in the environment. It did not come as a shock, even though only fragmentary information has been available from the news computer since project Aquarius was taken over by the International Ocean Commission in 1988.

The project was an outgrowth of experiments conducted during the 1960s to determine if land mammals could survive by breathing a highly oxygenated water solution. After limited success in subsequent experiments conducted on low order mammals and finally on humans, complete success was achieved in the early 80s by use of AWBA (Assisted Water-Breathing Apparatus), an electrically operated device that amplifies the action of the diaphragm and thoracic muscles. The program was taken over by the IOC because no country would accept complete responsibility for continuing the project. It was agreed that if complete implementation were to be achieved, massive genetic modification of the human system would be required. Under the project management of IOC, work continued and is still underway in the United States, Japan and Russia. The water-breathing *Homo Aquarius* are being given longer and longer exposures and appear to be surviving; prognosis sets their life spans at only a little less than that of their air-breathing counterparts. From the scanty information available, the aquarians' responses to the experience have been positive with two exceptions—they have been unable to make the transition from air to water without an initial psychological shock upong coming out of medical "hybernation" in the water environment, and secondly, they object to the underwater communication method, direct brain stimulation, because it infringes on the human prerogatives of private thought.

Bibliography to Part 4

Booda, Larry L. 1971. More dollars for marine sciences and technology in FY 1972. *Undersea Technology* vol. 12, no. 12.

Craven, J.P. 1970. Marine technology and the urban crisis. *Undercurrents* vol. 3, no. 3.

Dinnen, J. 1969. Capt. Mitchell reviews Navy salvage and diving operations. *Data* vol. 14, no. 11.

Dubai's offshore facility: a preview of future systems: *Ocean Industry* vol. 6, no. 3.

Commission on Marine Science, Engineering and Resources. 1969. *Our nation and the sea; a plan for action*. Washington: U.S. Government Printing Office.

Hardach, J. 1968. *Harvest of the sea*. New York: Harper & Row.

Kahn, H. and Weiner, A. 1967. *The year 2000, a framework for speculation on the next thirty-three years*. London: MacMillan.

Kenny, J.E. 1971*a*. Diver communications 1971. *Ocean Industry* vol. 6, no. 2.

___ . 1971*b*. Survey of buying habits; 100 southern California skindivers. San Diego: Ocean Market Consultants.

Kylstra, J.A. 1969. The feasibility of liquid-breathing and artificial gills. In *The physiology and medicine of diving and compressed air work*. Bennett, P.B. and Elliott, D.H., eds. London: Bailliere Tindall and Cassell.

Lindberg, Jon M. 1971. What direction for men in the sea? *Undercurrents* vol. 3, no. 2.

Mishler, H.W., and Randall, M.D. 1969. Underwater joining and cutting. *Battelle Research Outlook* vol. 1, no. 1, pp. 17-22.

National Petroleum Council. 1969. *Petroleum resources under the ocean floor*. Washington: U.S. Government Printing Office.

Navy reorganization emphasis diving. *Undercurrents* vol. 3, no. 3.

Paganelli, C.V., Bateman, N. and Rahn, H. 1966. Artificial gills for gas exchange in water. *Proceedings, third symposium on underwater physiology*. Lambertsen, C.J., ed. Baltimore: Williams & Wilkins.

Research Dept., Peterson Publishing Co. 1970. *1970 skindiver survey*. Los Angeles: Peterson Publishing.

Taylor, D.M. 1969. Potential of the sea. *Ocean Industry* vol. 4, no. 6.

Glossary

ambient the environment surrounding a body.

anoxia (anoxemia) the absence of oxygen; an abnormal condition produced by breathing air which is deficient in oxygen.

antisubmarine warfare operations conducted against submarines, their supporting forces and operating bases.

aseptic bone necrosis (ABN) damage to bone; thought to be caused by repeated, rapid decompression procedures.

assault diver Swimmer diver used for military assault operations (U.S. Navy).

atmosphere pressure equal to normal atmospheric pressure at sea level (14.7 psi).

atmospheres absolute (ATA) total pressure at a depth underwater expressed as a multiple of normal atmospheric pressure.

baralyme carbon dioxide absorbent chemical used in closed and semi-closed circuit breathing apparatus.

base line a surveyed line established with more than usual care, to which surveys are referred for coordination and correlation.

bathyscaphe (bathyscaph, bathyscap) a free, manned vehicle designed for exploring the deep ocean.

bathysphere a spherical pressure proof chamber in which persons are lowered for observation and study of ocean depths.

bends see decompression sickness.

bio-diver scientific diver working in the field of biology.

biosphere the transition zone between earth and atmosphere within which most forms of terrestrial life are commonly found.

bivalve one of a class of mollusks generally sessile or burrowing into soft sediment, rock, wood, or other materials. Bivalves possess a hinged shell and a hatchet-shaped foot, which some use for digging.

blowup diving accident in which the diver becomes positively buoyant and makes a rapid, uncontrolled ascent to the surface.

bounce dive a rapid dive with a very short bottom time to minimize decompression time.

breathing bag part of the semi-closed circuit breathing apparatus used to mix gas and assure low breathing resistance.

British thermal unit (BTU) a unit of energy defined as the heat required to raise the temperature of one pound of water one degree Fahrenheit.

buoyancy that property of an object that enables it to float on the surface of a liquid or ascend through and remain freely suspended in a compressible fluid such as the atmosphere.

caisson disease see decompression sickness.

carbon dioxide excess a partial pressure of carbon dioxide in equipment or the body above normal human tolerance.

carrier gas see inert gas.

center of gravity a point at which the mass of the entire body may be regarded as being concentrated.

cetacean a marine mammal of the order *Cetacea,* which includes the whales, dolphins and porpoises.

chaffing gear heavy canvas or nylon overgarment worn to protect diving suit.

chemical oceanography the study of the chemical composition of the dissolved solids and gases, material in suspension, and acidity of ocean waters and their variability both geographically and temporally in relationship to the adjoining domains, namely, the atmosphere and the ocean bottom.

clearance diver explosive ordinance disposal diver (Royal Navy).

closed center hydraulic tool system in which oil flows only when power is being used.

closed circuit a life-support system in which the gas is continually recycled, carbon dioxide removed and oxygen periodically added.

coast the general region of indefinite width that extends from the sea inland to the first major change in terrain features.

Coastal area the land and sea area bordering the shoreline.

compression the raising of the ambient gas pressure to a specific water depth equivalent.

compression chamber see decompression chamber.

condensation the physical process by which a vapor becomes a liquid or solid.

cone shell a tropical marine snail of the family *Conidae* possessing a venom injecting apparatus used to subdue its prey. Several Indo-Pacific species have been implicated in human fatalities.

conduction the transfer of energy within and through a conductor by means of internal particle or molecular activity, and without any net external motion.

continental shelf a zone adjacent to a continent or around an island, and extending from the low water line to the depth at which there is usually a marked increase of slope to greater depth.

coral the hard calcareous skeleton of various anthrozoans and a few hydrozoans (the *millepores*), or the stony solidified mass of a number of such skeletons. In warm water, colonial coral forms extensive reefs of limestone. In cool or cold water, coral usually appears in the form of isolated solitary individuals.

coral head a massive mushroom or pillar-shaped coral growth.

coral reef a ridge or mass of limestone built up of detrital material deposited around a framework of the skeletal remains of mollusks, colonial coral and massive calcareous algae. Coral may constitute less than half of the reef material.

crustacean one of a class (*Crustacea*) of arthropods which breathe by means of gills or branchiae and with the body commonly covered by a hard shell or crust. The group includes the barnacles, crabs, shrimps and lobsters.

current a horizontal movement of water.

cyanosis bluish skin due to lack of oxygen in the blood.

datum any numerical or geometrical quantity or set of such quantities which may serve as a reference or base for other quantities.

dead air space spaces in diving equipment and in the human respiratory system that receive minimum ventilation.

decibel a value that expresses the comparison of sound of two different intensities. This value is defined as ten times the common logarithm of the two *sound intensities*.

deck decompression chamber (DDC) a decompression chamber of one or more compartments designed as a portable or permanent installation on the deck of a work boat or barge.

deck decompression complex (DDG) two or more decompression chambers with separate pressure controls for compressing and decompressing teams of divers during saturation diving activity.

decompression the process of gradually lowering elevated ambient pressure at a rate that permits the body to release inert gas without any detrimental effects.

decompression chamber a pressure vessel of one or more interconnected chambers that can be pressurized with air or mixed gas to simulate ocean depths. Used in hyperbaric experimentation, standard and saturated diving techniques and medical treatment of diving accidents.

decompression sickness (bends, caisson disease, compressed-air illness)   a condition resulting from the formation of gas bubbles in the blood or tissues of divers during ascent. Depending on their number, size and location, these bubbles may cause a wide variety of symptoms including pain, paralysis, unconsciousness and occasionally death.

deep scattering layer   the stratified population(s) of organisms in most oceanic waters which scatter sound.

deep sea dress   see standard dress.

demand system   diver life-support equipment in which gas flows only during diver's inhalation and exhalation.

depth   the vertical distance from a specified sea level to the sea floor.

differential pressure (P)   a difference in absolute pressure on the two sides of an interface.

diffusion   the spreading or scattering of matter under the influence of a concentration gradient with movement from the stronger to the weaker.

dilutent gas   see inert gas.

diving bell   a cylindrical or spherical enclosure used to lower and raise divers at a work site.

diving chamber   see diving bell.

dolphin   a member of the cetacean suborder *Odontoceti*. The name is used interchangeably with porpoise.

drilling rig   floating platform or ship used primarily for drilling offshore oil wells.

dry submersible   a small submarine in which the occupants are maintained in a dry environment at near atmospheric conditions.

dry suit   protective diving garment which is completely sealed to prevent water entry.

dry welding   arc, gas or plasma welding performed in an underwater habitat with a gas environment at ambient pressure.

dump   expel gas into the water.

ecology (ecosystem)   the natural environment of the earth in which organisms can live in natural relationships between themselves and the environment.

EEG   electroencephalogram.

elasmobranch   any of numerous cartilaginous fishlike vertebrates belonging to the subclass *Elasmobranchii*, which includes the sharks, skates and rays.

environment   the sum of all the external conditions which may affect an organism, community, material or energy if brought under the influence of these external conditions.

equilibrium   in thermodynamics, any state of a system which would not undergo change if the system were to be isolated.

ergometric   level of muscular work.

estuary   a tidal bay formed by submergence or drowning of the lower portion of a river valley and containing a measurable quantity of sea salt.

excursion dive   short period diving to deeper depths during long saturation diving operations.

explosive ordnance disposal units (EODU)   Navy diving unit whose primary mission is disarmament of underwater ordnance devices.

experimental diving unit (EDU)   U.S. Navy diving research center at Washington, D.C.

extrapolation   the extension of a relationship between two or more variables beyond the range covered by knowledge, or the calculation of a value outside that range.

face mask (face plate)   rubber frame with clear flat lens used to seal all or a portion of the diver's face from the water environment.

face squeeze   differential pressure between internal volume of face mask and water. Caused by improper diving technique.

fathom   the common unit of depth in the ocean for countries using the English system of units, equal to 6 feet (1.83 meters).

fault   a fracture or fracture zone in rock along which one side has been displaced in relation to the other side. The intersection of the fault surface with any designated surface, such as the sea bottom, is called a fault line.

fins   semi-rigid, paddle-like extensions worn on the feet to increase swimming propulsion power.

flora   the plant population of a particular location, region, or period.

fouling   the mass of living and nonliving bodies and particles attached to or lying on the surface of a submerged man-made or introduced object; more commonly considered to be only the living or attached bodies.

free air space   all air spaces in the human body that contain air and are normally connected to the atmosphere. Includes pulmonary system, cranial sinuses and middle ear.

free diving (skin diving) diving, using breath-holding techniques and no life-support equipment.

free flow system continuous flow life-support system with flow rate independent of divers breathing.

gage pressure pressure above atmospheric indicated by gage that reads zero pressure at sea level.

gas console station for monitoring and controlling flow of gas to divers on mixed gas operations.

geological oceanography the study of the floors and margins of the oceans, including description of submarine relief features, chemical and physical composition of bottom materials, interaction of sediments and rocks with air and sea water.

gill a platelike or filamentous outgrowth; respiratory organ of aquatic animals.

gradient the rate of decrease of one quantity with respect to another, for example, the rate of decrease of temperature with depth.

half time time required to reach 50% saturation for a specific partial pressure.

heat conduction the transfer of heat from one part of a body to another, or from one body to another in physical contact with it without displacement of the particles of the body.

Heliox diving mixed gas diving using oxygen with helium as the inert diluent.

helmet protective enclosure for the entire head that forms part of the diver's life-support system.

high pressure nervous syndrome (HPNS) a term used to describe symptoms caused by high partial pressures of helium near 40 AtA.

hot water suit a loose-fitting wet suit through which hot water is circulated to maintain thermal equilibrium in extreme cold water exposure.

hyperventilation repeated forced exhalation leading to abnormally low partial pressure of carbon dioxide in the body.

hyperoxia partial pressure of oxygen in the body above that normal at sea level.

hypocapnia low partial pressure of carbon dioxide in the body

hypothermia low body temperature caused by exposure and thermal stress.

inert gas (carrier gas, diluent gas) that part of the breathing medium that serves as a transport for oxygen and is not used by the body as a life-support agent.

inert gas narcosis the narcotic effect of inert gas on the body at elevated partial pressures.

in-situ a Latin term meaning "in place"; in the natural or original position.

interface a surface separating two media, across which there is a discontinuity of some property.

jump dive.

joint juice lubricating fluid in the bone joints.

kelp one of an order (laminariales) of usually large, blade-shaped, or vinelike brown algae.

kluge line auxiliary gas line used in conjunction with a pneumofathometer for diver depth measurement.

lightweight gear all diving systems less complex than the standard dress, particularly shallow water diving gear.

limnology the physics and chemistry of fresh water bodies and ecology of the organisms living in them.

littoral the benthic zone between high and low water marks.

marine biology the study of the plants and animals living in the sea.

marine ecology the science which embraces all aspects of the interrelations of marine organisms and their environment and the interrelations between the organisms themselves.

marine diving technician diver with high degree of technical skills.

mixed gas breathing medium consisting of oxygen and one or more inert gases synthetically mixed.

nearshore zone pertaining to the zone extending seaward from the shore to an indefinite distance beyond the surf zone.

nitrogen narcosis (rapture of the deep) an intoxicating or narcotic effect of gaseous nitrogen, produced in divers breathing air at depth. Usually the effect first becomes noticeable at a depth of 100 feet or more, although individuals vary in their susceptibility.

no joint juice pain in the joints; common during rapid compression to depths greater than 500 feet.

oceanography the study of the sea, embracing and integrating all knowledge pertaining to the sea's physical boundaries, the chemistry and physics of sea water, and marine biology.

open center hydraulic system in which oil is continuously cycled.

open circuit   diving life-support system in which the diver's exhalation is completely vented to the water.

outcrop   naturally protruding, or erosionally exposed or uncovered part of a rock, bed or formation, most of which is covered by over-lying material.

oxygen high pressure (OHP)   higher than normal partial pressure of oxygen within the body.

oxygen low pressure (OLP)   lower than normal partial pressure of oxygen within the body.

oxygen toxicity   condition of elevated partial pressure of oxygen in the body causing un-desirable reactions. Extremely dangerous in diving operations because it can result in loss of consciousness and subsequent drowning.

oxyarc burning   a process for cutting ferrous metal using an electrode and stream of oxygen.

paradoxical shivering (helium tremors)   uncon-trollable shivering under high helium partial pressure with a concomitant subjective feeling of warmth.

parameter   in general, any quantity of a problem that is not an independent variable.

partial pressure   that portion of the total pressure exerted by a specific constituent of a gas mixture.

personnel transfer capsule (PTC)   a diving bell used to transport divers to a work site or sea-floor habitat.

physio-diver   scientific diver working in the field of underwater physiology or pyschology.

physical oceanography   the study of the physical aspects of the ocean, such as its density, temperature, ability to transmit light and sound, and sea ice; the movements of the sea, such as tides, currents and waves.

pinger   a battery powered acoustic device equip-ped with a transducer that transmits sound waves.

platform   any man-made structure (aircraft, ship, buoy or tower) from or on which ocean-ographic instruments are suspended or in-stalled, or from which resource recovery operations are conducted.

pneumofathometer (neumo, pneumo, kluge)   diver depth-indicating system based on air pressure necessary to reach steady flow condi-tion in an open tube terminating at the diver's equipment.

pressurize   to increase the internal pressure of a closed vessel.

pressure proof   able to structurally resist ambient pressure without leakage.

quadrant   a specific area marked off on the sea floor and used by biologists for long term animal counts.

recompression   increasing the ambient pressure on a diver for the primary purpose of medical treatment.

recompression chamber   see decompression chamber.

refraction   the process in which the direction of energy propagation is changed as the energy passes through the interface representing a density discontinuity between two media.

saturation   the condition in which the partial pressure of any fluid constituent is equal to its maximum possible partial pressure under the existing environmental conditions.

saturation diving   technique of maintaining divers in a condition of maximum inert gas absorp-tion for long periods of time.

saturation diving system (Sat system)   a system consisting of one or more deck decompression chambers, personnel transfer capsule and either a surface or sea-floor habitat. The sys-tem is capable of supporting several teams of divers at elevated pressure for long periods of time.

SCAL   skin-diver contact air lens.

SCUBA   self-contained underwater breathing apparatus.

scrubber   unit for removing carbon dioxide from breathing material by using chemical absorp-tion, cryogenics or molecular screen.

sea   all seas (except inland seas) are physically interconnected parts of the earth's total salt water system.

sea floor (sea bed, sea bottom)   the bottom of the ocean where there is a generally smooth, gentle gradient.

sea-floor habitat   undersea living quarters for aquanauts in saturated condition.

SEAL (Sea-Air-Land)   military assault force with underwater capability.

sediment   particulate organic and inorganic matter which accumulates in a loose, uncon-solidated form.

semi-closed circuit   a life-support system in which the gas is partially vented, the remainder is re-cycled, purified and reoxygenated.

sensor   a technical means, usually electronic, to extend man's natural senses by means of energy emitted or reflected.

shallow water gear   diving system consisting of full face mask, hose and compressor.

ship's diver   SCUBA diver trained for bottom inspection and minor repair (Royal Navy).

signature   the characteristic frequency pattern of the target displayed by detection and classification equipment.

sonar   an acronym derived from the expression "sound navigation and ranging." The method or the equipment for determining by underwater sound techniques the presence, location or nature of objects in the sea.

spectrum   the distribution of the intensity of energy dispersion of a given kind as a function of its wavelength, energy, frequency, momentum, mass or any related quantity.

squeeze (barotrauma)   a type of injury occurring in divers, usually during descent, which comes about because of inability to equalize pressure between a closed air space, such as the middle ear, and outside water pressure.

standard dress   diving system consisting of brass diving helmet, breastplate, heavy dry suit, weighted shoes, weight belt, hose, compressor and communications.

submersible decompression chamber (SDC)   a decompression chamber which the diver enters underwater. After being sealed, the chamber is hoisted to the surface, where the diver undergoes the decompression process.

sugar sand   very fine sand common to delta areas.

surface habitat   a large decompression chamber forming part of a saturation diving system. The habitat serves as pressurized living quarters.

swimmer delivery vehicle (SDV)   a wet submersible used for underwater transport of divers.

technical divers   deep sea and SCUBA divers (Federal German Navy).

telemetry   the technique involved in measuring a variety of quantities in place, transmitting the values to a station, and there interpreting, indicating or recording the quantities.

tethered diving   use of an umbilical hose to connect the diver to his gas supply.

thermal balance   a condition where sufficient heat is provided to the diver to avoid a drop in body core temperature during cold water diving.

thermal stress   exposure of the body to environmental temperatures that cause changes in body functions and alteration of body core temperature.

thermocline   a vertical negative temperature gradient in some layer of a body of water which is appreciably greater than the gradients above and below it.

thermogenesis   the ability to generate heat.

thoracic squeeze   development of a differential pressure between environment and free air spaces of the body. Lower pressure in lungs results in hemorrhage of pulmonary capillaries.

transverse section (TRANSEC)   line of survey used in underwater biological, military and geological surveys.

trawl   a bag or funnel-shaped net to catch bottom fish by dragging along the bottom.

turbidity   reduced water clarity resulting from the presence of suspended matter.

U.C.T.   underwater construction teams (U.S. Navy).

U.D.T.   underwater demolition teams (U.S. Navy).

umbilical   normally a composite hose/cable for supplying breathing gas and communications to a diver.

ventilate   procedure in which the diver increases gas flow to ventilate or flush the life-support system.

water breathing   use of a highly oxygenated water solution in the lungs instead of gas for life support.

weapons divers   mine clearance divers and battle swimmers (Federal German Navy).

wet suit   closed cell neoprene rubber diving suit that provides a thermal barrier by insulating a thin layer of body-warmed water next to the diver's skin.

wet submersible   small free-flooding underwater vehicle used for diver transport.

wet welding   underwater welding performed without the use of a protective habitat.

wiping   clearing expired gas from the diver's helmet.

working pressure   the pressure to which a high pressure container can be repeatedly pressurized without causing structural fatigue.

Acknowledgments

This book could not have been written without the generous help and cooperation of the many individuals, organizations and companies that make up the business of diving. I would like to particularly thank:

Ackley, Tool Company
Clackamas, Oregon

Advanced Diving Equipment Company
Gretna, Louisiana

Applied Oceanographics Inc.
San Diego, California

Battelle Memorial Institute
Columbus, Ohio

Bio-Marine Industries
Devon, Pennsylvania

Burnett Electronics Laboratory
San Diego, California

Coastal School of Diving
Oakland, California

COMEX
Marseilles, France

Dillingham Corporation
HydroProducts Division
San Diego, California

Drägerwerk
Lubeck, Germany

Dr. Alan Grant
Wheaton, Maryland

Mr. Dave Woodward
International Underwater Explorers Society
Grand Bahama Island

Marine Services Inc.
San Diego, California

Marine Technology Society
Man's Underwater Activity Committee
San Diego, California

Dr. William Vaughn and Birger Andersen
Oceanautics Inc.
San Diego, California

Perry Submarine Builders
Riviera Beach, Florida

Reading and Bates Offshore Drilling Company
Tulsa, Oklahoma

Mr. Ramsey Parks
Santa Barbara City College
Santa Barbara, California

Scott Aviation Corp.
Lancaster, New York

Smithsonian Institution
Washington, DC

Societe Industrielle Des Establissments Piel
Etamps, France

Mr. Charles Turner
California Department of Fish and Game
Terminal Island, California

Drs. Gershon Weltman and Glen Egstrom
University of California, Los Angeles

Dr. Harry Hollien
University of Florida
Gainesville, Florida

Mr. Lee Sommers
University of Michigan
Ann Arbor, Michigan

Mr. Tommy Thompson
U.S. Divers Company
Santa Anna, California

United States Navy:
Deep Submergence Project Office
Chevy Chase, Maryland

Naval Civil Engineering Laboratory
Port Hueneme, California

Naval Ship Research and Development Center
Panama City, Florida

Naval Special Warfare Group, Pacific
Coronado, California

Naval Submarine Medical Center
Groton, Connecticut

Naval Underwater Research and Development Center
San Diego, California

Office of Naval Research
Washington, DC

Underwater Swimmers School
Key West, Florida

Westinghouse Ocean Research Laboratory
San Diego, California

Wilson Marine Systems
Houston, Texas

In addition, I also express my appreciation to my wife, Edna, for several years of encouragement and performance of the seemingly never ending chores necessary to make this book a reality. To Julia Fletcher, my thanks for months of manuscript assistance.

Index